LANDAU FERMI-LIQUID THEORY

Concepts and Applications

GORDON BAYM
University of Illinois at Urbana-Champaign

and

CHRISTOPHER PETHICK
Nordita, Copenhagen
and
University of Illinois at Urbana-Champaign

A Wiley-Interscience Publication

JOHN WILEY & SONS, INC.

New York • Chichester • Brisbane • Toronto • Singapore

Library of Congress Cataloging in Publication Data:

Baym, Gordon.

Landau Fermi-liquid theory: concepts and applications/Gordon Baym and Christopher Pethick.

p. cm.

Includes bibliographical references.

ISBN 0-471-82418-6

1. Fermi liquid theory. 2. Fermi liquids. I. Pethick, Christopher. II. Title.

QC174.85.F47B39 1991

530.1′3--dc20 91-16606

ISBN 0-471-82418-6 CIP

Printed in the United States of America

10 9 8 7 6 5 4 3 2 1

CONTENTS

PREFACE

Landau's theory of Fermi liquids, advanced in 1956, stands out as one of the high points of modern theoretical physics, a theory whose profundity goes beyond mere phenomenology. Although strongly interacting many-particle systems are very difficult to describe in general, Landau saw, with great insight, that at low temperature an exact simplicity arises, enabling one to describe the properties of extended systems in terms of a dilute collection of elementary excitations. These elementary excitations, or "quasiparticles," are not merely the motion of single particles, but represent the motion of many particles of the system simultaneously. As Landau realized, one can keep track of the thermodynamic and transport properties, that is, do the bookkeeping on conserved quantities and their currents, in terms of the elementary excitations; the detailed proof of this picture requires extensive use of field theory, including conservation laws, associated Ward identities, and renormalization theory. At the same time that the Landau theory has deep conceptual roots, it is a useful and practical tool for accounting for the low temperature properties of Fermi liquids in terms of a rather small number of parameters. Beyond being a vital part of a condensed matter physicist's culture, the theory is a necessary skill for the practitioner.

The major emphasis in this book is on the practical development and application of the theory, rather than on the description of its microscopic derivation. The first chapter delineates the basic theory of the thermodynamic and nonequilibrium properties of Fermi liquids, with application to liquid ^{3}He in its normal state. The second chapter discusses the important and tractable extension of the Fermi liquid theory to dilute solutions of ^{3}He in superfluid ^{4}He, an experimentally rich system where, in addition to Fermi excitations, boson degrees of freedom—of the ^{4}He—play a role.

The basis of this book, two review articles on the Landau Fermi-liquid theory and the application to dilute solutions of ^{3}He in superfluid ^{4}He, were originally written as part of a collection of articles in the books, *The Physics of Liquid and Solid Helium*, edited by K. H. Bennemann and J. B. Ketterson (Wiley, Part 1, 1976, and Part 2, 1978). In retrospect, because of the general utility of the Landau Fermi-liquid theory, it appeared to be valuable to make them more widely available in the present format.

Despite the passing of over a decade, the two review articles remain timely, having been written to emphasize the fundamental ideas of the theory. Experiments on normal ^{3}He in the intervening years have continued to confirm the general theory, and provide information on its extension to higher temperatures. In addition, the theory has found further application in other areas, such as nuclear and neutron star matter, superfluid ^{3}He, and modern problems in superconductivity. To avoid resetting type for the entire book, we have chosen to give a brief summary of the developments in the physics of Fermi liquids since the writing of the original text in an added final Chapter 3, and at the same time have made minor modifications to the introductory sections of the two original chapters. Citations in Chapter 3 references in earlier chapters are given in the format "[Chap. 1: 46, 102]."

We would like to take this opportunity to thank our colleagues in Urbana and Copenhagen for the many interesting and provocative conversations over the years that have helped to shape this monograph. The supportive environment of Nordita, and the continued funding by the U.S. National Science Foundation of our research in low temperature physics, have both contributed greatly to our writing of the present volume.

We dedicate this book to the memory of our friend and colleague, John Wheatley, whose virtuosity in extracting the physics of Fermi liquids, through both his imaginative experiments and his challenging and probing interactions with theorists, made the field for us vital and exciting.

GORDON BAYM
CHRISTOPHER PETHICK

Urbana, Illinois
Copenhagen, Denmark

LANDAU FERMI-LIQUID THEORY

1

LANDAU FERMI-LIQUID THEORY AND LOW TEMPERATURE PROPERTIES OF NORMAL LIQUID ^{3}He

INTRODUCTION

Landau advanced his theory of normal Fermi liquids [1, 2] in 1956, and by the early 1960s, experiments indicated that the predictions of the Landau theory were satisfied, at least qualitatively, by liquid ^{3}He. During the past three decades, precision measurements of the properties of liquid ^{3}He have been carried out at temperatures below 100 mK, where the Landau theory enables one to calculate. More recent theoretical and experimental attention has turned to properties at higher temperatures, and phenomena that vary rapidly in space and time. We shall focus our attention here chiefly on theoretical developments and related experiments. A detailed account of experimental work on Fermi liquid properties of ^{3}He is given by Wheatley in his review article [3]. The main body of the theory of Fermi liquids is concerned with the low temperature thermodynamic properties, such as specific heat, compressibility, and magnetic susceptibility; transport properties, including spin diffusion, viscosity, and thermal conduction; and collective properties at long wavelengths and low frequencies, such as zero sound and spin waves. For discussions of Landau theory at the early stages of its development, we refer to Abrikosov and Khalatnikov's review [4], Wheatley's review at the Sussex Conference on Quantum Fluids [5], and to the books by Pines and Nozières [6], and Wilks [7]. The Landau theory has been derived from microscopic first principles, confirming Landau's intuition; we will not review that work here, but refer the reader instead to Nozières book [8] for an introduction to this subject.

Landau Fermi-liquid theory, in its core, is the application to normal—that is, nonsuperfluid—Fermi systems of the concept of elementary excitations, introduced by Landau, which has proved so fruitful in clarifying the physics of strongly interacting quantum systems at low temperatures. For a Fermi system, these elementary excitations are the fermion quasiparticles, the fully dressed single-particle-like motions. Other applications of the elementary excitation concept are to superfluid ^{4}He, where the elementary excitations are the boson phonons and rotons (see the review of Khalatnikov [120]), and to dilute solutions of ^{3}He in superfluid ^{4}He, where both the phonons and rotons and ^{3}He quasiparticles comprise the elementary excitations. This latter system is described in the following chapter of this book. Considerable strides have been made in the direct microscopic calculations of properties of strongly interacting Fermi systems (see, e.g., [121]); however, these calculations are difficult, and are subject to significant uncertainties. Landau theory provides a way of parametrizing the low temperature properties of Fermi systems, thus enabling one to relate experimentally measured quantities to each other directly. The results of Landau theory follow mainly from rather general arguments, based on conservation laws and general quantum mechanical symmetry principles, and do not depend on the details of the microscopic correlations in the liquid. Despite the rather general nature of most of the results of Landau theory, we shall see that one can give a consistent account of the measured low temperature properties of normal liquid ^{3}He using a rather small number of parameters.

The exciting discovery [9] that liquid ^{3}He becomes superfluid at extremely low temperatures (< 3 mK) has stimulated considerable work which has led not only to a good understanding of the superfluid phases, but also to a much improved characterization of the normal phase. The experimental properties of superfluid ^{3}He has been reviewed by Wheatley [112], and by Lee and Richardson [122], while the basic theory has been reviewed by Leggett [113], and Anderson and Brinkman [123]. Even though the superfluid phases go beyond the normal structure assumed for the Landau theory, the properties of the quasiparticle interaction deduced from measurements in the normal phase above the superfluid transition temperature enable one to estimate the interaction between the elementary excitations in the superfluid phases.

1.1. STATIC PROPERTIES

1.1.1. Quasiparticles

Let us consider first the noninteracting Fermi gas. The quantum states of the individual particles are specified by a momentum **p** and a spin quantum number σ (generally the projection of the spin along the z axis). The energy eigenstates of the system as a whole may be specified by giving the number of particles $N_{\mathbf{p}\sigma}(= 0, 1)$ in each of the single-particle states. In particular, the ground state of the system is one in which all single-particle states with momenta less than the Fermi momentum, p_f, are occupied ($N_{\mathbf{p}\sigma} = 1$) and all other single-particle states are empty ($N_{\mathbf{p}\sigma} = 0$). Excited states of the system may be created either by adding particles with momentum greater than p_f, or by removing particles with momenta less than p_f. The elementary fermion-like excitations of the system created by adding a particle to the system are referred to as particle-like excitations, or just particles, while those created by removing a particle are termed hole-like excitations or just holes. Any excited energy eigenstate of the system may be constructed by creating a number of these elementary excitations.

Now let us consider interacting Fermi systems. Landau's theory of normal Fermi liquids applies to Fermi systems whose spectrum of elementary excitations is similar to that of a free Fermi gas. To be more precise, it is assumed that there exists a one-to-one correspondence between the states of the free Fermi gas and those of the interacting system; that is, if one takes a non-interacting Fermi gas in a particular state and adiabatically turns on the interaction between particles, one obtains a state of the interacting system. Thus, states of the interacting system may be classified by the distribution of particles, $N_{\mathbf{p}\sigma}$, in the corresponding state of the free Fermi gas. The elementary excitations of the interacting system correspond to the particle and hole excitations of the free Fermi gas, and are referred to as *quasiparticles* and *quasiholes* respectively. The distribution of particles in the free Fermi gas $N_{\mathbf{p}\sigma}$ is referred to as the *quasiparticle distribution function* for the corresponding state of the interacting Fermi system. There can clearly be at most one quasiparticle present in a state $\mathbf{p}\sigma$; quasiparticles obey the exclusion principle.

The distribution function $N_{\mathbf{p}\sigma}$ of a general energy eigenstate is a highly discontinuous function of $\mathbf{p}$, jumping back and forth between zero and one. To describe the macroscopic properties of a Fermi liquid, it is sufficient to describe the states in terms of a mean or *smoothed* quasiparticle distribution function, denoted by $n_{\mathbf{p}\sigma}$, that is an average of $N_{\mathbf{p}\sigma}$ over a group of neighboring single-particle states. In the microscopic states of the system, $n_{\mathbf{p}\sigma}$ is a smooth function of $\mathbf{p}$.

The assumption of a one-to-one correspondence between states of the free Fermi gas and those of the interacting system clearly fails if bound states appear when the interaction is turned on. For example, the ground state of a superconductor is not related in a direct way to any one state of the free Fermi gas, but rather to a coherent superposition of a large number of states of the free Fermi system; for such a system Landau's basic assumption is not valid. In particular, the assumption fails for the superfluid phases of liquid ^{3}He recently observed at temperatures of a few mK. However, there will still be a range of temperatures given by $T_C \ll T \ll T_F$, where T_C is the superfluid transition temperature and T_F is the Fermi temperature, in which the superfluid correlations are unimportant and Landau theory is applicable.

In general, a Fermi system also has collective modes, such as zero sound; states containing collective modes have no analogs in the free Fermi gas and therefore the assumption of one-to-one correspondence is not strictly true. However, at low temperatures the phase space available for collective modes is very small, and they may generally be neglected in calculations of bulk properties.

The description of the excited states in terms of superpositions of quasiparticle excitations should be reasonable so long as the quasiparticle states excited have sufficiently long lifetimes. As we shall see in Section 1.2 the width, $\hbar/\tau$, of a quasiparticle state whose excitation energy [measured with respect to the chemical potential μ] is $(\varepsilon_p - \mu)$ varies as $(\varepsilon_p - \mu)^2$, and therefore for sufficiently small excitation energies, or, equivalently, low enough temperatures, the quasiparticle states have well-defined energies.*

There is no way of knowing a priori whether or not a given system of fermions is a normal Fermi liquid; the only way to decide this is to measure the properties of the system experimentally and see if the results are consistent with the predictions of Landau theory.

1.1.2. Quasiparticle Energy and Interactions

Since the low lying states of a Fermi liquid can be characterized by the quasiparticle distribution function $N_{\mathbf{p}\sigma}$, the energy (per unit volume) of each state may be regarded as a functional $E\{N_{\mathbf{p}'\sigma'}\}$ of the distribution function. If we add a quasiparticle to an unoccupied quasiparticle state, $\mathbf{p}\sigma$, then the total energy of the system will increase by an amount $\varepsilon_{\mathbf{p}\sigma}$, called the *quasiparticle energy*, which is itself a functional of the distribution function. In general we may assume that $\varepsilon_{\mathbf{p}\sigma}$ is a smooth function of $\mathbf{p}$, in which case we can equivalently regard the energy (per unit volume) as a functional $E\{n_{\mathbf{p}'\sigma'}\}$

*There exist calculations that define real "statistical quasiparticle energies," and which show that Landau theory is exact at arbitrary temperatures, and is not limited by the lifetime of the quasiparticle states. One example of such a calculation is that by Balian and De Dominicis [10], in which references to other work along these lines may be found. These calculations do not take into account collective modes and therefore give correct results only at low temperatures; they do not, for example, give the correct result for the T^3 term in the specific heat. In calculations of the thermodynamic properties of Fermi systems that do not use statistical quasiparticle energies, quasiparticle lifetime effects give rise to contributions of the same order in the temperature as the collective modes, so that at present it is not clear that the statistical quasiparticle energy type of calculations can be applied over a wider range of temperatures than the other calculations.

of the smoothed distribution function, $n_{\mathbf{p}\sigma}$. Then the quasiparticle energy $\varepsilon_{\mathbf{p}\sigma}\{n_{\mathbf{p}'\sigma'}\}$ is defined as the variation of E with respect to $n_{\mathbf{p}\sigma}$:

$$\delta E = \frac{1}{V}\sum_{\mathbf{p}\sigma} \varepsilon_{\mathbf{p}\sigma}\delta n_{\mathbf{p}\sigma}, \qquad (1.1.1)$$

where V is the volume.

In a macroscopic state of thermal equilibrium the smoothed distribution function $n_{\mathbf{p}\sigma}$ may be determined from the fact that for *any* variation about thermodynamic equilibrium at finite temperature

$$\delta E = T\delta s + \mu\delta n, \qquad (1.1.2)$$

where δs is the variation of the entropy density, δn is the variation of the particle density, T the temperature, and μ the chemical potential. The entropy of the state of the system characterized by a particular smoothed distribution function is given purely by combinatorial considerations and, since the states of the system are in one-to-one correspondence with the states of a free Fermi gas, the entropy density must have the same *form* as for a free Fermi gas:

$$s = -\frac{\kappa}{V}\sum_{\mathbf{p}\sigma} [n_{\mathbf{p}\sigma} \ln (n_{\mathbf{p}\sigma}) + (1 - n_{\mathbf{p}\sigma}) \ln (1 - n_{\mathbf{p}\sigma})], \qquad (1.1.3)$$

where κ is Boltzmann's constant. Since the interaction between particles conserves particle number, the total number of particles in the state of the interacting system must be the same as in the noninteracting system; further, the number of quasiparticles in the state of the interacting system is the same as the number of particles in the corresponding state of the free Fermi gas, and, therefore, the density of particles is given by

$$n = \frac{1}{V}\sum_{\mathbf{p}\sigma} n_{\mathbf{p}\sigma}. \qquad (1.1.4)$$

Consider, then, a variation $\delta n_{\mathbf{p}\sigma}$ of the quasiparticle distribution function; from (1.1.3) and (1.1.4) we have

$$\delta s = -\frac{\kappa}{V}\sum_{\mathbf{p}\sigma} \delta n_{\mathbf{p}\sigma} \ln \frac{n_{\mathbf{p}\sigma}}{1 - n_{\mathbf{p}\sigma}} \qquad (1.1.5)$$

and

$$\delta n = \frac{1}{V}\sum_{\mathbf{p}\sigma} \delta n_{\mathbf{p}\sigma}. \qquad (1.1.6)$$

Using (1.1.1), (1.1.5), and (1.1.6) in (1.1.2) and equating identically the coefficients of $\delta n_{\mathbf{p}\sigma}$ on both sides, we find $\varepsilon_{\mathbf{p}\sigma} = \mu + \kappa T \ln (n_{\mathbf{p}\sigma}^{-1} - 1)$, or

$$n_{\mathbf{p}\sigma} = \frac{1}{e^{(\varepsilon_{\mathbf{p}\sigma}-\mu)/\kappa T} + 1}, \tag{1.1.7}$$

the usual Fermi–Dirac distribution function. Note, however, that $\varepsilon_{\mathbf{p}\sigma}$ itself depends on the entire quasiparticle distribution function; (1.1.7) is therefore in fact a rather complicated implicit equation for $n_{\mathbf{p}\sigma}$.

At $T = 0$ the distribution function (1.1.7) takes the familiar form

$$n^0_{\mathbf{p}\sigma} = \begin{cases} 1, & \varepsilon_{\mathbf{p}\sigma} < \mu \\ 0, & \varepsilon_{\mathbf{p}\sigma} > \mu, \end{cases} \tag{1.1.8}$$

of a Fermi sea occupied up to a given momentum p_f, the Fermi momentum. Under the assumption that $\varepsilon_{p\sigma}$ is an continuous function of p (at least in the neighborhood of p_f), we see from (1.1.8) that μ equals the quasiparticle energy $\varepsilon^0_{p_f}$ at the Fermi surface. (In the presence of a magnetic field the Fermi momentum for up-spin particles is different from that for down-spin particles.)

For slight perturbations about zero temperature equilibrium, the quasiparticle distribution function varies only in the neighborhood of the Fermi surface. Consider, then, the state produced by adding a quasiparticle ($\mathbf{p}\sigma$) to the ground state; its energy measured relative to the ground state is given by

$$\varepsilon^0_{\mathbf{p}\sigma} = \varepsilon_{\mathbf{p}\sigma}\{n^0_{\mathbf{p}'\sigma'}\},$$

where the superscript 0 denotes the ground state. The velocity of a quasiparticle at the Fermi surface (the Fermi velocity) is given by*

$$v_f = \left(\frac{\partial \varepsilon^0_{p\sigma}}{\partial p}\right)_{p=p_f}. \tag{1.1.9}$$

The quasiparticle effective mass m^* is then defined by writing

$$v_f = \frac{p_f}{m^*}. \tag{1.1.10}$$

In the neighborhood of the Fermi surface the quasiparticle energy takes the form

$$\varepsilon^0_{\mathbf{p}\sigma} = \mu + v_f(p - p_f). \tag{1.1.11}$$

It is convenient also to introduce the density of quasiparticle states at the Fermi surface, defined by

$$N(0) = \frac{1}{V}\sum_{\mathbf{p}\sigma}\delta(\varepsilon^0_{\mathbf{p}\sigma} - \mu) = -\frac{1}{V}\sum_{\mathbf{p}\sigma}\frac{\partial}{\partial \varepsilon_{\mathbf{p}\sigma}} n^0_{\mathbf{p}\sigma}. \tag{1.1.12}$$

*The reader is cautioned to distinguish the italicized Roman letter v as in (1.1.9) from the Greek letter ν, as in (1.1.71).

Replacing the sum over p by an integral and taking $\varepsilon_{\mathbf{p}\sigma}$ as the variable of integration, we find (for zero magnetic field)

$$N(0) = \frac{m^* p_f}{\pi^2 \hbar^3}. \tag{1.1.13}$$

Because of the interactions between quasiparticles, the quasiparticle energy $\varepsilon_{\mathbf{p}\sigma}$ depends on the entire quasiparticle distribution function $n_{\mathbf{p}'\sigma'}$. The interaction energy of two quasiparticles is defined as the amount $f_{\mathbf{p}\sigma,\mathbf{p}'\sigma'}/V$ that the energy of one ($\mathbf{p}\sigma$) changes due to the presence of the other ($\mathbf{p}'\sigma'$); a variation of the distribution function produces a variation of $\varepsilon_{\mathbf{p}\sigma}$ given by

$$\delta\varepsilon_{\mathbf{p}\sigma} = \frac{1}{V}\sum_{\mathbf{p}'\sigma'} f_{\mathbf{p}\sigma,\mathbf{p}'\sigma'}\delta n_{\mathbf{p}'\sigma'}. \tag{1.1.14}$$

We see from (1.1.1) and (1.1.14) that f is a second variation of the total energy:

$$f_{\mathbf{p}\sigma,\mathbf{p}'\sigma'} = V^2 \frac{\delta^2 E}{\delta n_{\mathbf{p}\sigma}\delta n_{\mathbf{p}'\sigma'}}; \tag{1.1.15}$$

clearly $f_{\mathbf{p}\sigma,\mathbf{p}'\sigma'} = f_{\mathbf{p}'\sigma',\mathbf{p}\sigma}$. The quasiparticle interaction energy f is itself a functional of the distribution function, though rarely is it necessary to consider this dependence explicitly. Unless otherwise mentioned, f shall denote $f\{n^0_{\mathbf{p}''\sigma''}\}$, the ground state value.

The variation of the energy due to a variation $\delta n_{\mathbf{p}\sigma}$ of the distribution function from its ground state form can then be written as

$$E = E_0 + \frac{1}{V}\sum_{\mathbf{p}\sigma} \varepsilon^0_{\mathbf{p}\sigma}\delta n_{\mathbf{p}\sigma} + \frac{1}{2}\frac{1}{V^2}\sum_{\mathbf{p}\sigma,\mathbf{p}'\sigma'} f_{\mathbf{p}\sigma,\mathbf{p}'\sigma'}\delta n_{\mathbf{p}\sigma}\delta n_{\mathbf{p}'\sigma'} + \ldots, \tag{1.1.16}$$

where E_0 is the ground state energy per unit volume; the quasiparticle energy can be written as

$$\varepsilon_{\mathbf{p}\sigma} = \varepsilon^0_{\mathbf{p}\sigma} + \frac{1}{V}\sum_{\mathbf{p}'\sigma'} f_{\mathbf{p}\sigma,\mathbf{p}'\sigma'}\delta n_{\mathbf{p}'\sigma'} + \cdots. \tag{1.1.17}$$

So far we have, in considering the quasiparticle distribution function and energies, assumed that the spin quantization axes for all quasiparticle states were the same (generally the z axis), that is, all quasiparticles were assumed to be in eigenstates of σ_z. One can also consider the more general situation of a distribution of quasiparticles whose spins are not all quantized along the same axis. In such a case we must treat the quasiparticle distribution function as a 2×2 density matrix $(n_{\mathbf{p}})_{\alpha\bar{\alpha}}$ in spin space. The observables are then also represented by 2×2 matrices in spin space. If we write $(\tau_i)_{\alpha\bar{\alpha}}$ to denote the ith 2×2 Pauli spin matrix,* then the expectation value of the spin polarization in the ith direction will be given by

*We use $\boldsymbol{\tau}$ for Pauli spin matrices and $\boldsymbol{\sigma}$ for expectation values of the spin polarization.

$$\sigma_i = \sum_{\mathbf{p}} \sum_{\alpha\bar{\alpha}} (\tau_i)_{\alpha\bar{\alpha}} (n_{\mathbf{p}})_{\bar{\alpha}\alpha}. \tag{1.1.18}$$

The distribution function can be equivalently specified in terms of the four quantities

$$n_{\mathbf{p}} = \tfrac{1}{2} \sum_{\alpha} (n_{\mathbf{p}})_{\alpha\alpha} \tag{1.1.19}$$

$$\boldsymbol{\sigma}_{\mathbf{p}} = \tfrac{1}{2} \sum_{\alpha\bar{\alpha}} (\boldsymbol{\tau})_{\alpha\bar{\alpha}} (n_{\mathbf{p}})_{\bar{\alpha}\alpha}, \tag{1.1.20}$$

as

$$(n_{\mathbf{p}})_{\alpha\bar{\alpha}} = n_{\mathbf{p}} \delta_{\alpha\bar{\alpha}} + \boldsymbol{\sigma}_{\mathbf{p}} \cdot \boldsymbol{\tau}_{\alpha\bar{\alpha}}. \tag{1.1.21}$$

The energy of a quasiparticle is also represented by a 2×2 matrix $(\epsilon_{\mathbf{p}})_{\alpha\bar{\alpha}}$ defined by the first order variation of the total energy as $(n_{\mathbf{p}})$ is varied [cf. (1.1)]

$$\delta E = \frac{1}{V} \sum_{\mathbf{p}} \sum_{\alpha\bar{\alpha}} (\epsilon_{\mathbf{p}})_{\alpha\bar{\alpha}} (\delta n_{\mathbf{p}})_{\bar{\alpha}\alpha}. \tag{1.1.22}$$

Since any 2×2 matrix can be expressed as a linear combination of the Pauli spin matrices τ_i and the unit matrix, $(\epsilon_{\mathbf{p}})_{\alpha\bar{\alpha}}$ can always be written in the form

$$(\epsilon_{\mathbf{p}})_{\alpha\bar{\alpha}} = \varepsilon_{\mathbf{p}} \delta_{\alpha\bar{\alpha}} + \mathbf{h}_{\mathbf{p}} \cdot \boldsymbol{\tau}_{\alpha\bar{\alpha}}; \tag{1.1.23}$$

$\varepsilon_{\mathbf{p}}$ is a mean quasiparticle energy. $\mathbf{h}_{\mathbf{p}}$ is proportional to an effective magnetic field; it includes a term $\mathbf{h}_{\mathbf{p}}^0 \equiv -\frac{1}{2}\gamma\hbar\boldsymbol{\mathcal{H}}$ representing the coupling to an external magnetic field (where γ is the gyromagnetic ratio), plus terms that represent the internal fields acting on the quasiparticle because of its interaction with other quasiparticles.

The second variation of the energy is of the form

$$\delta^2 E = \frac{1}{V^2} \sum_{\mathbf{pp'}} \sum_{\alpha\bar{\alpha}\alpha'\bar{\alpha}'} f_{\mathbf{p}\alpha\bar{\alpha},\, \mathbf{p'}\alpha'\bar{\alpha}'} (\delta n_{\mathbf{p}})_{\bar{\alpha}\alpha} (\delta n_{\mathbf{p'}})_{\bar{\alpha}'\alpha'}. \tag{1.1.24}$$

Such an $f_{\mathbf{pp'}}$ enables us to describe the interactions of quasiparticles with spins quantized along different axes. The most general form for $f_{\mathbf{p}\alpha\bar{\alpha},\, \mathbf{p'}\alpha'\bar{\alpha}'}$ having the correct rotational properties in spin space is a linear combination of terms of the form $\delta_{\alpha\bar{\alpha}}\delta_{\alpha'\bar{\alpha}'}$, $(\sigma_i)_{\alpha\bar{\alpha}}\delta_{\alpha'\bar{\alpha}'}$, $\delta_{\alpha\bar{\alpha}}(\sigma_i)_{\alpha'\bar{\alpha}'}$, and $(\sigma_i)_{\alpha\bar{\alpha}}(\sigma_j)_{\alpha'\bar{\alpha}'}$ with coefficients dependent on $\mathbf{p}$ and $\mathbf{p'}$. In the absence of a net spin polarization one can see from isotropy that the resultant combination must be a scalar under simultaneous spatial rotation of both the momenta and spins. In the further absence of explicit spin-orbit coupling in the Hamiltonian we require f to be a scalar under independent rotations of the spatial momenta, or the spins. In this case f must be of the form:

$$f_{\mathbf{p}\alpha\bar{\alpha},\, \mathbf{p'}\alpha'\bar{\alpha}'} = f^s_{\mathbf{pp'}} \delta_{\alpha\bar{\alpha}} \delta_{\alpha'\bar{\alpha}'} + f^a_{\mathbf{pp'}} \boldsymbol{\tau}_{\alpha\bar{\alpha}} \cdot \boldsymbol{\tau}_{\alpha'\bar{\alpha}'} \tag{1.1.25}$$

or in matrix notation

$$f_{\mathbf{pp}'} = f^s_{\mathbf{pp}'} + f^a_{\mathbf{pp}'}\boldsymbol{\tau}\cdot\boldsymbol{\tau}'. \tag{1.1.26}$$

The superscripts s and a stand for *symmetric* and *antisymmetric*. When there is spin-orbit coupling, as in nuclear or neutron matter, $f_{\mathbf{pp}'}$ will contain terms of the form $(\mathbf{p}\cdot\boldsymbol{\tau})(\mathbf{p}'\cdot\boldsymbol{\tau}')$, etc. In liquid ^{3}He the dipole–dipole interaction between nuclear spins does not conserve total spin and cannot be written in the form (1.1.26). Because of coherence effects the dipole–dipole interaction plays an important role in the superfluid phases, but has little effect on the properties of normal ^{3}He.

By using (1.1.21), (1.1.24), and (1.1.25), we may write, in analogy to (1.1.17), an expression for the quasiparticle energy to first order in the deviation of the quasiparticle distribution from its value in the (unpolarized) ground state

$$\varepsilon_{\mathbf{p}} = \varepsilon^0_{\mathbf{p}} + \frac{2}{V}\sum_{\mathbf{p}'} f^s_{\mathbf{pp}'}\delta n_{\mathbf{p}'} \tag{1.1.27}$$

$$\mathbf{h}_{\mathbf{p}} = \mathbf{h}^0_{\mathbf{p}} + \frac{2}{V}\sum_{\mathbf{p}'} f^a_{\mathbf{pp}'}\delta\boldsymbol{\sigma}_{\mathbf{p}'}. \tag{1.1.28}$$

Here $\mathbf{h}^0_{\mathbf{p}} = -\frac{1}{2}\gamma\hbar\boldsymbol{\mathscr{H}}$ is the coupling to the external magnetic field $\boldsymbol{\mathscr{H}}$.

When dealing with quasiparticles whose spins are all eigenstates of the spin along a given direction, say z, the quasiparticle interaction assumes the simpler form:

$$\begin{aligned} f_{\mathbf{p}\uparrow,\mathbf{p}'\uparrow} &= f_{\mathbf{p}\downarrow,\mathbf{p}'\downarrow} = f^s_{\mathbf{pp}'} + f^a_{\mathbf{pp}'} \\ f_{\mathbf{p}\uparrow,\mathbf{p}'\downarrow} &= f_{\mathbf{p}\downarrow,\mathbf{p}'\uparrow} = f^s_{\mathbf{pp}'} - f^a_{\mathbf{pp}'}, \end{aligned} \tag{1.1.29}$$

where the arrows denote the eigenstates of σ_z. The functions f^s and f^a are the same as in the more general case (1.1.25).

We need be concerned only with the interactions of quasiparticles both basically on the Fermi surface. It is useful then to expand $f^s_{\mathbf{pp}'}$ and $f^a_{\mathbf{pp}'}$ in terms of the angle θ between $\mathbf{p}$ and $\mathbf{p}'$:

$$\begin{aligned} f^s_{\mathbf{pp}'} &= \sum_{l=0}^{\infty} f^s_l P_l(\cos\theta) \\ f^a_{\mathbf{pp}'} &= \sum_{l=0}^{\infty} f^a_l P_l(\cos\theta), \end{aligned} \tag{1.1.30}$$

where the P_l are Legendre polynomials. The f_l are given in terms of the $f_{\mathbf{pp}'}$ by the inverse relations

$$f^s_l = \frac{2l+1}{2}\int_{-1}^{1} d\mu P_l(\mu)\,\frac{f_{\mathbf{p}\uparrow,\mathbf{p}'\uparrow} + f_{\mathbf{p}\uparrow,\mathbf{p}'\downarrow}}{2} \tag{1.1.31}$$

$$f_l^a = \frac{2l+1}{2}\int_{-1}^{1} d\mu P_l(\mu)\,\frac{f_{\mathbf{p}\uparrow,\mathbf{p}'\uparrow} - f_{\mathbf{p}\uparrow,\mathbf{p}'\downarrow}}{2}, \tag{1.1.32}$$

where μ is the cosine of the angle between $\mathbf{p}$ and $\mathbf{p}'$ (assumed to be on the Fermi surface).

The conventional Landau parameters, defined by

$$F_l^s \equiv N(0) f_l^s, \qquad F_l^a \equiv N(0) f_l^a, \tag{1.1.33}$$

provide useful dimensionless measures of the strengths of the interactions between quasiparticles on the Fermi surface. The Landau parameters F_l and Z_l generally used in the Russian literature are related to these by

$$F_l = F_l^s, \qquad Z_l = 4F_l^a. \tag{1.1.34}$$

1.1.3. Equilibrium Properties

(a) Entropy and Specific Heat. Quite generally the low temperature specific heat of a Fermi liquid is linear in the temperature, with the coefficient given in terms of the effective mass of the quasiparticles at the Fermi surface. To see this, we calculate the first variation of the quasiparticle entropy per unit volume (1.1.3) as we vary the temperature by δT; on using (1.1.7) for $n_{\mathbf{p}\sigma}$ and (1.1.5) we find

$$\delta s = \frac{1}{TV}\sum_{\mathbf{p}\sigma} (\varepsilon_{\mathbf{p}\sigma} - \mu)\,\delta n_{\mathbf{p}\sigma}. \tag{1.1.35}$$

Now from (1.1.7) we see that

$$\delta n_{\mathbf{p}\sigma} = \frac{\partial n_{\mathbf{p}\sigma}}{\partial \varepsilon_{\mathbf{p}\sigma}}\left[-\frac{\varepsilon_{\mathbf{p}\sigma} - \mu}{T}\delta T + \delta\varepsilon_{\mathbf{p}\sigma} - \delta\mu\right]. \tag{1.1.36}$$

As we shall see in Section 1.4, the term involving $\delta\varepsilon_{\mathbf{p}\sigma} - \delta\mu$ leads to terms of order at least $T^3 \ln T$ in the entropy. The first contribution to the entropy is due to the explicit δT, and is

$$\delta s = -\frac{1}{V}\sum_{\mathbf{p}\sigma} \frac{\partial n_{\mathbf{p}\sigma}}{\partial \varepsilon_{\mathbf{p}\sigma}}(\varepsilon_{\mathbf{p}\sigma} - \mu)^2 \frac{\delta T}{T^2}. \tag{1.1.37}$$

The $\partial n_{\mathbf{p}\sigma}/\partial \varepsilon_{\mathbf{p}\sigma}$ restricts the momenta in the sum to those within $\sim \kappa T/v_f$ of the Fermi surface. Replacing the sum by an integral over energy, we then have

$$\begin{aligned}\delta s &= -\sum_\sigma \int p^2 \frac{dp}{d\varepsilon}\frac{4\pi}{(2\pi\hbar)^3} d\varepsilon \frac{\partial}{\partial\varepsilon}\left[\frac{1}{e^{(\varepsilon-\mu)/\kappa T} + 1}\right]\left(\frac{\varepsilon - \mu}{T}\right)^2 \delta T \\ &= -\kappa^2 N(0)\int_{-\infty}^{\infty} dx \frac{\partial}{\partial x}\left(\frac{1}{e^x + 1}\right) x^2 \delta T, \end{aligned} \tag{1.1.38}$$

where again $N(0)$ is the $(T = 0)$ density of states at the Fermi surface (summed over the two spins). The integral in (1.1.38) has the value $\pi^2/3$, and the entropy is thus

$$s = \frac{\pi^2}{3} N(0)\kappa^2 T. \tag{1.1.39}$$

The specific heat at constant volume is then

$$c_V = T\left(\frac{\partial s}{\partial T}\right)_V = s = \frac{m^* p_f}{3\hbar^3}\kappa^2 T. \tag{1.1.40}$$

It is also convenient to express c_V in terms of the Fermi temperature, $T_f \equiv p_f^2/2m^*\kappa$, and the total particle density n as

$$c_V = \frac{\pi^2}{2} n\kappa \frac{T}{T_f}, \tag{1.1.41}$$

which is simply a factor $\frac{1}{3}\pi^2 T/T_f$ times the classical specific heat. It should be noted that since the first temperature correction to the chemical potential is proportional to T^2, the specific heat at constant pressure equals that at constant volume, to lowest order in T.

The first temperature variation (at constant volume) of the free energy $F = E - Ts$ equals $-s\delta T$, so that at low temperature

$$F = E_0 - \frac{\pi^2}{4} n\kappa \frac{T^2}{T_f}, \tag{1.1.42}$$

where E_0 is the ground state energy density.

Finally, we calculate the first correction to the chemical potential by using $\mu = -(\partial F/\partial n)_T$; at low T we have

$$\mu(n, T) = \mu(n, 0) - \frac{\pi^2}{4}\kappa\left(\frac{1}{3} + \frac{n}{m^*}\frac{\partial m^*}{\partial n}\right)\frac{T^2}{T_f}. \tag{1.1.43}$$

(b) Compressibility. The zero temperature compressibility of a Fermi liquid,

$$\mathscr{K} = -\frac{1}{V}\frac{\partial V}{\partial P} = \frac{1}{n^2}\frac{\partial n}{\partial \mu} \tag{1.1.44}$$

can be calculated by varying μ in (1.1.7) for $n_{\mathbf{p}\sigma}$ and then using the relation (1.1.6) to determine δn. From (1.1.7) we find

$$\delta n_{\mathbf{p}\sigma} = \frac{\partial n_{\mathbf{p}\sigma}}{\partial \varepsilon_{\mathbf{p}\sigma}}(\delta\varepsilon_{\mathbf{p}\sigma} - \delta\mu). \tag{1.1.45}$$

The quasiparticle energies depend on μ through their dependence on $n_{\mathbf{p}'\sigma'}$, as in (1.1.14). Since $\partial n/\partial\varepsilon$, and therefore $\delta n_{\mathbf{p}\sigma}$, vanishes except at the Fermi surface, both $\mathbf{p}$ and $\mathbf{p}'$ in (1.1.14) are at the Fermi surface. Furthermore, a variation of μ must produce a variation of $n_{\mathbf{p}\sigma}$ that is isotropic and independent of spin; thus, from (1.1.31),

$$\delta\varepsilon_{\mathbf{p}\sigma} = f_0^s \frac{1}{V} \sum_{\mathbf{p}'\sigma'} \delta n_{\mathbf{p}'\sigma'} = f_0^s \delta n. \tag{1.1.46}$$

Summing (1.1.45) over $\mathbf{p}\sigma$ we then find

$$\delta n = N(0)(\delta\mu - f_0^s \delta n), \tag{1.1.47}$$

or equivalently

$$\frac{\partial n}{\partial \mu} = \frac{N(0)}{1 + F_0^s}; \tag{1.1.48}$$

the compressibility (1.1.44) is therefore

$$\mathscr{K} = \frac{1}{n^2} \frac{N(0)}{1 + F_0^s}. \tag{1.1.49}$$

This is simply the result for a free Fermi gas of particles of mass m^*, divided by the factor $1 + F_0^s$. When F_0^s is large and positive (as in ^{3}He where, for example, at $P = 0$, $F_0^s = 10.8$), the liquid is much stiffer than a free gas of the same effective mass.

(c) Spin Susceptibility. To determine the spin magnetic susceptibility we compute the magnetization produced by an external magnetic field applied to the system. We assume that the field is applied in the z direction. Then the quasiparticle energies are changed by an explicit amount $-\frac{1}{2}\hbar\gamma\sigma_z\mathscr{H}$, where γ is the gyromagnetic ratio and $\sigma_z = \pm 1$; in addition, the quasiparticle energies change because of the change in the distribution function. Thus

$$\delta\varepsilon_{\mathbf{p}\sigma} = -\frac{\hbar}{2}\gamma\sigma_z\mathscr{H} + \frac{1}{V}\sum_{\mathbf{p}'\sigma'} f_{\mathbf{p}\sigma,\mathbf{p}'\sigma'} \delta n_{\mathbf{p}'\sigma'}. \tag{1.1.50}$$

The change in the distribution function is given again by (1.1.45). It is clear, however, that the variation in μ cannot depend on the direction of $\mathscr{H}$, and hence must be proportional to $\mathscr{H}^2$; thus $\delta\mu$ can be neglected in the calculation of the linear susceptibility. Since $\delta n_{\mathbf{p}\sigma}$ is proportional to $\delta\varepsilon_{\mathbf{p}\sigma}$, we see from (1.1.50) that $\delta\varepsilon_{\mathbf{p}\sigma}$ and $\delta n_{\mathbf{p}\sigma}$ are independent of the direction of $\mathbf{p}$ and of opposite sign for spin up and spin down particles. Also $\delta n_{\mathbf{p}\sigma}$ is nonzero only at the Fermi surface. The sum in (1.1.50) thus becomes, from (1.1.32),

$$2f_0^a \delta n_\sigma = \sigma_z f_0^a(\delta n_\uparrow - \delta n_\downarrow) \tag{1.1.51}$$

for p on the Fermi surface, where δn_σ is the change in the total number of particles per unit volume of spin σ (corresponding to $\sigma_z = \pm 1$). Substituting this result in (1.1.45) and summing over $\mathbf{p}$ we find

$$\delta n_\sigma = \frac{1}{2}N(0)\left(\frac{\hbar}{2}\gamma\sigma_z\mathscr{H} - 2f_0^a\delta n_\sigma\right). \tag{1.1.52}$$

The net spin polarization is thus

$$\delta n_\uparrow - \delta n_\downarrow = \frac{\hbar}{2}\gamma\frac{N(0)\mathscr{H}}{1 + F_0^a}, \tag{1.1.53}$$

and the total magnetization is given by

$$\gamma\frac{\hbar}{2}(\delta n_\uparrow - \delta n_\downarrow) = \frac{\left(\gamma\frac{\hbar}{2}\right)^2 N(0)\mathscr{H}}{1 + F_0^a}. \tag{1.1.54}$$

The spin susceptibility,

$$\chi = \frac{\hbar^2}{4}\frac{\gamma^2 N(0)}{1 + F_0^a} \tag{1.1.55}$$

is that of a free Fermi gas of effective mass m^*, divided by $1 + F_0^a$. Negative values of F_0^a enhance the induced spin alignment over that of a free gas.

(d) Effective Mass and Galilean Invariance. In a single component system that is Galilean invariant, such as ^{3}He (but not solutions of ^{3}He in ^{4}He), we can write a simple relation between the effective mass at the Fermi surface and the Landau parameters:

$$\frac{m^*}{m} = 1 + \tfrac{1}{3}F_1^s, \tag{1.1.56}$$

where m is the bare mass of the particles. To derive (1.1.56) we consider the results of a Galilean transformation on the system. If we observe the system from a frame (denoted by a prime) moving with velocity $\mathbf{u}$, then the Hamiltonian in the primed frame is related to that in the lab frame by

$$H' = H - \mathbf{P}\cdot\mathbf{u} + \tfrac{1}{2}Mu^2, \tag{1.1.57}$$

where $\mathbf{P}$ is the momentum operator in the lab frame and $M = Nm$ is the total mass of the system. Thus the total energy and momentum of a state in the primed frame are given by

$$\mathscr{E}' = \mathscr{E} - \mathbf{P}\cdot\mathbf{u} + \tfrac{1}{2}Mu^2 \tag{1.1.58}$$

$$\mathbf{P}' = \mathbf{P} - M\mathbf{u} \tag{1.1.59}$$

where $\mathbf{P}$ is the momentum of the state in the lab frame.

Consider now the change in energy due to adding a quasiparticle of momentum $\mathbf{p}$ in the lab frame. This increases the total mass of the system by m, the *bare* particle mass, since the addition of a quasiparticle involves the addition of one bare particle. In the lab frame the momentum of the system increases by $\mathbf{p}$, and the energy increases by $\varepsilon_{\mathbf{p}}$. (We suppress the spin index here.) From (1.1.59), we see that in the primed frame the momentum increases by $\mathbf{p} - m\mathbf{u}$ while the energy increases by $\varepsilon_{\mathbf{p}} - \mathbf{p}\cdot\mathbf{u} + \frac{1}{2}mu^2$. Thus the quasiparticle energy, in the primed frame, is given by

$$\varepsilon'_{\mathbf{p}-m\mathbf{u}} = \varepsilon_{\mathbf{p}} - \mathbf{p}\cdot\mathbf{u} + \tfrac{1}{2}mu^2, \tag{1.1.60}$$

or

$$\varepsilon'_{\mathbf{p}} = \varepsilon_{\mathbf{p}+m\mathbf{u}} - \mathbf{p}\cdot\mathbf{u} - \tfrac{1}{2}mu^2. \tag{1.1.61}$$

This relation is the fundamental consequence of Galilean invariance for the quasiparticle spectrum.

To derive (1.1.56) we expand both sides of (1.1.61) to order $\mathbf{u}$, assuming $\mathbf{p}$ to be on the Fermi surface. The right side becomes

$$\varepsilon_{\mathbf{p}} + \frac{m - m^*}{m^*}\mathbf{p}\cdot\mathbf{u}. \tag{1.1.62}$$

As seen from the primed frame, the ground state distribution function is a filled Fermi sea centered on $\mathbf{p} = -m\mathbf{u}$, that is,

$$\begin{aligned} n'_{\mathbf{p}} &= n^0_{\mathbf{p}+m\mathbf{u}} \\ &= n^0_{\mathbf{p}} + m\mathbf{u}\cdot\nabla_{\mathbf{p}} n^0_{\mathbf{p}} + \dots . \end{aligned} \tag{1.1.63}$$

The quasiparticle energy in this frame is simply

$$\varepsilon'_{\mathbf{p}} = \varepsilon_{\mathbf{p}}\{n'_{\mathbf{p}'}\} = \varepsilon_{\mathbf{p}}\{n^0_{\mathbf{p}'+m\mathbf{u}}\}; \tag{1.1.64}$$

the only way the quasiparticle energy depends on $\mathbf{u}$ is through the shifting of the Fermi sea. In a multicomponent system, such as a solution of ^{3}He in ^{4}He, $\varepsilon'_{\mathbf{p}}$ would depend explicitly on the ^{4}He superfluid velocity as well as the ^{3}He quasiparticle distribution function, and the present argument would be invalid.

Expanding (1.1.64) to first order in $\mathbf{u}$, and writing

$$\nabla_{\mathbf{p}'} n_{\mathbf{p}'} = \nabla_{p'}\varepsilon_{\mathbf{p}'}\frac{\partial n^0_{p'}}{\partial \varepsilon_{p'}},$$

we find

$$\begin{aligned} \varepsilon'_{\mathbf{p}} &= \varepsilon_{\mathbf{p}} + \frac{1}{V}\sum_{\mathbf{p}'\sigma'} f^s_{\mathbf{p}\mathbf{p}'}\, m\mathbf{u}\cdot\frac{\mathbf{p}'}{m^*}\frac{\partial n^0_{p'}}{\partial \varepsilon_{p'}} \\ &= \varepsilon_{\mathbf{p}} - \frac{F^s_1}{3}\frac{m}{m^*}\mathbf{p}\cdot\mathbf{u}, \end{aligned} \tag{1.1.65}$$

for $\mathbf{p}$ on the Fermi surface. Comparing (1.1.65) with (1.1.62), we find the desired relation (1.1.56).

(e) Thermodynamic Stability. The requirement that the ground state

energy be a minimum, and not simply stationary, restricts the possible values of the Landau parameters to the rangé

$$F_l^s > -(2l+1)$$
$$F_l^a > -(2l+1). \tag{1.1.66}$$

To see this we consider the effects on the ground state energy caused by a distortion of the Fermi sea. We characterize the distortion by a direction dependent Fermi momentum $p_f(\theta, \sigma)$; then the distorted distribution function is given by

$$n_{\mathbf{p}\sigma} = \Theta(p_f(\theta, \sigma) - p),$$

where $\Theta(x) = 1, x > 0$ and $\Theta(x) = 0, x < 0$; σ is the spin orientation and θ is the polar angle of **p**. (Consideration of a more general distribution dependent on the azimuthal angle leads to no new results.)

Stability requires that $E - \mu n$ be an absolute minimum for the undistorted distribution function. (The μn term allows for possible changes in the particle density caused by the distortion.) To second order the change in $E - \mu n$ produced by the distortion takes the form

$$(E - \mu n) - (E - \mu n)_0 = \frac{1}{V} \sum_{\mathbf{p}\sigma} (\varepsilon_p^0 - \mu)\delta n_{\mathbf{p}\sigma} + \frac{1}{2V^2} \sum_{\mathbf{p}\sigma, \mathbf{p}'\sigma'} f_{\mathbf{p}\sigma, \mathbf{p}'\sigma'} \delta n_{\mathbf{p}\sigma} \delta n_{\mathbf{p}'\sigma'}, \tag{1.1.67}$$

where

$$\delta n_{\mathbf{p}\sigma} = n_{\mathbf{p}\sigma} - n_{\mathbf{p}}^0 = (\delta p_f)\delta(p_f - p) - \tfrac{1}{2}(\delta p_f)^2 \frac{\partial}{\partial p} \delta(p_f - p), \tag{1.1.68}$$

and $\delta p_f(\theta, \sigma) \equiv p_f(\theta, \sigma) - p_f^0$ is the change in the Fermi momentum.

We now calculate (1.1.67) to second order in δp_f. The first order term vanishes identically. The first term on the right side of (1.1.67) becomes

$$\frac{1}{4} N(0)\, \nu_f^2 \sum_{\sigma} \int_{-1}^{1} \frac{d\cos\theta}{2} [\delta p_f(\theta, \sigma)]^2, \tag{1.1.69}$$

while the second term on the right becomes

$$\frac{1}{8} [N(0)\nu_f]^2 \sum_{\sigma\sigma'} \int_{-1}^{1} \frac{d\cos\theta}{2} \int_{-1}^{1} \frac{d\cos\theta'}{2} f_{\mathbf{p}\sigma, \mathbf{p}'\sigma'} \delta p_f(\theta, \sigma) \delta p_f(\theta', \sigma') \tag{1.1.70}$$

to second order in δp_f. If we expand δp_f in Legendre polynomials as

$$\nu_f \delta p_f(\theta, \sigma) = \sum_l \nu_{l\sigma} P_l(\cos\theta) \tag{1.1.71}$$

then the sum of (1.1.69) and (1.1.70) becomes

$$\delta E - \mu\delta n = \sum_l \frac{N(0)}{8(2l+1)}\left[(\nu_{l\uparrow} + \nu_{l\downarrow})^2\left(1 + \frac{F_l^s}{2l+1}\right)\right.$$
$$\left. + (\nu_{l\uparrow} - \nu_{l\downarrow})^2\left(1 + \frac{F_l^a}{2l+1}\right)\right]. \quad (1.1.72)$$

This sum is positive definite only if the conditions (1.1.66) are obeyed.

The reader will discover, on closer examination, that the condition for the $l = 1$ symmetric term is

$$N(0)\left(1 + \frac{F_1^s}{3}\right) = \frac{m^* p_f}{\pi^2\hbar^3}\left(1 + \frac{F_1^s}{3}\right) > 0.$$

However, the relation (1.1.56) in a single component Galilean invariant system implies that this condition is always satisfied, whatever the value of F_1^s. Nevertheless, one must still have $1 + F_1^s/3 > 0$ for such systems in order for the effective mass at the Fermi surface to be positive.

1.2. NONEQUILIBRIUM PROPERTIES

1.2.1. Quasiparticle Energies and Interaction

In the previous section we studied the equilibrium properties of a homogeneous Fermi liquid; we now consider the behavior of a Fermi liquid in nonequilibrium and inhomogeneous situations that differ little from the equilibrium state of the homogeneous liquid. In such situations it would be simplest to describe the state of the liquid by specifying the quasiparticle distribution $n_{\mathbf{p}\sigma}(\mathbf{r}, t)$ as a function of position and time. The problem is the uncertainty principle. We know that quantum mechanically if a quasiparticle has a definite momentum $\mathbf{p}$, then its position is completely uncertain. To what extent are the quasiparticles localized in position and momentum space? If the spatial inhomogeneity of the system occurs over a characteristic length λ, then the particles are localized in space only within a distance λ. At temperature T the distribution function varies in momentum space over a characteristic momentum $\kappa T/v_f$, and so use of the distribution function does not require any localization of the quasiparticles in momentum space to less than $\Delta p = \kappa T/v_f$. Thus, as long as $\lambda\Delta p \gg \hbar$, or

$$\lambda \gg \frac{\hbar v_f}{\kappa T}, \quad (1.2.1)$$

the Heisenberg uncertainty principle causes no trouble, and we may legitimately use the "classical" distribution function $n_{\mathbf{p}\sigma}(\mathbf{r}, t)$ to specify the density of quasiparticles of momentum $\mathbf{p}$ at space point $\mathbf{r}$.

More generally, the quasiparticle distribution must be described by a Wigner distribution function $\mathcal{N}(\mathbf{r}_1\sigma_1, \mathbf{r}_2\sigma_2, t)$ that is essentially the amplitude for removing at time t a quasiparticle of spin σ_1 at point $\mathbf{r}_1$, then immediately adding one of spin σ_2 at point $\mathbf{r}_2$ and thereby returning to the initial state of the system. In terms of the "second quantized" operators $a^\dagger_{\mathbf{p}\sigma}$ and $a_{\mathbf{p}\sigma}$ that create and annihilate quasiparticles, we may write

$$\mathcal{N}(\mathbf{r}_1\sigma_1, \mathbf{r}_2\sigma_2, t) = \int \frac{d^3p_1}{(2\pi\hbar)^3} e^{i\mathbf{p}_1\cdot\mathbf{r}_1/\hbar} \frac{d^3p_2}{(2\pi\hbar)^3} e^{-i\mathbf{p}_2\cdot\mathbf{r}_2/\hbar} \langle a^\dagger_{\mathbf{p}_2\sigma_2} a_{\mathbf{p}_1\sigma_1}\rangle, \tag{1.2.2}$$

where the expectation value is in the state of the system at time t. The Fourier transform of the Wigner distribution function:

$$\begin{aligned}\Big(n_\mathbf{p}(\mathbf{r},t)\Big)_{\sigma\sigma'} &\equiv \int d^3r' e^{-i\mathbf{p}\cdot\mathbf{r}'/\hbar}\mathcal{N}\Big(\mathbf{r} + \frac{\mathbf{r}'}{2}, \sigma; \mathbf{r} - \frac{\mathbf{r}'}{2}, \sigma', t\Big)\\ &= \int \frac{d^3q}{(2\pi\hbar)^3} e^{i\mathbf{q}\cdot\mathbf{r}/\hbar}\langle a^\dagger_{\mathbf{p}-\mathbf{q}/2,\sigma'}\, a_{\mathbf{p}+\mathbf{q}/2,\sigma}\rangle,\end{aligned} \tag{1.2.3}$$

a 2×2 matrix in spin space, is the quantum mechanical generalization to spatially inhomogeneous situations of the homogeneous distribution function $(n_\mathbf{p})_{\sigma\sigma'}$ [cf. (1.1.21)]. For example, the total quasiparticle density at $\mathbf{r}$ is given by

$$\sum_\sigma \int \frac{d^3p}{(2\pi\hbar)^3}\Big(n_\mathbf{p}(\mathbf{r}, t)\Big)_{\sigma\sigma} = \sum_\sigma \mathcal{N}(\mathbf{r}\sigma, \mathbf{r}\sigma), \tag{1.2.4}$$

while the total number of quasiparticles of momentum $\mathbf{p}$ in the system is given by

$$\sum_\sigma \int d^3\mathbf{r}\Big(n_\mathbf{p}(r, t)\Big)_{\sigma\sigma} = \sum_\sigma \int d^3r_1 d^3r_2 e^{-i\mathbf{p}\cdot(\mathbf{r}_1-\mathbf{r}_2)/\hbar}\mathcal{N}(\mathbf{r}_1\sigma, \mathbf{r}_2\sigma). \tag{1.2.5}$$

When $\lambda \gg \hbar v_f/\kappa T$, the distribution function $(n_\mathbf{p}(\mathbf{r}, t))_{\sigma\sigma'}$ has the usual interpretation as a classical distribution function in momentum and position space (and a density matrix in spin space).

Consider now an inhomogeneous system, possibly in the presence of external space and time dependent forces, and not necessarily in equilibrium. Let $\mathscr{E}(t)$ be the total energy of the system, including interactions with the external forces. We can then define the energy $\varepsilon_{\mathbf{p}\sigma}(\mathbf{r}, t)$ of a quasiparticle at $\mathbf{r}$ by the variation of $\mathscr{E}(t)$ with respect to the quasiparticle distribution function at $\mathbf{r}$:

$$\delta\mathscr{E}(t) = \int d^3r\, \delta E(\mathbf{r}, t) = \sum_\sigma \int d^3r \int \frac{d^3p}{(2\pi\hbar)^3} \varepsilon_{\mathbf{p}\sigma}(\mathbf{r}, t)\delta n_{\mathbf{p}\sigma}(\mathbf{r}, t), \tag{1.2.6}$$

where $E(\mathbf{r}, t)$ is the local energy density. This is the generalization of (1.1.1). Similarly the variation of $\varepsilon_{\mathbf{p}\sigma}(\mathbf{r}, t)$ with respect to $n_{\mathbf{p}'\sigma'}(\mathbf{r}, t)$ defines, as in the homogeneous case, the effective quasiparticle interaction:

$$\delta\varepsilon_{\mathbf{p}\sigma}(\mathbf{r}, t) = \sum_{\sigma'} \int d^3r' \int \frac{d^3p'}{(2\pi\hbar)^3} f_{\mathbf{p}\sigma,\mathbf{p}'\sigma'}(\mathbf{r}, \mathbf{r}', t)\delta n_{\mathbf{p}'\sigma'}(\mathbf{r}', t). \tag{1.2.7}$$

For a neutral Fermi system the range of interaction between quasiparticles is on the order of a typical microscopic length, for example, $\hbar/p_f$. Therefore, when $\delta n_{\mathbf{p}'\sigma'}(\mathbf{r}', t)$ varies little over such distances (as we shall assume), the $\delta n_{\mathbf{p}'\sigma'}(\mathbf{r}', t)$ in (1.2.7) can be replaced by $\delta n_{\mathbf{p}'\sigma'}(\mathbf{r}, t)$, and the interaction can be regarded as effectively a local one, $f_{\mathbf{p}\sigma, \mathbf{p}'\sigma'}(\mathbf{r}, t) \equiv \int d^3r' f_{\mathbf{p}\sigma, \mathbf{p}'\sigma'}(\mathbf{r}, \mathbf{r}', t)$. In the first order variation of $\varepsilon_{\mathbf{p}\sigma}(\mathbf{r}, t)$ from its value for uniform equilibrium at $T = 0$, $f_{\mathbf{p}\sigma, \mathbf{p}'\sigma'}(\mathbf{r}, t)$ reduces simply to the $f_{\mathbf{p}\sigma, \mathbf{p}'\sigma'}$ of the equilibrium system (1.1.14).

If there are long range Coulomb forces between particles, then $f_{\mathbf{p}\sigma, \mathbf{p}'\sigma'}(\mathbf{r}, \mathbf{r}', t)$ is essentially a sum of a long range plus a short range part:

$$f_{\mathbf{p}\sigma, \mathbf{p}'\sigma'}(\mathbf{r}, \mathbf{r}', t) = \frac{e^2}{|\mathbf{r} - \mathbf{r}'|} + f_{\mathbf{p}\sigma, \mathbf{p}'\sigma'}\delta(\mathbf{r} - \mathbf{r}'), \tag{1.2.8}$$

where e is the particle charge.

1.2.2. The Kinetic Equation

Let us first consider the case when $n_{\mathbf{p}\sigma}(\mathbf{r}, t)$ may be regarded as a classical distribution function. The space and time dependence of the quasiparticle distribution function is determined by a kinetic equation, which, in the absence of quasiparticle collisions, takes the form of the continuity equation for the distribution function in **p** and **r** space. For simplicity, we shall at first neglect forces that rotate spins, and suppress the spin index. Then the changes of the distribution function obey

$$\frac{\partial n_{\mathbf{p}}(\mathbf{r}, t)}{\partial t} + \nabla_r \cdot [\mathbf{v}_{\mathbf{p}}(\mathbf{r}, t) n_{\mathbf{p}}(\mathbf{r}, t)] + \nabla_p \cdot [\mathbf{f}_{\mathbf{p}}(\mathbf{r}, t) n_{\mathbf{p}}(\mathbf{r}, t)] = 0, \tag{1.2.9}$$

where $\mathbf{v}_{\mathbf{p}}(\mathbf{r}, t)$ is the quasiparticle velocity in position space $(d\mathbf{r}/dt)$, and $\mathbf{f}_{\mathbf{p}}(\mathbf{r}, t)$ is the time rate of change of the quasiparticle momentum $(d\mathbf{p}/dt)$. The basic assumption of the quasiparticle kinetic theory is that $\varepsilon_{\mathbf{p}}(\mathbf{r}, t)$ plays the role of the quasiparticle Hamiltonian, so that

$$\mathbf{v}_{\mathbf{p}}(\mathbf{r}, t) = \nabla_p \varepsilon_{\mathbf{p}}(\mathbf{r}, t) \tag{1.2.10}$$

$$\mathbf{f}_{\mathbf{p}}(\mathbf{r}, t) = -\nabla_r \varepsilon_{\mathbf{p}}(\mathbf{r}, t). \tag{1.2.11}$$

Equation 1.2.10 is simply the statement that $\mathbf{v}_{\mathbf{p}}$ is the quasiparticle group velocity, while (1.2.11) is the classical statement that the rate of change of momentum, the effective force, is the negative spatial gradient of the particle energy. It is essentially the kinetic generalization of the concept that in a nonequilibrium thermodynamic system the gradient of the chemical potential acts as an effective force on the system. The last two terms on the left side of (1.2.9) describe the changes in the quasiparticle distribution function due to continuous changes in the quasiparticle positions and momenta. In the presence of quasiparticle collisions, in which the quasiparticles undergo sudden changes in momenta, to the right side of (1.2.9) must be added a collision integral $I[n_{\mathbf{p}'}]$, whose form we shall specify later. Then substituting (1.2.10) and (1.2.11) in (1.2.9), we find the Landau kinetic equation

$$\frac{\partial n_{\mathbf{p}}(\mathbf{r}, t)}{\partial t} + \nabla_p \varepsilon_{\mathbf{p}}(\mathbf{r}, t) \cdot \nabla_r n_{\mathbf{p}}(\mathbf{r}, t) - \nabla_r \varepsilon_{\mathbf{p}}(\mathbf{r}, t) \cdot \nabla_p n_{\mathbf{p}}(\mathbf{r}, t) = I[n_{\mathbf{p}'}], \quad (1.2.12a)$$

or

$$\frac{\partial n_{\mathbf{p}}(\mathbf{r}, t)}{\partial t} - [\varepsilon_{\mathbf{p}}(\mathbf{r}, t), n_{\mathbf{p}}(\mathbf{r}, t)]_{\text{P.B.}} = I[n_{\mathbf{p}'}], \quad (1.2.12b)$$

where $[\ \ ,\ \ ]_{\text{P.B.}}$ denotes the Poisson bracket.

The Landau kinetic equation is considerably richer than the usual Boltzmann equation used to describe weakly interacting gases. There are two additional physical features included here. First, the quasiparticle velocity $\nabla_p \varepsilon_{\mathbf{p}}(\mathbf{r}, t)$ can depend on position and time; this effect only arises in nonlinear deviations from homogeneous equilibrium. Second, the force term $\nabla_r \varepsilon_{\mathbf{p}}$ includes effective field contributions. To be specific, let us assume that an external scalar potential $U(\mathbf{r}, t)$ is applied to the system; this adds a term $\int d^3r U(\mathbf{r}, t) n(\mathbf{r}, t)$ to the total energy, increasing the quasiparticle energies by $U(\mathbf{r}, \mathbf{t})$. The $\mathbf{r}$ dependence of $\varepsilon_{\mathbf{p}}(\mathbf{r}, t)$ is due both to the $\mathbf{r}$ dependence of U and to the fact that $\varepsilon_{\mathbf{p}}$ depends on the entire quasiparticle distribution function (near $\mathbf{r}$), which itself is $\mathbf{r}$ dependent. Thus we can write

$$\nabla_r \varepsilon_{\mathbf{p}}(\mathbf{r}, t) = \nabla_r U(\mathbf{r}, t) + \int \frac{d^3p'}{(2\pi\hbar)^3} f_{\mathbf{p}\mathbf{p}'} \nabla_r n_{\mathbf{p}'}(\mathbf{r}, t). \quad (1.2.13)$$

The first term, present in the dilute gas case, is the force due directly to the external field. The second term is new; it represents the force acting on the quasiparticle because of all the other quasiparticles near it. This effective field term is responsible for much of the structure of transport phenomena in Fermi liquids.

In physical situations where the distribution function must be taken to be the Wigner function, rather than a classical-like distribution, the kinetic equation (1.2.12a) is replaced by a quantum mechanical equation of motion, which is essentially the classical equation (1.2.12b) with the Poisson bracket replaced by $(i\hbar)^{-1}$ times the commutator [11]. In this equation the quasiparticles are assumed to move independently, with the effective Hamiltonian given by the position dependent quasiparticle energy $(\epsilon_{\mathbf{p}}(\mathbf{r}))_{\alpha\alpha'}$. (Later we shall need the resulting equation for the case when the full spin dependence is taken into account, so we shall retain the spin indices $\alpha\alpha'$ in the present discussion.) In the position representation we may write the variation of the total energy with respect to a variation of the Wigner distribution function (1.2.2) as

$$\delta\mathscr{E} = \sum_{\alpha\alpha'} \int d^3r d^3r' (\epsilon(\mathbf{r}, \mathbf{r}'))_{\alpha\alpha'} (\delta\mathscr{N}(\mathbf{r}', \mathbf{r}))_{\alpha'\alpha} \qquad (1.2.14)$$

where the quasiparticle energy $(\epsilon(\mathbf{r}, \mathbf{r}'))_{\alpha\alpha'}$ corresponds to a nonlocal "potential."

Then

$$(\epsilon_{\mathbf{p}}(\mathbf{r}))_{\alpha\alpha'} = \int d^3r' e^{-i\mathbf{p}\cdot\mathbf{r}'/\hbar} (\epsilon(\mathbf{r} + \frac{\mathbf{r}'}{2}, \mathbf{r} - \frac{\mathbf{r}'}{2}))_{\alpha\alpha'}. \qquad (1.2.15)$$

The equation of motion for $\mathscr{N}$ is then simply

$$\frac{\partial(\mathscr{N}(\mathbf{r},\mathbf{r}'))_{\alpha\alpha'}}{\partial t} - \frac{1}{i\hbar}\sum_{\bar{\alpha}} \int d^3\bar{r} \{((\epsilon(\mathbf{r}, \bar{\mathbf{r}}))_{\alpha\bar{\alpha}} (\mathscr{N}(\bar{\mathbf{r}}, \mathbf{r}'))_{\bar{\alpha}\alpha'} - (\mathscr{N}(\mathbf{r}, \bar{\mathbf{r}}))_{\alpha\bar{\alpha}} (\epsilon(\bar{\mathbf{r}}, \mathbf{r}'))_{\bar{\alpha}\alpha'}\} = I_Q[\mathscr{N}],$$

where I_Q is the quantum mechanical collision integral, which is generally different from the classical one [12]. Writing this equation in terms of $n_{\mathbf{p}}(\mathbf{r}, t)$ and $\epsilon_{\mathbf{p}}(\mathbf{r}, t)$ we find

$$\frac{\partial(n_{\mathbf{p}}(\mathbf{r}, t))_{\alpha\alpha'}}{\partial t} - \frac{1}{i\hbar}\sum_{\bar{\alpha}} \int \frac{d^3p'}{(2\pi)^3} \frac{d^3q}{(2\pi)^3} \int d^3r' d^3\tau e^{i[(\mathbf{p}'-\mathbf{p})\cdot\boldsymbol{\tau}+\mathbf{q}\cdot(\mathbf{r}-\mathbf{r}')]}$$
$$\times \left\{ (\epsilon_{\mathbf{p}'+\hbar\mathbf{q}/2}(\mathbf{r}' + \frac{\hbar\boldsymbol{\tau}}{2}, t))_{\alpha\bar{\alpha}} (n_{\mathbf{p}'}(\mathbf{r}', t))_{\bar{\alpha}\alpha'} - (n_{\mathbf{p}'}(\mathbf{r}', t))_{\alpha\bar{\alpha}} (\epsilon_{\mathbf{p}'-\hbar\mathbf{q}/2}(\mathbf{r}' - \frac{\hbar\boldsymbol{\tau}}{2}, t))_{\bar{\alpha}\alpha'} \right\}$$
$$= I_Q[n_{\mathbf{p}''}]. \qquad (1.2.16)$$

If one expands the integrand in powers of $\hbar$ and retains only terms that do not vanish as $\hbar \to 0$, one recovers from (1.2.16) the classical result (1.2.12) when n and ϵ are proportional to the unit matrix $\delta_{\alpha\alpha'}$.

We shall often be interested in situations in which the distribution function differs by only a very small amount $\delta n_{\mathbf{p}}$ from its value in uniform equilibrium. In these situations we shall only need the kinetic equation (1.2.12) linearized in $\delta n_{\mathbf{p}}$. In (1.2.12) the term $\nabla_r n_{\mathbf{p}}$ is explicitly first order in $\delta n_{\mathbf{p}}$; the term $\nabla_r \varepsilon_{\mathbf{p}}$ equals

$$\nabla_r U + \int \frac{d^3p'}{(2\pi\hbar)^3} f_{\mathbf{p}\mathbf{p}'} \nabla_r \delta n_{\mathbf{p}'} = \nabla_r \delta\varepsilon_{\mathbf{p}}, \qquad (1.2.17)$$

to first order in $\delta n_{\mathbf{p}}$ and U, and therefore we may write the linearized kinetic equation as

$$\frac{\partial \delta n_{\mathbf{p}}}{\partial t} + \mathbf{v}_{\mathbf{p}} \cdot \nabla_r \left(\delta n_{\mathbf{p}} - \frac{\partial n_p^0}{\partial \varepsilon_p} \delta\varepsilon_{\mathbf{p}} \right) = I[n_{\mathbf{p}'}]. \qquad (1.2.18)$$

Here $\mathbf{v}_{\mathbf{p}}$ and $\partial n_p^0/\partial\varepsilon_p$ are the equilibrium functions; we have used the relation that in equilibrium $\nabla_p n_p = (\nabla_p \varepsilon_p) \partial n_p^0/\partial\varepsilon_p$. To complete the linearization we must also linearize the collision integral; we defer this until Section 1.2.4. It is convenient to introduce the quantity

$$\delta\bar{n}_{\mathbf{p}\sigma} \equiv \delta n_{\mathbf{p}\sigma} - \frac{\partial n_p^0}{\partial \varepsilon_p}\,\delta\varepsilon_{\mathbf{p}\sigma},$$
$$= \delta n_{\mathbf{p}\sigma} - \frac{\partial n_p^0}{\partial \varepsilon_p}\sum_{\sigma'}\int \frac{d^3p'}{(2\pi\hbar)^3} f_{\mathbf{p}\sigma,\,\mathbf{p}'\sigma'}\delta n_{\mathbf{p}'\sigma'}, \tag{1.2.19}$$

which is the linear deviation of the distribution function from the value $\{\exp[(\varepsilon_{\mathbf{p}}(\mathbf{r}, t) - \mu)/\kappa T] + 1\}^{-1}$ it would have for a quasiparticle of energy $\varepsilon_{\mathbf{p}}(\mathbf{r}, t)$ in a system in overall equilibrium. Equation 1.2.18 then has the simple form

$$\frac{\partial \delta n_{\mathbf{p}}}{\partial t} + \mathbf{v}_{\mathbf{p}}\cdot\nabla_r \delta\bar{n}_{\mathbf{p}} = I[n_{\mathbf{p}'}]. \tag{1.2.20}$$

Equation 1.2.18, Fourier transformed in space and time, becomes

$$(\omega - \mathbf{q}\cdot\mathbf{v}_{\mathbf{p}})\delta n_{\mathbf{p}}(\mathbf{q}, \omega) - \mathbf{q}\cdot\mathbf{v}_{\mathbf{p}}\frac{\partial n_p^0}{\partial \varepsilon_p}\,\delta\varepsilon_{\mathbf{p}}(\mathbf{q}, \omega) = iI[n_{\mathbf{p}'}]. \tag{1.2.21}$$

The analogous result obtained from the linearized version of the quantum-kinetic equation (1.2.16), assuming that ϵ and n are proportional to the unit matrix in spin space is

$$\left(\omega - \frac{1}{\hbar}(\varepsilon_{\mathbf{p}+\hbar\mathbf{q}/2} - \varepsilon_{\mathbf{p}-\hbar\mathbf{q}/2})\right)\delta n_{\mathbf{p}}(\mathbf{q}, \omega) - \frac{1}{\hbar}(n_{\mathbf{p}+\hbar\mathbf{q}/2} - n_{\mathbf{p}-\hbar\mathbf{q}/2})\delta\varepsilon_{\mathbf{p}}(\mathbf{q}, \omega)$$
$$= iI_Q[n_{\mathbf{p}}']. \tag{1.2.22}$$

This equation, valid even when (1.2.1) is not satisfied, reduces to (1.2.21) in the limit $\hbar \to 0$. In almost all circumstances with which we shall deal, the classical Boltzmann equation and the quantum-kinetic equation give identical results; the only case where this is not true is in the calculation of the attenuation of collective modes when $\hbar\omega \gtrsim \kappa T$.

1.2.3. The Conservation Laws

As the first application of the kinetic equation we derive, from the full nonlinear kinetic equation (1.2.12a), the local conservation laws of particle number, momentum, and energy. (The spin conservation law is discussed in Section 1.3.2(c).) These laws are useful for later derivations of the transport modes of a Fermi liquid, as well as for telling us how to write the currents of conserved quantities in terms of the quasiparticles. We shall assume that in quasiparticle collisions, described by the collision term $I[n_{\mathbf{p}'}]$, quasiparticle number, momentum, energy (sums of the $\varepsilon_{\mathbf{p}'}$), and spin are locally conserved. We shall also assume that an external scalar potential $U(\mathbf{r}, t)$ is applied to the system.

If we sum both sides of (1.2.12) over $\mathbf{p}$ and σ [it is easiest to use the form for the left side in (1.2.9) directly], we find

$$\frac{\partial}{\partial t} n(\mathbf{r}, t) + \nabla \cdot \mathbf{j}(\mathbf{r}, t) = 0 \tag{1.2.23}$$

where the quasiparticle, and particle, density is given by

$$n(\mathbf{r}, t) = \sum_{\sigma} \int \frac{d^3p}{(2\pi\hbar)^3} n_{\mathbf{p}\sigma}(\mathbf{r}, t), \tag{1.2.24}$$

and the particle current is given by

$$\mathbf{j}(\mathbf{r}, t) = \sum_{\sigma} \int \frac{d^3p}{(2\pi\hbar)^3} \nabla_p \varepsilon_{\mathbf{p}\sigma}(\mathbf{r}, t) n_{\mathbf{p}\sigma}(\mathbf{r}, t). \tag{1.2.25}$$

The sum over the collision term vanished by local conservation of quasiparticles in collisions.

The linearized particle current can be written as

$$\begin{aligned} \mathbf{j}(\mathbf{r}, t) &= \sum_{\sigma} \int \frac{d^3p}{(2\pi\hbar)^3} \left[\left(\nabla_p \varepsilon_p^0\right) \delta n_{\mathbf{p}\sigma}(\mathbf{r}, t) + n_p^0 \nabla_p \delta\varepsilon_{\mathbf{p}}(\mathbf{r}, t) \right] \\ &= \sum_{\sigma} \int \frac{d^3p}{(2\pi\hbar)^3} \left(\nabla_p \varepsilon_p^0\right) \delta \bar{n}_{\mathbf{p}\sigma}(\mathbf{r}, t), \end{aligned} \tag{1.2.26}$$

where the superscript 0 refers to equilibrium values; to derive the second line we integrate the final term in the first line by parts. If we assume that $\delta n_{\mathbf{p}}$ is nonzero only near the Fermi surface, then from the latter form for $\delta \bar{n}_{\mathbf{p}}$ in (1.2.19) we find

$$\begin{aligned} \mathbf{j}(\mathbf{r}, t) &= \sum_{\sigma} \int \frac{d^3p}{(2\pi\hbar)^3} (\nabla_p \varepsilon_p^0)\, \delta n_{\mathbf{p}\sigma}(\mathbf{r}, t) \left(1 + \frac{F_1^s}{3}\right) \\ &= \frac{1 + F_1^s/3}{m^*} \sum_{\sigma} \int \frac{d^3p}{(2\pi\hbar)^3} \mathbf{p}\, \delta n_{\mathbf{p}\sigma}(\mathbf{r}, t). \end{aligned} \tag{1.2.27}$$

This result is valid even if the system is not Galilean invariant.

In a Galilean invariant system $1 + F_1^s/3 = m^*/m$, so that $\mathbf{j}$ becomes just the local momentum density, divided by m, the bare particle mass, that is,

$$\mathbf{j}(\mathbf{r}, t) = \sum_{\sigma} \int \frac{d^3p}{(2\pi\hbar)^3} \frac{\mathbf{p}}{m} n_{\mathbf{p}\sigma}(\mathbf{r}, t). \tag{1.2.28}$$

This result is valid even when one does not linearize (1.2.25) for $\mathbf{j}$; it follows from the fact that for any isolated system the velocity of the center of mass is the total momentum divided by the total mass. To see how the equivalence of (1.2.28) and (1.2.25) arises in detail in the nonlinear case requires a lengthy calculation. However, we can gain some insight into the equivalence by studying a weakly interacting Fermi gas in which the quasiparticle energy is given explicitly by

$$\varepsilon_{\mathbf{p}} = \frac{p^2}{2m} + \int \frac{d^3p'}{(2\pi\hbar)^3} v_{\mathbf{p}\mathbf{p}'} n/n_{\mathbf{p}'}, \tag{1.2.29}$$

where $v_{\mathbf{pp}'}$ is independent of the distribution function; this is the appropriate form in the Hartree–Fock approximation and also for dilute systems with short-range interactions. Here $v_{\mathbf{pp}'}$ plays the role of the $f_{\mathbf{pp}'}$. Then

$$\int \frac{d^3p}{(2\pi\hbar)^3}\left[\nabla_p \varepsilon_{\mathbf{p}} n_{\mathbf{p}} - \frac{\mathbf{p}}{m} n_{\mathbf{p}}\right] = \int \frac{d^3p}{(2\pi\hbar)^3} \frac{d^3p'}{(2\pi\hbar)^3} (\nabla_p v_{\mathbf{pp}'}) n_{\mathbf{p}} n_{\mathbf{p}'}. \tag{1.2.30}$$

In a Galilean invariant system $v_{\mathbf{pp}'}$ depends only on $\mathbf{p} - \mathbf{p}'$. Thus, $\nabla_p v_{\mathbf{pp}'} = \frac{1}{2}(\nabla_p - \nabla_{p'}) v_{\mathbf{pp}'}$ is antisymmetric in $\mathbf{p}$ and $\mathbf{p}'$, and the integral on the right side of (1.2.30) vanishes by symmetry. When the system lacks Galilean invariance, as in the case of a ^{3}He Fermi liquid dissolved in ^{4}He, the current is still given by (1.2.25), but not by (1.2.28).

Returning to the general case, we find the local momentum conservation law by multiplying both sides of (1.2.12) by $\mathbf{p}$ and summing over $\mathbf{p}$ and σ. From conservation of momentum in collisions, the right side vanishes, and we find

$$\frac{\partial}{\partial t} g_i(\mathbf{r}, t) + \nabla_j T_{ij}(\mathbf{r}, t) + \sum_\sigma \int \frac{d^3p}{(2\pi\hbar)^3} \frac{\partial \varepsilon_{\mathbf{p}\sigma}}{\partial r_i} n_{\mathbf{p}\sigma} = 0, \tag{1.2.31}$$

where

$$\mathbf{g}(\mathbf{r}, t) = \sum_\sigma \int \frac{d^3p}{(2\pi\hbar)^3} \mathbf{p} n_{\mathbf{p}\sigma}(\mathbf{r}, t) \tag{1.2.32}$$

is the momentum density (equal to $m\mathbf{j}$ in a Galilean invariant system); a sum over the repeated index j is understood, and

$$T_{ij} = \sum_\sigma \int \frac{d^3p}{(2\pi\hbar)^3} p_i \frac{\partial \varepsilon_{\mathbf{p}\sigma}}{\partial p_j} n_{\mathbf{p}\sigma}. \tag{1.2.33}$$

In the dilute gas case T_{ij} would be the stress tensor and $\partial\varepsilon_{\mathbf{p}\sigma}/\partial r_i$ in (1.2.31) would be just $\partial U/\partial r_i$; the last term in (1.2.31) would be the (negative) force per unit volume produced by the external potential. We can in general write the last term in (1.2.31) as a total divergence by first writing

$$\frac{\partial \varepsilon_{\mathbf{p}\sigma}}{\partial r_i} n_{\mathbf{p}\sigma} = \frac{\partial}{\partial r_i}(\varepsilon_{\mathbf{p}\sigma} n_{\mathbf{p}\sigma}) - \varepsilon_{\mathbf{p}\sigma} \frac{\partial n_{\mathbf{p}\sigma}}{\partial r_i}. \tag{1.2.34}$$

The last term, when summed over $\mathbf{p}$ and σ gives the spatial variation of E due to its dependence on n [cf. (1.1.1) and (1.2.6)], but not its explicit spatial dependence due to U; it equals $\nabla_i E(\mathbf{r}, t) - n(\mathbf{r}, t)\nabla_i U(\mathbf{r}, t)$. Thus (1.2.31) takes the form

$$\frac{\partial g_i(\mathbf{r}, t)}{\partial t} + \nabla_j \Pi_{ij}(\mathbf{r}, t) + n(\mathbf{r}, t)\nabla_i U(\mathbf{r}, t) = 0, \tag{1.2.35}$$

where the total stress tensor Π_{ij} is given by

$$\Pi_{ij} = T_{ij} + \delta_{ij}\left(\sum_\sigma \int \frac{d^3p}{(2\pi\hbar)^3} \varepsilon_{\mathbf{p}\sigma} n_{\mathbf{p}\sigma} - E \right). \tag{1.2.36}$$

The derivation of the energy conservation law is similar. We multiply both sides of (1.2.12) by $\varepsilon_{\mathbf{p}\sigma}$ and sum over $\mathbf{p}$ and σ. Again, from conservation of total quasiparticle energy in collisions, the right side vanishes and we find after a little algebra:

$$\sum_\sigma \int \frac{d^3p}{(2\pi\hbar)^3} \varepsilon_{\mathbf{p}\sigma} \frac{\partial n_{\mathbf{p}\sigma}}{\partial t} + \nabla \cdot \sum_\sigma \int \frac{d^3p}{(2\pi\hbar)^3} (\nabla_p \varepsilon_{\mathbf{p}\sigma}) \varepsilon_{\mathbf{p}\sigma} n_{\mathbf{p}\sigma} = 0. \tag{1.2.37}$$

The first term can be written as

$$\frac{\partial E}{\partial t} - n\frac{\partial U}{\partial t} = \frac{\partial}{\partial t}(E - nU) + U\frac{\partial n}{\partial t} = \frac{\partial}{\partial t}(E - nU) - U\nabla \cdot \mathbf{j} \tag{1.2.38}$$

since it describes the time variation of E due to its dependence on n, but not its explicit time dependence due to U. We define the energy current by

$$\mathbf{j}_E = \sum_\sigma \int \frac{d^3p}{(2\pi\hbar)^3} (\nabla_p \varepsilon_{\mathbf{p}\sigma})(\varepsilon_{\mathbf{p}\sigma} - U) n_{\mathbf{p}\sigma}; \tag{1.2.39}$$

this current represents the transport of the energy density $E - Un$ of the system. The conservation law (1.2.37) then becomes

$$\frac{\partial}{\partial t}(E - Un) + \nabla \cdot \mathbf{j}_E = -\mathbf{j} \cdot \nabla U. \tag{1.2.40}$$

The source term on the right is the rate at which the energy density $E - Un$ is changed by the interaction of the system with the external field.

The energy current $\mathbf{j}_E$ is a sum of terms due to the bulk motion of the fluid, which transports energy by convection, as well as a term $\mathbf{j}_T$ due to thermal conduction. For calculating the thermal conductivity, it is useful, to have an explicit expression for $\mathbf{j}_T$. This can be found if we realize that in the frame in which the bulk motion of the fluid vanishes locally (we denote this frame by a prime), then the entire energy transport at that point is due solely to thermal conduction, and $\mathbf{j}'_E = \mathbf{j}_T$. The primed frame is determined by the condition

$$\mathbf{j}' = 0. \tag{1.2.41}$$

Let us assume that this frame travels with velocity $\mathbf{u}$. Then, as in Section 1.1.3(d)

$$n_{\mathbf{p}\sigma} = n'_{\mathbf{p}-m\mathbf{u},\sigma} \tag{1.2.42}$$

$$\varepsilon_{\mathbf{p}\sigma} = \varepsilon'_{\mathbf{p}-m\mathbf{u},\sigma} + \mathbf{p}\cdot\mathbf{u} - \tfrac{1}{2}mu^2 \tag{1.2.43}$$

$$\mathbf{v}_{\mathbf{p}\sigma} = \mathbf{v}'_{\mathbf{p}-m\mathbf{u},\sigma} + \mathbf{u}. \tag{1.2.44}$$

Substituting (1.2.42) and (1.2.44) into (1.2.25) for the current (and replacing the integration variable $\mathbf{p}$ by $\mathbf{p} + m\mathbf{u}$) we see that

$$\mathbf{j} = \mathbf{j}' + n\mathbf{u}, \tag{1.2.45}$$

so that from (1.2.41), we have $\mathbf{u} = \mathbf{j}/n$. On substituting (1.2.42), (1.2.43) and (1.2.44) into (1.2.39) for the energy current, and using (1.2.41), we similarly find

$$j_{E_i} = j_{T_i} + \Pi'_{ji}u_j + u_i[E' + \mathbf{g}'\cdot\mathbf{u} + \tfrac{1}{2}mnu^2] - nUu_i, \tag{1.2.46}$$

where Π'_{ji} is the stress tensor and $\mathbf{g}'$ the momentum density in the primed frame, and $\mathbf{j}_T = \mathbf{j}'_E$. However, from (1.1.58) and (1.1.59) we see that the term in square brackets in (1.2.46) is simply E. Thus, the thermal (or heat) current is given by

$$j_{T_i} = j_{E_i} - [(E - nU)\delta_{ji} + \Pi'_{ji}]u_j \tag{1.2.47}$$

The stress tensor Π'_{ji} is generally of the form

$$\Pi'_{ji} = P\delta_{ji} - \sigma'_{ji} \tag{1.2.48}$$

where P is the pressure, and σ_{ji} is the dissipative part of the stress tensor. As usual, the term $(E - nU)\mathbf{u}$ is the transport of energy by convection, while the $\boldsymbol{\Pi}'\boldsymbol{u}$ term represents the flow of energy resulting from work done by the stresses as the fluid is transported.

For small deviations from a system at rest and in local equilibrium, $\Pi'_{ji} = P\delta_{ji}$ and the expression for the thermal current becomes

$$\mathbf{j}_T = \sum_\sigma \int \frac{d^3p}{(2\pi\hbar)^3} n_{\mathbf{p}\sigma}\left(\varepsilon_{\mathbf{p}\sigma} - \frac{E + P}{n}\right)\mathbf{v}_{\mathbf{p}\sigma}. \tag{1.2.49}$$

1.2.4. Transport Coefficients

Corresponding to the five conservation laws of number, momentum (three components), and energy are five transport modes of the system, which are first or zero sound (counted as two modes corresponding to frequencies $\omega = \pm cq$), where c is the velocity of the mode and q the wavenumber, transverse viscous transport of momentum (doubly degenerate), and thermal conduction. In addition to the conservation laws of the three components of spin, there correspond modes of spin diffusion or spin waves. In this section we calculate the coefficients of viscosity and thermal conductivity, as well as simple spin diffusion, deferring consideration of sound and spin modes to Section 1.3.

(a) Collision Integral. In order to give explicit calculations of the transport coefficients we must first specify the details of the collision integral. At low

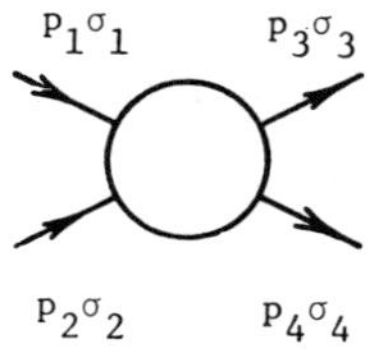

FIGURE 1.1. Scattering of two quasiparticles.

temperatures the density of thermally excited quasiparticles is low, and it is only necessary to consider collisions of two quasiparticles at a time, as shown in Figure 1.1. This process assumes conservation of quasiparticles. Let us write the amplitude for this process as

$$\langle \mathbf{p}_3\sigma_3, \mathbf{p}_4\sigma_4 | t | \mathbf{p}_1\sigma_1, \mathbf{p}_2\sigma_2 \rangle; \tag{1.2.50}$$

this includes all exchange contributions. The amplitudes for transitions to the physically indistinguishable states $|\mathbf{p}_3\sigma_3, \mathbf{p}_4\sigma_4\rangle$ and $|\mathbf{p}_4\sigma_4, \mathbf{p}_3\sigma_3\rangle$ differ by a minus sign. We assume the quasiparticle states to be normalized in a box of volume V. Then (1.2.50) is proportional to $\delta_{\mathbf{p}_1+\mathbf{p}_2, \mathbf{p}_3+\mathbf{p}_4}$, expressing conservation of momentum. The matrix element also contains conservation of the total quasiparticle spin; when all spins are referred to the same quantization axis (1.2.50) simply contains a factor $\delta_{\sigma_1+\sigma_2, \sigma_3+\sigma_4}$. The rate at which this process occurs in the system is

$$\frac{2\pi}{\hbar} |\langle 34|t|12\rangle|^2 \delta(\varepsilon_1 + \varepsilon_2 - \varepsilon_3 - \varepsilon_4) n_1 n_2 (1 - n_3)(1 - n_4), \tag{1.2.51}$$

where we use 1 to denote $\mathbf{p}_1\sigma_1$, and so on. The factors $n_1(\mathbf{r}, t)$ and $n_2(\mathbf{r}, t)$ are the probabilities of occupation of the incident quasiparticle states, while the factors $1 - n_3$ and $1 - n_4$, the probabilities that the final states are unoccupied, are required by the exclusion principle. The rate for the inverse process $3,4 \to 1,2$ is given by the same expression as (1.2.51) with $n_1 n_2 (1 - n_3)(1 - n_4)$ replaced by $(1 - n_1)(1 - n_2) n_3 n_4$. We define a transition probability $W(12;34)$ by writing

$$\frac{2\pi}{\hbar} |\langle 34|t|12\rangle|^2 = \frac{1}{V^2} W(12;34) \delta_{\mathbf{p}_1+\mathbf{p}_2, \mathbf{p}_3+\mathbf{p}_4} \delta_{\sigma_1+\sigma_2, \sigma_3+\sigma_4}. \tag{1.2.52}$$

Then the net rate $I_1[n_\mathbf{p}]$ at which collisions increase the occupation of state $\mathbf{p}_1\sigma_1$ is the rate for the inverse process $3,4 \to 1,2$ minus the rate for the process $1,2 \to 3,4$, summed over all states 2 and distinguishable final states 3 and 4:

$$I_1[n_\mathbf{p}] = \frac{1}{V^2} \sum_{\mathbf{p}_2\sigma_2,} \sum_{\substack{\mathbf{p}_3\sigma_3 \\ \mathbf{p}_4\sigma_4}}{}' W(12;34) \delta_{\mathbf{p}_1+\mathbf{p}_2, \mathbf{p}_3+\mathbf{p}_4} \delta_{\sigma_1+\sigma_2, \sigma_3+\sigma_4} \delta(\varepsilon_1 + \varepsilon_2 - \varepsilon_3 - \varepsilon_4)$$
$$\times [n_3 n_4 (1 - n_1)(1 - n_2) - n_1 n_2 (1 - n_3)(1 - n_4)]. \tag{1.2.53}$$

The prime denotes that the sum is only over distinguishable final states. The reader may wish to verify explicitly that the contributions of this collision integral to the local conservation laws do in fact vanish.

In most applications of the kinetic equation, the system is almost in local thermodynamic equilibrium at each point, with a local temperature $T(\mathbf{r}, t)$, chemical potential $\mu(\mathbf{r}, t)$ and fluid velocity $\mathbf{u}(\mathbf{r}, t)$. We would like then to linearize the collision term, at each point, about the deviations of the distribution function from the local equilibrium distribution

$$n_{\mathbf{p}\sigma}^{\text{l.e.}}(\varepsilon_{\mathbf{p}\sigma}^{\text{l.e.}}, \mathbf{r}, t) = \left[\exp\left(\frac{\varepsilon_{\mathbf{p}\sigma}^{\text{l.e.}} - \mathbf{p}\cdot\mathbf{u}(\mathbf{r}, t) - \mu(\mathbf{r}, t)}{\kappa T(\mathbf{r}, t)}\right) + 1\right]^{-1}. \tag{1.2.54}$$

Here the $\varepsilon_{\mathbf{p}}^{\text{l.e.}}$ are the local equilibrium quasiparticle energies that are determined self-consistently by

$$\varepsilon_{\mathbf{p}\sigma}^{\text{l.e.}} = \varepsilon_{\mathbf{p}\sigma}\{n_{\mathbf{p}'\sigma'}^{\text{l.e.}}\} \tag{1.2.55}$$

$$n_{\mathbf{p}'\sigma'}^{\text{l.e.}} = n_{\mathbf{p}'\sigma'}^{\text{l.e.}}(\varepsilon_{\mathbf{p}'\sigma'}^{\text{l.e.}}, \mathbf{r}, t). \tag{1.2.56}$$

The parameters μ, T, and $\mathbf{u}$ in the local equilibrium function are chosen to give the same local density, velocity, and energy as the true distribution function, that is,

$$\delta n = \sum_{\sigma} \int \frac{d^3p}{(2\pi\hbar)^3} \delta n_{\mathbf{p}\sigma}^{\text{l.e.}} \tag{1.2.57}$$

$$\delta E = \sum_{\sigma} \int \frac{d^3p}{(2\pi\hbar)^3} \varepsilon_{\mathbf{p}\sigma} \delta n_{\mathbf{p}\sigma}^{\text{l.e.}} \tag{1.2.58}$$

and

$$\delta \mathbf{g} = \sum_{\sigma} \int \frac{d^3p}{(2\pi\hbar)^3} \mathbf{p}\, \delta n_{\mathbf{p}\sigma}^{\text{l.e.}} \tag{1.2.59}$$

where $\delta n_{\mathbf{p}\sigma}^{\text{l.e.}}$ is the deviation of the local equilibrium distribution function from *global* equilibrium, and δn, δE and δg are the actual deviations of the density, energy density, and momentum density from their values in global equilibrium.

At first sight one would expect the collision term to vanish when $n_{\mathbf{p}\sigma} = n_{\mathbf{p}\sigma}^{\text{l.e.}}$. However, this is not true, because the energy conservation condition involves the true quasiparticle energies, $\varepsilon_{\mathbf{p}\sigma}\ \{n_{\mathbf{p}'\sigma'}\}$ and not the local equilibrium energies $\varepsilon_{\mathbf{p}\sigma}^{\text{l.e.}}$; hence, $\varepsilon_1^{\text{l.e.}} + \varepsilon_2^{\text{l.e.}} \neq \varepsilon_3^{\text{l.e.}} + \varepsilon_4^{\text{l.e.}}$. Since the collision term will vanish when

$$n_{\mathbf{p}\sigma} = n_{\mathbf{p}\sigma}^{\text{l.e.}}(\varepsilon_{\mathbf{p}\sigma}, \mathbf{r}, t), \tag{1.2.60}$$

we can linearize the collision integral, without explicitly expanding the energies in the delta function, by writing, to first order,

$$n_i = n_i^{\text{l.e.}}(\varepsilon_i) + \delta \bar{n}_i^{\text{l.e.}}, \tag{1.2.61}$$

where $\delta \bar{n}_i^{\text{l.e.}}$ is a small correction term, and $i = 1,2,3,4$. If we define a deviation from local equilibrium by writing

$$n_i = n_i^{\text{l.e.}}(\varepsilon_i^{\text{l.e.}}) + \delta \tilde{n}_i, \tag{1.2.62}$$

then

$$\delta \bar{n}_i^{\text{l.e.}} = \delta \tilde{n}_i - \frac{\partial n^{\text{l.e.}}(\varepsilon_i^{\text{l.e.}})}{\partial \varepsilon_i^{\text{l.e.}}} \delta \tilde{\varepsilon}_i, \tag{1.2.63}$$

where

$$\delta \tilde{\varepsilon}_i = \varepsilon_i - \varepsilon_i^{\text{l.e.}}. \tag{1.2.64}$$

One should compare (1.2.63) with (1.2.19). Now, after making liberal use of the identity

$$[n_1 n_2 (1 - n_3)(1 - n_4) - (1 - n_1)(1 - n_2) n_3 n_4]\, \delta(\varepsilon_1 + \varepsilon_2 - \varepsilon_3 - \varepsilon_4) = 0, \tag{1.2.65}$$

where here $n_i = n_i^{\text{l.e.}}(\varepsilon_i)$, we can write the linearized collision integral as

$$I_1[n_{\mathbf{p}}] = -\frac{1}{\kappa T V^2} \sum_2 \sum_{34}{}' n_1 n_2 (1 - n_3)(1 - n_4) W(12;34) \delta_{\mathbf{p}_1+\mathbf{p}_2,\, \mathbf{p}_3+\mathbf{p}_4} \times \delta_{\sigma_1+\sigma_2,\, \sigma_3+\sigma_4} \delta(\varepsilon_1 + \varepsilon_2 - \varepsilon_3 - \varepsilon_4)\, [\bar{\Phi}_1 + \bar{\Phi}_2 - \bar{\Phi}_3 - \bar{\Phi}_4], \tag{1.2.66}$$

where all n_i, ε_i, and W have their true local equilibrium values, and

$$\delta \bar{n}_i^{\text{l.e.}} \equiv -\frac{\partial n_i^0}{\partial \varepsilon_i} \bar{\Phi}_i. \tag{1.2.67}$$

The sums over states 2,3, and 4 proceed as follows. Let us consider the case in which σ_1 is [illegible] Then if σ_2 is up, both σ_3 and σ_4 must also be up. We write $W(12;34)$ for this case as $W_{\uparrow\uparrow}(\mathbf{p}_1\mathbf{p}_2, \mathbf{p}_3\mathbf{p}_4)$ and the sum over 3 and 4 becomes

$$\frac{1}{2} \sum_{\mathbf{p}_3\mathbf{p}_4} W_{\uparrow\uparrow}(\bar{\Phi}_{1\uparrow} + \bar{\Phi}_{2\uparrow} - \bar{\Phi}_{3\uparrow} - \bar{\Phi}_{4\uparrow}), \tag{1.2.68}$$

where the sum is over *all* $\mathbf{p}_3$ and $\mathbf{p}_4$. In the case when σ_1 is up but σ_2 is down, then one of the final spins σ_3 and σ_4 is up and the other is down; we choose 3 to be the particle with the same spin as 1, and write $W(12;34)$ as $W_{\uparrow\downarrow}(\mathbf{p}_1\mathbf{p}_2, \mathbf{p}_3\mathbf{p}_4)$. The sum over 3 and 4 then becomes

$$\sum_{\mathbf{p}_3\mathbf{p}_4} W_{\uparrow\downarrow}(\bar{\Phi}_{1\uparrow} + \bar{\Phi}_{2\downarrow} - \bar{\Phi}_{3\uparrow} - \bar{\Phi}_{4\downarrow}), \tag{1.2.69}$$

where again the sum is over all $\mathbf{p}_3$ and $\mathbf{p}_4$. Results similar to (1.2.68) and (1.2.69) obtain when σ_1 is down. In the absence of a magnetic field $W_{\downarrow\downarrow} = W_{\uparrow\uparrow}$ and $W_{\uparrow\downarrow} = W_{\downarrow\uparrow}$.

It is also interesting to note that the Landau kinetic equation with the collision term (1.2.53) obeys an H-theorem; one can show, by use of (1.2.12), that the total entropy of the system

$$S = -\kappa \sum_{\sigma} \int d^3r \frac{d^3p}{(2\pi\hbar)^3} \{n_{\mathbf{p}\sigma}(\mathbf{r}, t) \ln n_{\mathbf{p}\sigma}(\mathbf{r}, t) + [1 - n_{\mathbf{p}\sigma}(\mathbf{r}, t)] \ln [1 - n_{\mathbf{p}\sigma}(\mathbf{r}, t)]\} \tag{1.2.70}$$

always increases in time, except when $n_{\mathbf{p}\sigma}(\mathbf{r}, t)$ has the form of a local equilibrium distribution $n_{\mathbf{p}\sigma} = n_{\mathbf{p}\sigma}^{\text{l.e.}} (\varepsilon_{\mathbf{p}\sigma}(\mathbf{r}, t))$; in this case $dS/dt = 0$. We record, for future use, the form for dS/dt, evaluated to second order in the deviations $\delta\bar{n}_i^{\text{l.e.}}$:

$$\frac{dS}{dt} = \frac{1}{4\kappa T^2 V^3} \sum_{12} \sum_{34}' \int d^3r\, n_1 n_2 (1 - n_3)(1 - n_4) W(12;34) \delta_{\mathbf{p}_1 + \mathbf{p}_2, \mathbf{p}_3 + \mathbf{p}_4}$$
$$\times\, \delta_{\sigma_1 + \sigma_2, \sigma_3 + \sigma_4} \delta(\varepsilon_1 + \varepsilon_2 - \varepsilon_3 - \varepsilon_4) \left(\bar{\Phi}_1 + \bar{\Phi}_2 - \bar{\Phi}_3 - \bar{\Phi}_4\right)^2, \tag{1.2.71}$$

where again the n_i and ε_i denote local equilibrium distributions and quasiparticle energies.

In a number of problems it is too difficult to treat the full energy dependence of the collision integral; under these circumstances it is common to make a *relaxation time approximation* in which the collision integral is replaced by the approximate form

$$(I_1[n_{\mathbf{p}}])_{\text{R.T.}} = -\frac{\delta\bar{n}_1^{\text{l.e.}}}{\tau}, \tag{1.2.72}$$

where the relaxation time τ is independent of the energy of quasiparticle $\mathbf{p}_i$. A somewhat more general form of the relaxation time approximation in which the relaxation time depends on the angular behavior of the deviation of the distribution function from local equilibrium is

$$(I_1[n_{\mathbf{p}}])_{\text{R.T.}} = -\sum_{lm} \frac{((\delta\bar{n}^{\text{l.e.}}(p_1))_{lm}}{\tau_l} Y_{lm}(\hat{\mathbf{p}}_1), \tag{1.2.73}$$

where

$$(\delta\bar{n}^{\text{l.e.}}(p_1))_{lm} = \int d\Omega_1 \delta\bar{n}_{\mathbf{p}_1}^{\text{l.e.}} Y_{lm}^*(\hat{\mathbf{p}}_1), \tag{1.2.74}$$

and the Y_{lm} are spherical harmonics. The deviation from local equilibrium, not the deviation from global equilibrium, that occurs in these approximate forms of the collision integral automatically ensures that they are consistent with the conservation laws.

(b) Thermal Conductivity. In a Fermi liquid in which there is a small temperature gradient, the thermal current $\mathbf{j}_T$ has the form

$$\mathbf{j}_T = -K\nabla T, \tag{1.2.75}$$

where K is the thermal conductivity. To calculate K we assume that the system is in a steady state. For sufficiently small temperature gradients the distribution function $n_{\mathbf{p}}$ has the form of a local equilibrium distribution (1.2.54). In calculating thermal conductivity we shall assume that $\mathbf{u}$, the local fluid velocity, vanishes everywhere.

In (1.2.49) for the heat current we observe that if $n_{\mathbf{p}}$, $\varepsilon_{\mathbf{p}}$, and $\mathbf{v}_{\mathbf{p}}$ are given by their true local equilibrium values, then $\mathbf{j}_T$ vanishes. We then expand $\mathbf{j}_T$ to first order in $\delta\bar{n}_{\mathbf{p}}$ (1.2.62) and, after integrating by parts, find

$$\mathbf{j}_T = \sum_{\sigma} \int \frac{d^3p}{(2\pi\hbar)^3} \delta\bar{n}^{\text{l.e.}}_{\mathbf{p}\sigma}(\varepsilon_{\mathbf{p}\sigma} - \mu - Ts')\mathbf{v}_{\mathbf{p}\sigma}, \tag{1.2.76}$$

where we have used

$$E = Ts + \mu n - P \tag{1.2.77}$$

for a system at rest, and $s' = s/n$ as the entropy per particle.

We turn now to the linearized kinetic equation [(1.2.18) with (1.2.66)] to determine $\delta\bar{n}^{\text{l.e.}}_{\mathbf{p}}$. The linearized collision term (1.2.66) is of order $\delta\bar{n}^{\text{l.e.}}_{\mathbf{p}}/\tau$, where τ is a typical quasiparticle collision time. The left side of the kinetic equation is of order $v_f \delta n_p/L$, where $L^{-1} \sim \nabla T/T$. Therefore

$$\delta\bar{n}^{\text{l.e.}}_{\mathbf{p}} \sim \delta\bar{n}_{\mathbf{p}} \sim \frac{\lambda}{L}\delta n_{\mathbf{p}}, \tag{1.2.78}$$

where $\lambda = v_f\tau$ is a typical quasiparticle mean free path. It is only necessary then to evaluate the left side of (1.2.18) for $n_{\mathbf{p}} = n^{\text{l.e.}}_{\mathbf{p}}$; to first order it becomes $-(\partial n^0_p/\partial\varepsilon_p)\mathbf{v}_p\cdot[(\varepsilon_p - \mu)\nabla T/T + \nabla\mu]$ where all quantities, except ∇T and $\nabla\mu$, refer to global equilibrium.

$\nabla\mu$ can be evaluated by noting that, from (1.2.35), in a steady state $\partial g_i/\partial t = -\nabla_j \Pi_{ij} = 0$ (for $U = 0$). The stress tensor Π_{ij} evaluated in local equilibrium is simply $P\delta_{ij}$, where P is the pressure. Thus, $\nabla P = 0$, and from the Gibbs–Duhem relation

$$\nabla P = n\nabla\mu + s\nabla T, \tag{1.2.79}$$

we conclude

$$\nabla\mu = -s'\nabla T. \tag{1.2.80}$$

The kinetic equation then assumes the form

$$-\frac{\partial n^0_p}{\partial\varepsilon_p}(\varepsilon_p - \mu - Ts')\,\mathbf{v}_{\mathbf{p}}\cdot\frac{\nabla T}{T} = I[n'_{\mathbf{p}}]; \tag{1.2.81}$$

this equation determines $\delta\bar{n}^{\text{l.e.}}_{\mathbf{p}\sigma}$, contained in the right side, in terms of ∇T, on the left side.

Historically, the first approximate method of solving this equation for a low temperature Fermi liquid was that of Abrikosov and Khalatnikov [13], who calculated the limiting low temperature behavior of K, as well as the coefficient of viscosity η. Hone [14] later applied this method in an approximate calculation of the spin-diffusion coefficient D_σ in the low temperature limit. It was found though that these solutions were in serious disagreement with very low temperature measurements of K and D_σ in dilute solutions of ^{3}He in superfluid ^{4}He [15]. Using the variational method [16], with the Abrikosov–Khalatnikov and Hone solutions for $\delta n_{\mathbf{p}\sigma}$ as trial functions, Baym and Ebner [17] were able to provide a more consistent account of the transport coefficients of the dilute solutions. Subsequently, Dy and Pethick [18] provided exact numerical solutions of the transport equation, and Emery and Cheng [19] gave an improved analytic approximation to the solution. At about that time, Brooker and Sykes [20], and Højgaard Jensen, Smith, and Wilkins [21] supplied exact analytical solutions to the low temperature transport equations for K, η, and D_σ; a detailed account of the exact solutions may be found in the paper by Sykes and Brooker [22]. We shall describe here this method for solving the transport equation exactly, beginning with (1.2.81) to determine the thermal conductivity.

We first note that the factor $n_1n_2(1-n_3)(1-n_4)\,\delta(\varepsilon_1+\varepsilon_2-\varepsilon_3-\varepsilon_4)$ in the collision term (1.2.53) restricts the quasiparticles that appreciably scatter to those with energies $\lesssim \kappa T$ from the Fermi surface. We then simplify the linear collision term (1.2.66) by using momentum conservation to do the sum over $\mathbf{p}_4$ and converting the $\mathbf{p}_2$ and $\mathbf{p}_3$ sums to integrals. The d^3p_2 we write as $p_f^2dp_2 \sin\theta\, d\theta d\phi_2 = m^*p_f d\varepsilon_2 \sin\theta\, d\theta d\phi_2$, where the polar angle θ is defined as that between $\mathbf{p}_2$ and $\mathbf{p}_1$. Also we write d^3p_3 as $m^*p_f d\varepsilon_3 \sin\theta_3\, d\theta_3 d\phi$, where the polar axis is taken along $\mathbf{p}_1+\mathbf{p}_2(=\mathbf{p}_3+\mathbf{p}_4)$ and the azimuthal angle of $\mathbf{p}_3$ is measured with respect to the plane containing $\mathbf{p}_1$ and $\mathbf{p}_2$. These angles are illustrated in Figure 1.2. We next replace the θ_3 integral by one over ε_4, at constant ε_3; by inspecting Figure 1.3 we see that

$$\left(\frac{\partial p_4}{\partial \theta_3}\right)_{p_3} = p_f \sin\theta_{34}, \tag{1.2.82}$$

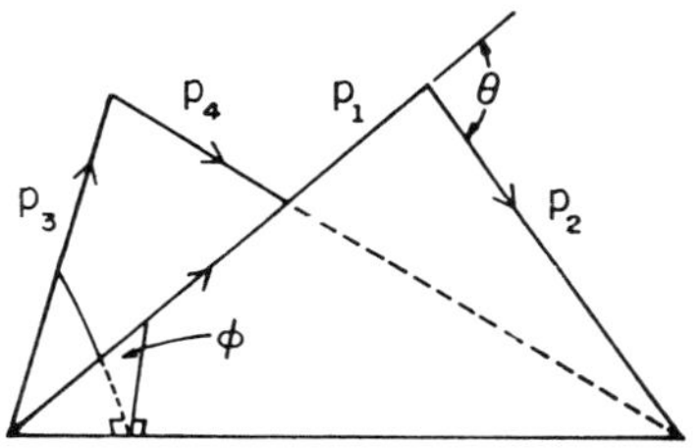

FIGURE 1.2. The relation of scattering angles θ and ϕ to the incident ($\mathbf{p}_1$ and $\mathbf{p}_2$) and final ($\mathbf{p}_3$ and $\mathbf{p}_4$) quasiparticle momenta.

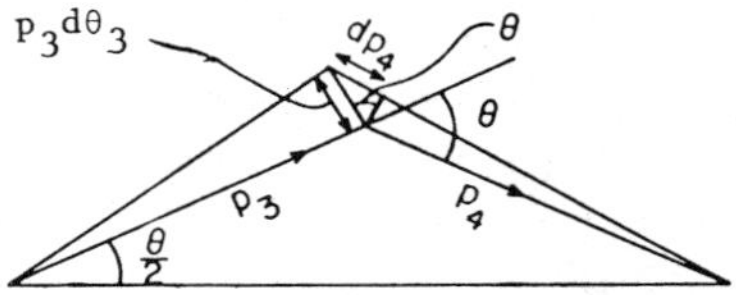

FIGURE 1.3. Diagram for calculating $(\partial\theta_3/\partial p_4)_{p_3}$. The angle between $\mathbf{p}_3$ and $\mathbf{p}_4$ may be put equal to θ, the angle between $\mathbf{p}_1$ and $\mathbf{p}_2$, since $\mathbf{p}_1$, $\mathbf{p}_2$, $\mathbf{p}_3$, and $\mathbf{p}_4$ are all very close to p_f.

where θ_{34} is the angle between $\mathbf{p}_3$ and $\mathbf{p}_4$. Since all four momenta are close to p_f, $\theta_{34} \simeq \theta$ in the low temperature limit, and we find

$$d\theta_3 = \frac{dp_4}{p_f \sin\theta} = \frac{m^*}{p_f^2}\frac{d\varepsilon_4}{\sin\theta}. \tag{1.2.83}$$

(To compute the finite temperature corrections to the transport coefficients, in Section 1.4, we shall have to take into account the slight deviations of the momenta from the Fermi surface.) From Figure 1.3 we see that $\theta_3 \simeq \theta/2$, so that finally

$$d^3p_2 d^3p_3 = \frac{m^{*3}}{2\cos(\theta/2)} d\varepsilon_2 d\varepsilon_3 d\varepsilon_4 \sin\theta\, d\theta\, d\phi\, d\phi_2. \tag{1.2.84}$$

Because of the $\delta(\varepsilon_1 + \varepsilon_2 - \varepsilon_3 - \varepsilon_4)$ and Fermi functions, the limits of the integrations over quasiparticle energies may be replaced by $\pm\infty$. Aside from the spin dependence, $W(12;34)$ depends only on the angles θ and ϕ. Note that interchange of $\mathbf{p}_3$ and $\mathbf{p}_4$ corresponds to the replacement of ϕ by $\phi + \pi$, leaving θ fixed.

The way to carry out the spin sum has been discussed in Section 1.2.4(a). In the absence of a magnetic field, $n_{\mathbf{p}\sigma}$ is independent of spin in a thermal conduction process, only the transition probability enters the spin sum, and the the result is

$$\sum_{\sigma_2\sigma_3\sigma_4}' W(12;34) = \frac{1}{2}W_{\uparrow\uparrow} + W_{\uparrow\downarrow} \equiv 2W(\theta, \phi). \tag{1.2.85}$$

$W(\theta, \phi)$ is the average scattering probability for σ_2 unpolarized with respect to σ_1. The transport equation therefore reduces to

$$\begin{aligned} n_1(1 - n_1)(\varepsilon_1 - \mu - Ts)\frac{\mathbf{v}_1\cdot\nabla T}{T} &= -\frac{m^{*3}}{(2\pi\hbar)^6}\int d\varepsilon_2 d\varepsilon_3 d\varepsilon_4 n_1 n_2(1 - n_3)(1 - n_4)\delta(\varepsilon_1 + \varepsilon_2 - \varepsilon_3 - \varepsilon_4) \\ &\times \int_0^\pi \sin\theta\, d\theta \int_0^{2\pi} d\phi \int_0^{2\pi} d\phi_2 \frac{W(\theta, \phi)}{\cos(\theta/2)}[\bar{\Phi}_1 + \bar{\Phi}_2 - \bar{\Phi}_3 - \bar{\Phi}_4]. \end{aligned} \tag{1.2.86}$$

The angular integration is in fact decoupled from the energy integrals in (1.2.86). To see this we note that since the driving term on the left is proportional to $\mathbf{v}_1 \cdot \nabla T$, and W depends only on the relative angles between the four momenta, the deviation of the distribution function, $\bar{\Phi}_i$, must be proportional to $\mathbf{v}_i \cdot \nabla T$ times a function of $\varepsilon_i - \mu$. From the addition theorem for spherical harmonics we can write the angular integrals over θ and ϕ in (1.2.86) as

$$\frac{1}{8\pi^2}\int_0^{2\pi} d\phi_2 \int d\Omega \frac{W(\theta, \phi)}{\cos(\theta/2)} \mathbf{v}_i \cdot \nabla T = \mathbf{v}_1 \cdot \nabla T \int \frac{d\Omega}{4\pi} \frac{W(\theta, \phi)}{\cos(\theta/2)} \hat{\mathbf{v}}_1 \cdot \hat{\mathbf{v}}_i$$
$$\equiv \langle W \rangle w_i \mathbf{v}_1 \cdot \nabla T, \tag{1.2.87}$$

where $d\Omega = \sin\theta \, d\theta \, d\phi$ and

$$\langle W \rangle \equiv \int \frac{d\Omega}{4\pi} \frac{W(\theta, \phi)}{\cos(\theta/2)}. \tag{1.2.88}$$

The resulting kinetic equation can be made dimensionless by writing

$$\bar{\Phi}_i = -\mathbf{v}_i \cdot \nabla(\kappa T) \tau \Psi(x_i) \tag{1.2.89}$$

where the dimensionless variable x_i is defined by

$$x_i \equiv \frac{\varepsilon_i - \mu}{\kappa T}, \tag{1.2.90}$$

and

$$\tau \equiv \frac{8\pi^4 \hbar^6}{m^{*3} \langle W \rangle (\kappa T)^2} \tag{1.2.91}$$

is a characteristic relaxation time. Note that τ behaves as T^{-2}. The kinetic equation, written in terms of $\Psi(x)$, is then

$$n_1(1 - n_1)(x_1 - s/\kappa) = \int dx_2 dx_3 dx_4 n_1 n_2 (1 - n_3)(1 - n_4)\delta(x_1 + x_2 - x_3 - x_4)$$
$$\times [\Psi(x_1) + w_2 \Psi(x_2) - w_3 \Psi(x_3) - w_4 \Psi(x_4)]. \tag{1.2.92}$$

Here $n_i = (e^{x_i} + 1)^{-1}$. At this point we can let $T \to 0$ in this equation; the only explicit occurrences of T are in the Ts/κ term and in the finite temperature correction to μ; both terms are of order T^2, and may be neglected as $T \to 0$. The left side of the resultant equation is odd in x_1. If we replace x_1 by $-x_1$, let $x_2 \to -x_2$, $x_3 \to -x_3$, $x_4 \to -x_4$, and observe that with the delta function $n_1 n_2 (1 - n_3)(1 - n_4) = (1 - n_1)(1 - n_2) n_3 n_4$, we see that $-\Psi(-x)$ obeys the same equation as $\Psi(x)$, and hence $\Psi(x)$ must be odd in x. We now replace the integration variables x_3 and x_4 in (1.2.92) by $-x_3$ and $-x_4$, and use the resultant symmetry in x_2, x_3, and x_4 to write

$$n_1(1 - n_1)x_1 = \int_{-\infty}^{\infty} dx_2 \int_{-\infty}^{\infty} dx_3 \int_{-\infty}^{\infty} dx_4 n_1 n_2 n_3 n_4$$
$$\times\ \delta(x_1 + x_2 + x_3 + x_4)[\Psi(x_1) + \lambda_K \Psi(x_2)], \tag{1.2.93}$$

where

$$\lambda_K \equiv w_2 + w_3 + w_4 = \frac{1}{\langle W \rangle} \int \frac{d\Omega}{4\pi} \frac{W(\theta, \phi)}{\cos(\theta/2)} (1 + 2\cos\theta). \qquad (1.2.94)$$

The integrals in (1.2.93) are readily evaluated in terms of the integral I_ν defined and evaluated in Appendix A; we find

$$\begin{aligned} n_1(1 - n_1)x_1 &= I_3(x_1)\Psi(x_1) + \lambda_K \int_{-\infty}^{\infty} dx_2 I_2(x_1 + x_2)\Psi(x_2) \\ &= \frac{x_1^2 + \pi^2}{2} n_1(1 - n_1)\Psi(x_1) \\ &\quad + \lambda_K n_1 \int_{-\infty}^{\infty} dx_2 \frac{(x_1 + x_2)n_2}{1 - e^{-(x_1+x_2)}} \Psi(x_2). \end{aligned} \qquad (1.2.95)$$

This equation can be brought to soluble form if we define the (odd) function ζ by

$$\Psi(x) = 2\cosh\left(\frac{x}{2}\right)\zeta(x); \qquad (1.2.96)$$

then (1.2.95) reduces to

$$\frac{x}{\cosh \frac{x}{2}} = (x^2 + \pi^2)\zeta(x) - \lambda_K \int_{-\infty}^{\infty} dx_2 \frac{(x - x_2)}{\sinh\left(\frac{x - x_2}{2}\right)} \zeta(x_2). \qquad (1.2.97)$$

This equation was first derived by Abrikosov and Khalatnikov [13] (though in somewhat different form).

It is useful, before solving (1.2.97), to go back and express the thermal conductivity in terms of $\zeta(x)$. From (1.2.76), (1.2.67), and (1.2.89) we have

$$\mathbf{j}_T = -2\frac{\tau}{T} \int \frac{d^3p}{(2\pi\hbar)^3} (\varepsilon - \mu - Ts)\mathbf{v}_\mathbf{p} n_\mathbf{p}^0(1 - n_\mathbf{p}^0)\mathbf{v}_\mathbf{p} \cdot \nabla T \Psi. \qquad (1.2.98)$$

The angular integration gives $v_f^2/3$, and at low T, the Ts term, as well as the finite temperature correction to μ, can be neglected. The integrand is nonzero only near the Fermi surface; using (1.2.96) we can then write $\mathbf{j}_T = -K\nabla T$, where

$$K = \frac{1}{2\pi^2} c_V v_f^2 \tau \int_{-\infty}^{\infty} dx \frac{x}{\cosh \frac{x}{2}} \zeta(x); \qquad (1.2.99)$$

here c_V is the heat capacity per unit volume (1.1.40).

The procedure for solving (1.2.97) is first to find the eigenfunctions $\phi_\nu(t)$ and eigenvalues $1/\alpha_\nu$ of the integral operator in (1.2.97); these obey the defining equation

$$\int_{-\infty}^{\infty} dx_2 \frac{x - x_2}{\sinh\left(\frac{x - x_2}{2}\right)} \phi_\nu^*(x_2) = \frac{1}{\alpha_\nu}(x^2 + \pi^2)\phi_\nu(x). \tag{1.2.100}$$

Because of the hermiticity of the kernel, the α_ν are real, and the eigenfunctions, with appropriate choice of normalization, obey the orthonormality condition

$$\int_{-\infty}^{\infty} dx(x^2 + \pi^2)\phi_\nu^*(x)\phi_\mu(x) = \delta_{\nu,\mu}. \tag{1.2.101}$$

The $\phi_\nu(x)$ are simply polynomials of order $\nu - 1$ in x, divided by 2cosh $(x/2)$. We then expand $\zeta(x)$ in terms of the ϕ_ν as

$$\zeta(x) = \sum_\nu c_\nu \phi_\nu(x). \tag{1.2.102}$$

Substituting (1.2.102) into (1.2.97) we have

$$\frac{x}{\cosh\frac{x}{2}} = (x^2 + \pi^2)\sum_\nu c_\nu\left(1 - \frac{\lambda_K}{\alpha_\nu}\right)\phi_\nu(x), \tag{1.2.103}$$

and using the orthogonality relation (1.2.101) we find

$$c_\nu = \frac{1}{1 - \lambda_K/\alpha_\nu}\int_{-\infty}^{\infty} dx \frac{x}{\cosh\frac{x}{2}} \phi_\nu^*(x). \tag{1.2.104}$$

This determines the solution $\zeta(x)$. The thermal conductivity, given by (1.2.99), is

$$K = \frac{1}{2\pi^2} c_V v_f^2 \tau \sum_\nu \frac{1}{1 - \lambda_K/\alpha_\nu}\left|\int_{-\infty}^{\infty} dx \frac{x}{\cosh\frac{x}{2}}\phi_\nu(x)\right|^2. \tag{1.2.105}$$

The problem now is to find the eigenfunctions ϕ_m; this is done by Fourier transforming (1.2.100). We define

$$\theta_\nu(k) = \int_{-\infty}^{\infty} dx\, e^{-ikx}\phi_\nu(x); \tag{1.2.106}$$

then from (1.2.100) we see that

$$F(k)\theta_\nu(\kappa) = \frac{1}{\alpha_\nu}\left(\pi^2 - \frac{\partial^2}{\partial k^2}\right)\theta_\nu(k), \tag{1.2.107}$$

where

$$F(k) = \int_{-\infty}^{\infty} dx \frac{x}{\sinh\frac{x}{2}} e^{-ikx} = i\frac{\partial}{\partial k}\int_{-\infty}^{\infty} dx \frac{e^{-ikx}}{\sinh\frac{x}{2}}. \tag{1.2.108}$$

The latter integral is closely related to the Fourier transform of the function $b(z)$ in Appendix A, so that from (A.5) we can conclude

$$F(k) = \frac{2\pi^2}{(\cosh \pi k)^2}. \tag{1.2.109}$$

The differential equation (1.2.107) can be brought to a standard form by introducing the variable $z = \tanh \pi k$ and letting

$$u_\nu(z) = \theta_\nu(k);$$

then

$$(1 - z^2)\frac{d^2u_\nu}{dz^2} - 2z\frac{du_\nu}{dz} + \left(2\alpha_\nu - \frac{1}{1 - z^2}\right)u_\nu = 0. \tag{1.2.110}$$

The solution of this equation is the associated Legendre polynomial [23] $P_\nu^1(z)$ with eigenvalue

$$2\alpha_\nu = \nu(\nu + 1), \qquad \nu = 1,2,3 \ldots . \tag{1.2.111}$$

Thus

$$\theta_\nu(k) = \frac{\sqrt{\nu + \frac{1}{2}}}{\nu(\nu + 1)} P_\nu^1(\tanh \pi k); \tag{1.2.112}$$

the coefficient is chosen so that the ϕ_ν obey the normalization condition (1.2.101). [The P_ν^1 satisfy $\int_{-1}^{1} dx[P_\nu^1(x)]^2 = 2\nu(\nu + 1)/(2\nu + 1)$.]

The integral in (1.2.105) for the thermal conductivity can be calculated directly in terms of the $\theta_\nu(k)$; the details are given in Appendix A. Using (A.16), noting that only even ν contribute, we find that the thermal conductivity is given by

$$K = \tfrac{1}{3} c_V v_f^2 \tau_K \tag{1.2.113a}$$

where

$$\frac{\tau_K}{\tau} = \frac{6}{\pi^2} \sum_{\nu=2,4,6\ldots} \frac{(2\nu + 1)}{\nu(\nu + 1)[\nu(\nu + 1) - 2\lambda_K]} \tag{1.2.113b}$$

$$= \frac{12 - \pi^2}{2\pi^2} + \frac{12}{\pi^2}\lambda_K \sum_{\nu=2,4,6\ldots} \frac{2\nu + 1}{\nu^2(\nu + 1)^2[\nu(\nu + 1) - 2\lambda_K]}. \tag{1.2.113c}$$

Given λ_K and τ, the thermal conductivity is easily evaluated numerically. Since $c_V \sim T$ and $\tau \sim T^{-2}$, K varies as T^{-1}.

The form (1.2.113c) is more useful for numerical calculations because the series converges rapidly. It may easily be obtained from (1.2.113b) by adding and subtracting the value of τ_K for $\lambda_K = 0$, and using the fact that $\sum_{\nu=2,4,6\ldots}(2\nu + 1)/\nu^2(\nu + 1)^2 = (1 - \pi^2/12)$. The result may also be derived more directly, by the method Højgaard Jensen, Smith, and Wilkins used in their calculations [21]. To solve (1.2.97) they cast it in the form of an equation for the difference between $\zeta(x)$ and $x/[(x^2+\pi^2)\cosh(x/2)]$, the value of $\zeta(x)$ for $\lambda_K = 0$, and then expanded the difference in terms of the eigenfunctions (1.2.100). This leads directly to (1.2.113c). Note that the parameters α used in Reference [21] are just 2λ.

As we see from (1.2.94) the parameter λ_K falls within the range $-1 \leq \lambda_K \leq 3$. The case $\lambda_K = 3$ occurs when the only scattering is for $\cos\theta = 0$, that is, between particles of parallel momenta. In this case the collisions conserve the heat current; this is reflected in the divergence of K for $\lambda_K = 3$. It is interesting to note that the usual variational calculation [17] of K yields simply the $\nu = 2$ term in the sum (1.2.113b).

$$(\tau_K)_{\text{var}} = \frac{5}{2\pi^2}\frac{\tau}{3-\lambda_K}. \tag{1.2.114}$$

This is because the trial function in the variational calculation is just $\Psi(x) \sim x$, or $\zeta(x) \sim x/\cosh(x/2)$, which is just the $\nu = 2$ eigenfunction of the collision operator. The original Abrikosov-Khalatnikov calculation of K gave the result $K_{AK} = 12K_{\text{var}}/5$. The variational solution becomes exact in the limit $\lambda_K \to 3$, while for $\lambda_K = 0$, (1.2.95) is trivial to solve for Ψ. These facts were used by Emery and Cheng [19] to construct an interpolative solution for Ψ, of the form

$$\Psi(x) = \frac{2x}{x^2+\pi^2} + \frac{\lambda_K}{3-\lambda_K}\frac{15}{2\pi^2}x; \tag{1.2.115}$$

this approximate form agrees with the exact result when $\lambda_K = 0$ and when $\lambda_K \to 3$. The result for the thermal conductivity time in Emery and Cheng's approximation is

$$(\tau_K)_{EC} = \tau\left(1 - \frac{\pi^2}{12} + \frac{5}{6\pi^2}\frac{\lambda_K}{3-\lambda_K}\right), \tag{1.2.116}$$

which is in fact the same as the exact result (1.2.113c), but with only the first term in the sum retained. For the values of λ_K encountered in most calculations, K_{EC} differs from the exact result by no more than a few percent.

The exact result for the thermal conductivity lies between the variational result and the Abrikosov–Khalatnikov value. As λ_K varies between -1 and $+3$, K/K_{var} varies from 1.35 to 1.

Let us summarize the general method of solution of the transport equation, since we employ essentially the same arguments in the following sections to calculate the other transport coefficients. We may quite generally expand $\delta\bar{n}_{\mathbf{p}}$, near the Fermi surface, in terms of the eigenfunctions $\phi_\nu(x)$ and spherical harmonics $Y_{lm}(\Omega_{\mathbf{p}})$, where $\Omega_{\mathbf{p}}$ denotes the spherical angles of $\mathbf{p}$; we write

$$\delta\bar{n}_{\mathbf{p}} = \sum_{lm\nu} \delta n_{lm\nu} Y_{lm}(\Omega_{\mathbf{p}}) \frac{\phi_\nu(x)}{2\cosh\dfrac{x}{2}}. \tag{1.2.117}$$

(More generally the δn also carry a spin index.) Comparing with (1.2.67), (1.2.89), (1.2.96), and (1.2.102), we see that in the case of thermal conductivity only $\delta n_{10\nu}$ is nonzero, assuming ∇T to be in the z direction, and

$$\delta n_{10\nu} Y_{10} = -\mathbf{v_p} \cdot \nabla T \frac{\tau}{T} c_\nu. \tag{1.2.118}$$

The effect of the collision operator acting on $\delta \bar{n}_{\mathbf{p}}$ [see (1.2.86)] can then be generally expressed in terms of the $\delta n_{lm\nu}$ and the eigenvalues $1/\alpha_\nu$ (1.2.100) as

$$I[n_{\mathbf{p}'}] = -\frac{x^2 + \pi^2}{2\tau} \sum_{lm\nu} \left[\delta n_{lm\nu} Y_{lm}(\Omega_{\mathbf{p}}) \frac{\phi_\nu(x)}{2\cosh \frac{x}{2}} \right] \left(1 - \frac{\lambda_{l\nu}}{\alpha_\nu}\right), \tag{1.2.119}$$

where

$$\lambda_{l\nu} = (-1)^\nu w_{2l} + w_{3l} + w_{4l}, \tag{1.2.120}$$

and

$$w_{il} = \frac{1}{\langle W \rangle} \int \frac{d\Omega}{4\pi} \frac{W(\theta, \phi)}{\cos(\theta/2)} P_l(\hat{v}_1 \cdot \hat{v}_i). \tag{1.2.121}$$

Note that w_i introduced earlier (1.2.87) is just w_{il}. The factor $(-1)^\nu$ in (1.2.120) results from the fact that the $\phi_\nu(x)$ have parity $(-1)^{\nu+1}$; in the case of thermal conductivity only the odd ϕ_ν(even ν) enter. The kinetic equation is then solved by using the orthornormality relations for the ϕ_ν (1.2.101) and Y_{lm} to determine the $\delta n_{lm\nu}$ in terms of the left side of the kinetic equation. The general eigenfunction expansion of the kinetic equation is discussed in considerable detail by Sykes and Brooker [22].

(c) Viscosity. To calculate the viscosity of a Fermi liquid we assume the system to be approximately in local equilibrium with a small spatially varying local fluid velocity $\mathbf{u}(\mathbf{r})$. Then the stress tensor Π_{ij} assumes the form

$$\Pi_{ij} = \Pi_{ij}^{\text{l.e.}} - \sigma_{ij}, \tag{1.2.122}$$

where the dissipative part is

$$\sigma_{ij} = \eta(\nabla_j u_i + \nabla_i u_j - \tfrac{2}{3}\delta_{ij}\nabla \cdot \mathbf{u}) + \zeta \delta_{ij} \nabla \cdot \mathbf{u}, \tag{1.2.123}$$

and $\Pi_{ij}^{\text{l.e.}} = P\delta_{ij}$, where P is the local pressure. The coefficients η and ζ are the first and second viscosities. As we shall see, in the low temperature limit $\zeta/\eta \sim T^4$, and only the first viscosity η is important.

A small deviation $\delta n_{\mathbf{p}}$ of the distribution function from global equilibrium, together with a small external potential U, produce a first order change in the stress tensor Π_{ij} [(1.2.36) and (1.2.33)] given by

$$\delta \Pi_{ij} = \sum_\sigma \int \frac{d^3p}{(2\pi\hbar)^3} p_i v_{\mathbf{p}j} \delta \bar{n}_{\mathbf{p}} - \delta_{ij} n U; \tag{1.2.124}$$

in deriving (1.2.124) it is necessary to do a partial integration in the $\delta \mathbf{v}_{\mathbf{p}}$ term. Let us expand $\delta n_{\mathbf{p}}$ about local equilibrium (1.2.54), as $\delta \bar{n}_{\mathbf{p}} = \delta n_{\mathbf{p}}^{\text{l.e.}}(\varepsilon_{\mathbf{p}}^{\text{l.e.}}) - \delta \varepsilon_{\mathbf{p}}^{\text{l.e.}}\, \partial n_p^0/\partial \varepsilon_p + \delta \bar{n}_{\mathbf{p}}^{\text{l.e.}}$, where $\bar{n}_{\mathbf{p}}^{\text{l.e.}}$ is given by (1.2.63), and substitute this in (1.2.124). The local equilibrium terms simply produce the variation of the pressure, while the deviation $\delta \bar{n}_{\mathbf{p}}^{\text{l.e.}}$ from local equilibrium produces the dissipative terms in Π_{ij}. Comparing with (1.2.122) we have

$$\sigma_{ij} = -\sum_{\sigma} \int \frac{d^3p}{(2\pi\hbar)^3}\, p_i v_{\mathbf{p}_j} \delta \bar{n}_{\mathbf{p}}^{\text{l.e.}}. \tag{1.2.125}$$

We calculate the first viscosity by assuming that $\mathbf{u}$ is in the x direction and varies only in the y direction. Then

$$\sigma_{xy} = \eta \partial u_x/\partial y = -\sum_{\sigma} \int \frac{d^3p}{(2\pi\hbar)^3}\, p_x (v_{\mathbf{p}})_y \delta \bar{n}_{\mathbf{p}}^{\text{l.e.}} \tag{1.2.126}$$

Clearly only terms with $l = 2$ in $\delta \bar{n}_{\mathbf{p}}^{\text{l.e.}}$ in (1.2.126) contribute to η. To find these terms we again assume a steady state and using the local equilibrium distribution in the left side of the kinetic equation (1.2.12) we find

$$-\frac{\partial n_p^0}{\partial \varepsilon_p} \mathbf{v}_{\mathbf{p}} \cdot \left[\frac{\nabla T}{T}(\varepsilon - \mu) + \nabla \mu - \nabla(\mathbf{p}\cdot\mathbf{u})\right] = I[n'_{\mathbf{p}}]. \tag{1.2.127}$$

The term with $l = 2$ symmetry on the left side is

$$\frac{\partial n_p^0}{\partial \varepsilon_p} (v_{\mathbf{p}})_y p_x \frac{\partial u_x}{\partial y}.$$

Multiplying both sides of the transport equation by $\phi_\nu^*(x)/(2\cosh(x/2))$ and integrating over x (1.2.90) we see that

$$\sum_m \delta n_{2m\nu} Y_{2m} = -\frac{\tau}{\kappa T}\left(\int \frac{\phi_\nu^*}{\cosh \frac{x}{2}}\, dx\right) \frac{(v_{\mathbf{p}})_y p_x \partial u_x/\partial y}{1 - \lambda_{2\nu}/\alpha_\nu}. \tag{1.2.128}$$

On substituting (1.2.128) into (1.2.117) and computing σ_{xy} in (1.2.126), we arrive at the exact result for the low temperature first viscosity:

$$\eta = \tfrac{1}{5} n p_f v_f \tau_\eta \tag{1.2.129a}$$

where

$$\frac{\tau_\eta}{\tau} = \frac{2}{\pi^2} \sum_{\nu=1,3,5\ldots} \frac{2\nu + 1}{\nu(\nu + 1)[\nu(\nu + 1) - 2\lambda_\eta]} \tag{1.2.129b}$$

$$= \frac{1}{6} + \frac{4}{\pi^2}\lambda_\eta \sum_{\nu=1,3,5\ldots} \frac{2\nu + 1}{\nu^2(\nu + 1)^2[\nu(\nu + 1) - 2\lambda_\eta]}. \tag{1.2.129c}$$

Equation 1.2.129c is again the form that Højgaard Jensen, Smith, and Wilkins [21] derive directly by expanding the difference between $\zeta(x)$ and the solution for $\lambda_\eta = 0$ in terms of eigenfunctions (cf. (1.2.113c) for K). In deriving (1.2.129) we have replaced the explicit factors $(v_{\mathbf{p}})_x^2 p_y^2$ by their values on the Fermi surface; their angular average gives a factor $v_f^2 p_f^2/15$. The integral over ϕ_ν is given by (A.15), and

$$\lambda_\eta = \lambda_{2\nu} = \frac{1}{\langle W \rangle} \int \frac{d\Omega}{4\pi} \frac{W(\theta, \phi)}{\cos(\theta/2)} \left(1 - 3 \sin^4 \frac{\theta}{2} \sin^2\phi\right). \qquad (1.2.130)$$

Note that

$$-2 \leqslant \lambda_\eta \leqslant 1; \qquad (1.2.131)$$

the case $\lambda_\eta = 1$, corresponding to scattering only between particles of parallel momenta, gives rise to a singularity in η, since such collisions do not alter the momentum flux.

The variational solution for the viscosity, which is given by just the $\nu = 1$ term in (1.2.129a), is $\frac{3}{4}$ of the Abrikosov–Khalatnikov value. For λ_η in the allowed range between -2 and $+1$, η/η_{var} lies between 1.23 and 1. Emery and Cheng's approximate solution is again given by taking only the first term in the exact solution (1.2.129c); its explicit form is

$$(\tau_\eta)_{EC} = \frac{2}{\pi^2} \tau \left[\frac{\pi^2}{12} + \frac{3}{4} \frac{\lambda_\eta}{(1 - \lambda_\eta)}\right], \qquad (1.2.132)$$

and it lies within a few percent of the exact result for the values of λ_η of physical interest.

We shall not calculate the second viscosity ζ in detail, but only indicate how the calculation is formulated. The complete calculation of ζ is given in [22]. From (1.2.123) we see that ζ is found from

$$\frac{1}{3} \sum_i \sigma_{ii} = \zeta \, \nabla \cdot \mathbf{u}. \qquad (1.2.133)$$

In order to calculate ζ then, we must assume that the liquid has a local velocity obeying $\nabla \cdot \mathbf{u} \neq 0$. But from the linearized continuity equation for a local equilibrium distribution:

$$\frac{\partial n}{\partial t} + n \, \nabla \cdot \mathbf{u} = 0, \qquad (1.2.134)$$

we see that $\partial n/\partial t \neq 0$. We must therefore, in solving the kinetic equation, retain the term $\partial n_{\mathbf{p}}^{\text{l.e.}}/\partial t$.

The calculation of $\sum_i \sigma_{ii}$ requires only the $l = 0$ terms in the kinetic equation. The $l = 0$ term in $\partial n_{\mathbf{p}}^{\text{l.e.}}/\partial t$ can be calculated if we regard n as the independent variable and write to first order

$$\frac{\partial n_{\mathbf{p}}^{\text{l.e.}}}{\partial t} = \frac{\partial n_p^0}{\partial \varepsilon_p} T \left[\frac{d}{dn}\left(\frac{\varepsilon_p - \mu}{T}\right)\right] \frac{\partial n}{\partial t}. \qquad (1.2.135)$$

[The $\mathbf{p}\cdot\mathbf{u}$ term contributes only to $l = 1$.] From the continuity equation, this term is proportional to $\nabla\cdot\mathbf{u}$. Similarly, the $l = 0$ component in the driving term on the left side of (1.2.127) is $(\partial n_p^0/\partial\varepsilon_p)\,(\mathbf{v_p}\cdot\mathbf{p})\,\nabla\cdot\mathbf{u}/3$, and the complete $l = 0$ component on the left side of the time dependent kinetic equation becomes

$$\frac{\partial n_p^0}{\partial\varepsilon_p}\left[-nT\frac{d}{dn}\left(\frac{\varepsilon_p-\mu}{T}\right)+\frac{1}{3}v_p p\right]\nabla\cdot\mathbf{u}. \tag{1.2.136}$$

Sykes and Brooker [22] show that the terms in the square bracket that are even in x are of order T^2 while those that are odd in x are of order T^3. One would expect that the terms even in x would lead to terms of order $T^2\tau$ in ζ; however, because of condition (1.2.57), these terms vanish, and the first contribution is of order $T^4\tau$. One would also expect that the terms odd in x would lead to terms of order $T^4\tau$; again, the condition that

$$\delta\varepsilon-\mu\delta n=\sum_\sigma\int\frac{d^3p}{(2\pi\hbar)^3}(\varepsilon_p-\mu)\,\delta n_{\mathbf{p}\sigma}^{\text{l.e.}} \tag{1.2.137}$$

causes these terms to vanish. The entire contribution comes then from odd ν eigenfunctions. Since for ν odd $\lambda_{l=0,\nu} = 1$ the sums over ν are straightforward, and Sykes and Brooker's result for ζ can be written as

$$\zeta=\frac{\pi^2}{18}n\tau p_f v_f\left(\frac{T}{T_f}\right)^4\left[\frac{p_f^2}{2m^*}\frac{\partial^2}{\partial\varepsilon_p^2}\left(\frac{pv_p}{3}+n\int\frac{d\Omega_{\mathbf{p}}}{4\pi}f^s_{\mathbf{pp}'}\right)_{p=p'=p_f}\right]^2. \tag{1.2.138}$$

It is interesting to note that ζ vanishes, to this order in T, for a free Fermi gas, since there $pv_p = 2\varepsilon_p$. The reason is that the second viscosity is determined by the extent to which the system, when uniformly compressed at a steady rate, is driven out of thermodynamic equilibrium. A uniform compression by a linear scale factor θ (so that $V\to\theta^{-3}V$) is equivalent to increasing all momenta by θ; as long as ε_p is a homogeneous function of p (of order n, say), any equilibrium distribution is transformed into a new equilibrium distribution with new temperature $T\theta^n$.

(d) Spin Diffusion Coefficient. In discussing the spin transport properties we first consider the situation in which an external magnetic field $\mathscr{H}_z(\mathbf{r},t)$ is applied only in the z direction (which we take to be the spin quantization axis). If we multiply both sides of (1.2.12a) by σ, sum over $\mathbf{p}$ and σ, and use the conservation of spin in quasiparticle collisions, we find the spin density continuity equation:

$$\frac{\partial\sigma_z(\mathbf{r},t)}{\partial t}+\nabla\cdot\mathbf{j}_{\sigma_z}(\mathbf{r},t)=0, \tag{1.2.139}$$

where the density of the z component of the spin is given by

$$\sigma_z(\mathbf{r}, t) = \sum_\sigma \int \frac{d^3p}{(2\pi\hbar)^3} \sigma n_{\mathbf{p}\sigma} \tag{1.2.140}$$

and the current of the z component of spin is

$$\mathbf{j}_{\sigma_z}(\mathbf{r}, t) = \sum_\sigma \int \frac{d^3p}{(2\pi\hbar)^3} \sigma \mathbf{v}_{\mathbf{p}\sigma} n_{\mathbf{p}\sigma}. \tag{1.2.141}$$

Because a single spin component of a Fermi liquid does not form a Galilean invariant system, there is no alternate expression for $\mathbf{j}_{\sigma_z}$ analogous to (1.2.28).

The linearized spin current is [cf. (1.2.26)]

$$\mathbf{j}_{\sigma_z}(\mathbf{r}, t) = \sum_\sigma \int \frac{d^3p}{(2\pi\hbar)^3} \sigma \mathbf{v}_{\mathbf{p}\sigma} \delta\bar{n}_{\mathbf{p}\sigma}. \tag{1.2.142}$$

If we assume that $\delta n_{\mathbf{p}\sigma}$ is nonzero only near the Fermi surface, then [cf. (1.2.27)]

$$\mathbf{j}_{\sigma_z}(\mathbf{r}, t) = \left(1 + \frac{F_1^a}{3}\right) \sum_\sigma \int \frac{d^3p}{(2\pi\hbar)^3} \sigma \mathbf{v}_\mathbf{p}^0 \, \delta n_{\mathbf{p}\sigma}(\mathbf{r}, t). \tag{1.2.143}$$

Because spin current is not conserved, due to collisions between quasiparticles of opposite spin, $\mathbf{j}_{\sigma_z}$ in the presence of a weak slowly varying $\mathscr{H}_z(x, y, t)$ (note $\nabla \cdot \mathscr{H} = 0$) is of the Fick form

$$\mathbf{j}_{\sigma_z}(\mathbf{r}, t) = -D_\sigma \nabla(\sigma_z(\mathbf{r}, t) - \sigma_z^0(\mathbf{r}, t)), \tag{1.2.144}$$

where σ_z^0 is the equilibrium spin density produced by the field $\mathscr{H}_z$, and D_σ is the spin diffusion coefficient.

We calculate D_σ by the now familiar technique; we take a static field $\mathscr{H}_z(x)$ and use the local equilibrium distribution function

$$n_{\mathbf{p}\sigma}^{\text{l.e.}} = \left\{\exp\left[\frac{\varepsilon_{\mathbf{p}\sigma} - \mu_\sigma(\mathbf{r}, t)}{\kappa T}\right] + 1\right\}^{-1} \tag{1.2.145}$$

in the left side of the kinetic equation (1.2.18). The $\varepsilon_{\mathbf{p}\sigma}$ includes the explicit coupling, $-(\gamma\hbar/2)\sigma\mathscr{H}_z$, to the magnetic field. The chemical potentials μ_σ, different for spin-up and spin-down particles, are chosen so that

$$n_\sigma = \int \frac{d^3p}{(2\pi\hbar)^3} n_{\mathbf{p}\sigma}^{\text{l.e.}}. \tag{1.2.146}$$

Then the left side of the kinetic equation becomes $-(\partial n_{pl}^0/\partial \varepsilon_p)\mathbf{v}_\mathbf{p} \cdot \nabla \mu_\sigma(\mathbf{r}, t)$. The spin sums in the collision term may be carried out according to (1.2.68) and (1.2.69); the result is

$$\sum_{\sigma_2\sigma_3\sigma_4}' W(\bar{\Phi}_1 + \bar{\Phi}_2 - \bar{\Phi}_3 - \bar{\Phi}_4) = \frac{1}{2} W_{\uparrow\uparrow}(\bar{\Phi}_{1\sigma} + \bar{\Phi}_{2\sigma} - \bar{\Phi}_{3\sigma} - \bar{\Phi}_{4\sigma}) + W_{\uparrow\downarrow}(\bar{\Phi}_{1\sigma} + \bar{\Phi}_{2,-\sigma} - \bar{\Phi}_{3\sigma} - \bar{\Phi}_{4,-\sigma}), \tag{1.2.147}$$

where σ is the direction of spin 1, and $-\sigma$ denotes the direction opposite to σ. The linearized spin current for a paramagnetic liquid involves only the difference between the up-spin and down-spin distribution functions. We write

$$\begin{aligned}\delta\bar{n}_i^a &= \delta\bar{n}_{i\uparrow} - \delta\bar{n}_{i\downarrow},\\ \bar{\Phi}_i^a &= \bar{\Phi}_{i\uparrow} - \bar{\Phi}_{i\downarrow},\end{aligned} \tag{1.2.148}$$

and take the difference of the kinetic equations for the two spin orientations, to find

$$\begin{aligned}-\frac{\partial n_p^0}{\partial \varepsilon_p}\mathbf{v}_\mathbf{p}\cdot\nabla(\mu_\uparrow - \mu_\downarrow) &= I[n_{\mathbf{p}'}]\\ &= -\frac{1}{V^2}\sum_{\mathbf{p}_2\mathbf{p}_3\mathbf{p}_4} n_1 n_2(1-n_3)(1-n_4)\delta_{\mathbf{p}_1+\mathbf{p}_2,\,\mathbf{p}_3+\mathbf{p}_4}\delta(\varepsilon_1+\varepsilon_2-\varepsilon_3-\varepsilon_4)\\ &\times\left[\frac{1}{2}W_{\uparrow\uparrow}(\bar{\Phi}_1^a+\bar{\Phi}_2^a-\bar{\Phi}_3^a-\bar{\Phi}_4^a)+W_{\uparrow\downarrow}(\bar{\Phi}_1^a-\bar{\Phi}_2^a-\bar{\Phi}_3^a+\bar{\Phi}_4^a)\right].\end{aligned} \tag{1.2.149}$$

The equation is solved by the method outlined at the end of Section 1.2.4(b). The left side has simply a $\cos\theta_p$ angular dependence, and is even in x; therefore, we can expand $\delta\bar{n}_p^a$ as

$$\delta\bar{n}_\mathbf{p}^a = -\mathbf{v}_\mathbf{p}\cdot\nabla(\mu_\uparrow - \mu_\downarrow)\frac{\tau}{T}\sum_{\nu_\text{odd}}\frac{d_\nu\phi_\nu(x)}{2\cosh\frac{x}{2}}. \tag{1.2.150}$$

Substituting this form in the right side of (1.2.149) we find, as in (1.2.119),

$$I[n_{\mathbf{p}'}] = \frac{x^2+\pi^2}{2}\,\frac{\mathbf{v}_\mathbf{p}\cdot\nabla(\mu_\uparrow-\mu_\downarrow)}{T}\sum_{\nu_\text{odd}}\frac{d_\nu\phi_\nu(x)}{2\cosh\frac{x}{2}}\left(1-\frac{\lambda_D}{\alpha_\nu}\right), \tag{1.2.151}$$

where

$$\begin{aligned}\lambda_D &= \frac{1}{\langle W\rangle}\int\frac{d\Omega}{4\pi}\left[\frac{1}{2}W_{\uparrow\uparrow}(-\hat{\mathbf{v}}_1\cdot\hat{\mathbf{v}}_2+\hat{\mathbf{v}}_1\cdot\hat{\mathbf{v}}_3+\hat{\mathbf{v}}_1\cdot\hat{\mathbf{v}}_4)\right.\\ &\quad\left.+W_{\uparrow\downarrow}(\hat{\mathbf{v}}_1\cdot\hat{\mathbf{v}}_2+\hat{\mathbf{v}}_1\cdot\hat{\mathbf{v}}_3-\hat{\mathbf{v}}_1\cdot\hat{\mathbf{v}}_4)\right]\Big/(2\cos(\theta/2))\end{aligned} \tag{1.2.152}$$

$$= 1-\frac{1}{\langle W\rangle}\int\frac{d\Omega}{4\pi}\,\frac{W_{\uparrow\downarrow}(1-\cos\theta)(1-\cos\phi)}{2\cos(\theta/2)} \tag{1.2.153}$$

We now multiply both sides of the kinetic equation by cosh x/2, and use the orthogonality relation (1.2.101) to determine d_ν:

$$d_\nu = \frac{1}{1-\lambda_D/\alpha_\nu}\int_{-\infty}^{\infty}\frac{dx\,\phi_\nu(x)}{\cosh x/2}. \tag{1.2.154}$$

The spin current is then given by

$$\mathbf{j}_{\sigma_z} = -\frac{1}{12}N(0)v_f^2\tau\nabla(\mu_\uparrow - \mu_\downarrow)\sum_\nu \frac{1}{1-\lambda_D/\alpha_\nu}\left(\int_{-\infty}^{\infty}\frac{dx\phi_\nu(x)}{\cosh x/2}\right)^2. \quad (1.2.155)$$

To determine the spin diffusion coefficient, we must relate changes in σ_z, the local spin density, to changes in $\mu_\uparrow - \mu_\downarrow$. This calculation is carried out in a fashion analogous to that of Section 1.1.3(b) for $\partial n/\partial \mu$, and the result is [cf. (1.1.48) and (1.1.52)]

$$d\sigma_z = \frac{N(0)}{2(1+F_0^a)}d[\mu_\uparrow - \mu_\downarrow + \hbar\gamma\mathscr{H}_z]. \quad (1.2.156)$$

In the presence of a weak magnetic field the equilibrium spin density is given by

$$\sigma_z^0 = \frac{\hbar\gamma N(0)}{2(1+F_0^a)}\mathscr{H}_z \quad (1.2.157)$$

and we have

$$N(0)d(\mu_\uparrow - \mu_\downarrow) = 2(1+F_0^a)d(\sigma_z - \sigma_z^0). \quad (1.2.158)$$

Combining (1.2.158), (1.2.155), and (1.2.144), and using (A.15) we find finally that the spin diffusion coefficient is given by

$$D_\sigma = \tfrac{1}{3}v_f^2(1+F_0^a)\tau_D \quad (1.2.159a)$$

where

$$\frac{\tau_D}{\tau} = \frac{2}{\pi^2}\sum_{\nu_{\text{odd}}}\frac{2\nu+1}{\nu(\nu+1)[\nu(\nu+1)-2\lambda_D]} \quad (1.2.159b)$$

$$= \frac{1}{6} + \frac{4\lambda_D}{\pi^2}\sum_{\nu_{\text{odd}}}\frac{2\nu+1}{\nu^2(\nu+1)^2[\nu(\nu+1)-2\lambda_D]}. \quad (1.2.159c)$$

The Emery and Cheng approximation is obtained by taking only the first term in (1.2.159c). The variational solution [17] for τ_D is given by the $\nu = 1$ term in (1.2.159b), and it depends on the transition probability only through the characteristic time $\tau(1-\lambda_D)^{-1}$; the net spin diffusion coefficient in that case depends only on $W_{\uparrow\downarrow}$. This is what one would expect in the simplest picture of spin diffusion in which only collisions between opposite spin quasiparticles can change the spin current. We see from (1.2.152) that

$$\langle W\rangle(1-\lambda_D) = \int\frac{d\Omega}{4\pi}W_{\uparrow\downarrow}\hat{\mathbf{v}}_1\cdot(\hat{\mathbf{v}}_1 - \hat{\mathbf{v}}_2 - \hat{\mathbf{v}}_3 + \hat{\mathbf{v}}_4). \quad (1.2.160)$$

In the up-down scattering, quasiparticles 1 and 3 have the same spin, opposite to that of 2 and 4; the factor in parentheses is thus proportional to the initial spin current, $\sim(\mathbf{v}_1 - \mathbf{v}_2)$. minus the final spin current, $\sim(\mathbf{v}_3 - \mathbf{v}_4)$. The exact solution for D_σ involves $W_{\uparrow\uparrow}$ as well as $W_{\uparrow\downarrow}$; collisions between parallel spin quasiparticles, while not directly affecting the spin current, do change the distribution function, and hence modify the way collisions between anti-parallel spin quasiparticles change the spin current.

One should note that λ_D lies between -3 and $+1$, and in this range $D_\sigma/D_{\sigma,\,\mathrm{var}}$ varies from 1.28 to 1. The original Hone solution for D_σ equals 4/3 of the variational result.

To complete the detailed calculations of the transport coefficients one further requires an expression for the scattering amplitude; we shall therefore defer the discussion of theoretical estimates for the transport coefficients in ^{3}He until we have considered approximations for the scattering amplitude in Section 1.4. We first discuss the collective effects in Fermi liquids.

1.3. COLLECTIVE EFFECTS

1.3.1. Sound in Fermi Liquids

(a) First Sound. At frequencies ω sufficiently small that $\omega\tau \ll 1$, where τ is a typical quasiparticle collision time, sound in Fermi liquids takes the form of ordinary hydrodynamic, or first sound. From the particle and momentum conservation laws, (1.2.23) and (1.2.31), together with the relation $\mathbf{g} = m\mathbf{j}$, we find the completely general linearized equation of motion for the density disturbance δn produced by a weak external field U:

$$m\frac{\partial^2 \delta n}{\partial t^2} - \nabla_i \nabla_j \delta \Pi_{ij} = n\nabla^2 U, \tag{1.3.1}$$

where $\delta\Pi_{ij}$ is given in terms of $\delta n_{\mathbf{p}}$ by (1.2.124). In the hydrodynamic limit $\delta\Pi_{ij} = \delta_{ij}\delta P - \sigma_{ij}$ where δP is the variation of the pressure, and σ_{ij} is given by (1.2.123). For $\omega\tau \ll 1$ the system is essentially in local thermodynamic equilibrium, and δP can be approximated by $n(\partial\mu/\partial n)_{T=0}\delta n = [(m^* v_f^2)/3](1 + F_0^s)\,\delta n$. [Cf. (1.1.48).] Finite temperature corrections to this yield terms of relative order T^2 in the first sound velocity and attenuation; also second viscosity can be neglected in σ_{ij}. If we use $n\nabla \cdot \mathbf{u} = -\partial\delta n/\partial t$, we then arrive at the equation for first sound at low temperatures

$$\left[m\frac{\partial^2}{\partial t^2} - m^*\frac{v_f^2}{3}(1 + F_0^s)\nabla^2 - \frac{4}{3}\frac{\eta}{n}\frac{\partial}{\partial t}\nabla^2\right]\delta n = n\nabla^2 U. \tag{1.3.2}$$

The first sound velocity is

$$c_1 = \frac{v_f}{\sqrt{3}}[(1 + F_1^s/3)(1 + F_0^s)]^{1/2}, \tag{1.3.3}$$

and the amplitude attenuation coefficient α, defined by the relation

$$q = \frac{\omega}{c_1} + i\alpha \tag{1.3.4}$$

between the frequency and complex wavevector q, is given by

$$\alpha = \frac{2\omega^2\eta}{3mnc_1^3} \propto \frac{\omega^2}{T^2}. \tag{1.3.5}$$

Inclusion of thermal conduction produces a correction to α of relative order T^2.

(b) Zero Sound. As the temperature is lowered, the mean quasiparticle scattering time increases as T^{-2} and, for fixed frequency, $\omega\tau$ increases. As $\omega\tau$ nears 1 the quasiparticles no longer have time to relax in one period of the sound; the liquid then no longer remains in local thermodynamic equilibrium, and the character of the sound propagation begins to change. When $\omega\tau \ll 1$ (the "hydrodynamic regime"), an element of the fluid that has an increased density and pressure pushes, through quasiparticle collisions, coherently on its neighboring elements. If there are no quasiparticle interactions, $f_{\mathbf{pp}'}$, then in the collisionless regime ($\omega\tau \gg 1$) the excess quasiparticles in a fluid element with increased density simply diffuse away, without driving the neighboring elements. However, as Landau observed, if there are quasiparticle interactions, then a local increase in the density of a fluid element can drive neighboring elements via the modification of the effective field $\Sigma_{\mathbf{p}'} f_{\mathbf{pp}'}\delta n_{\mathbf{p}'}(\mathbf{r}, t)$. Such restoring forces between neighboring elements can give rise to a sound-like collective mode of oscillation of the fluid, called *zero sound.* We turn now to the description of this mode.

We consider first the extreme collisionless regime, $\omega\tau \gg 1$, where the right side of the kinetic equation can be neglected. Then the linearized kinetic equation (1.2.18) becomes*

$$\left(\frac{\partial}{\partial t} + \mathbf{v}_{\mathbf{p}}\cdot\nabla\right)\delta n_{\mathbf{p}}(\mathbf{r}, t) - \left(\frac{\partial n_p^0}{\partial \varepsilon_p}\right)\mathbf{v}_{\mathbf{p}}\cdot\nabla\delta\varepsilon_{\mathbf{p}}(\mathbf{r}, t) = 0, \tag{1.3.6}$$

where $\delta\varepsilon_{\mathbf{p}}(\mathbf{r}, t)$, the effective field that drives the quasiparticles, is given by

$$\delta\varepsilon_{\mathbf{p}}(\mathbf{r}, t) = U(\mathbf{r}, t) + \sum_{\mathbf{p}'} f_{\mathbf{pp}'}\delta n_{\mathbf{p}'}(\mathbf{r}, t); \tag{1.3.7}$$

U is the external driving potential and the second term is the internal "molec-

*To be definite we use the "classical" kinetic equation (1.2.18), which is valid for wavenumbers q satisfying (1.2.1), $\hbar v_f\, q \ll \kappa T$. Otherwise, one must use the quantum kinetic equation (1.2.22). In the case that $\hbar q \ll p_f$, the quantum equation is the same as the classical equation with $\partial n_p^0/\partial\varepsilon_p$ replaced by $(n^0_{\mathbf{p}+\hbar q/2} - n^0_{\mathbf{p}-\hbar q/2})/(\varepsilon_{\mathbf{p}+\hbar q/2} - \varepsilon_{\mathbf{p}-\hbar q/2})$. Also the classical expression for the collision term is valid only if $\hbar\omega \ll \kappa T$.

ular" field. We may solve (1.3.6) by Fourier transforming; assuming that $U(\mathbf{r}, t) = Ue^{i(\mathbf{q}\cdot\mathbf{r}-\omega t)}$, we have $\delta n_{\mathbf{p}}(\mathbf{r}, t) = \delta n_{\mathbf{p}}(\mathbf{q}, \omega)e^{i(\mathbf{q}\cdot\mathbf{r}-\omega t)}$, and

$$(\omega - \mathbf{q}\cdot\mathbf{v}_{\mathbf{p}})\delta n_{\mathbf{p}} + \frac{\partial n_p^0}{\partial \varepsilon_p}\,\mathbf{q}\cdot\mathbf{v}_{\mathbf{p}}(U + \sum_{\mathbf{p}'} f_{\mathbf{pp}'}\delta n_{\mathbf{p}'}) = 0. \tag{1.3.8}$$

We assume ω to have an infinitesimal positive imaginary part corresponding to U being adiabatically turned on. If we write $\delta n_{\mathbf{p}}$ in the form

$$\delta n_{\mathbf{p}} \equiv -\frac{\partial n_p^0}{\partial \varepsilon_p}\nu_{\mathbf{p}}, \tag{1.3.9}$$

where $\nu_{\mathbf{p}}$ has the simple interpretation as the energy by which the zero sound wave shifts the quasiparticle distribution in the direction of $\hat{\mathbf{p}}$, then the $\nu_{\mathbf{p}}$ obey

$$\nu_{\mathbf{p}} + \frac{\mathbf{q}\cdot\mathbf{v}_{\mathbf{p}}}{\omega - \mathbf{q}\cdot\mathbf{v}_{\mathbf{p}}}\sum_{\mathbf{p}'} f_{\mathbf{pp}'}\frac{\partial n_{p'}^0}{\partial \varepsilon_{p'}}\nu_{\mathbf{p}'} = \frac{\mathbf{q}\cdot\mathbf{v}_{\mathbf{p}}}{\omega - \mathbf{q}\cdot\mathbf{v}_{\mathbf{p}}}U. \tag{1.3.10}$$

At low temperatures only $\mathbf{p}$ near the Fermi surface enter $\delta n_{\mathbf{p}}$ and we may assume that $\nu_{\mathbf{p}}$ is given by its value on the Fermi surface, and hence that it depends only on the direction of $\mathbf{p}$. We first consider the azimuthally symmetric case where $\nu_{\mathbf{p}}$ depends only on the angle θ between $\mathbf{p}$ and $\mathbf{q}$, and expand $\nu_{\mathbf{p}}$ in Legendre polynomials as

$$\nu_{\mathbf{p}} = \sum_l P_l(\cos\theta)\nu_l. \tag{1.3.11}$$

Then from the addition theorem for Legendre polynomials, we find that

$$\sum_{\mathbf{p}'} f_{\mathbf{pp}'}\frac{\partial n_{p'}^0}{\partial \varepsilon_{p'}}\nu_{\mathbf{p}'} = -\sum_l \frac{1}{2l+1}F_l^s P_l(\cos\theta_{\mathbf{p}})\nu_l, \tag{1.3.12}$$

and that the ν_l are given as the solutions of the algebraic equations:

$$\frac{\nu_l}{2l+1} + \sum_l \Omega_{ll'}(s)F_{l'}^s\frac{\nu_{l'}}{2l'+1} = -\Omega_{l0}(s)U, \tag{1.3.13}$$

where

$$s = \omega/qv_f, \tag{1.3.14}$$

and

$$\Omega_{ll'}(s) = \Omega_{l'l}(s) = \int_{-1}^{1}\frac{d\mu}{2}P_l(\mu)\frac{\mu}{\mu - s}P_{l'}(\mu). \tag{1.3.15}$$

In particular,

$$\begin{aligned}\Omega_{00}(s) &= 1 + \frac{s}{2}\ln\frac{s-1}{s+1} \\ &= 1 + \frac{s}{2}\ln\left|\frac{s-1}{s+1}\right| + \frac{i\pi}{2}s\Theta(1 - |s|)\end{aligned} \tag{1.3.16}$$

(the latter form valid for s having an infinitesimal positive imaginary part),

$$\Omega_{l1} = s\Omega_{l0} + \tfrac{1}{3}\,\delta_{l1}, \tag{1.3.17}$$

and

$$\Omega_{20} = \frac{1}{2} + \frac{3s^2 - 1}{2}\Omega_{00}, \qquad \Omega_{22} = \frac{1}{5} + \frac{3s^2 - 1}{2}\Omega_{20}. \tag{1.3.18}$$

For small positive $s - 1$

$$\Omega_{00}(s) \simeq 1 + \frac{1}{2}\ln\frac{s - 1}{2}, \tag{1.3.19}$$

while as $s \to \infty$

$$\Omega_{00}(s) = -\frac{1}{3s^2} - \frac{1}{5s^4} - \frac{1}{7s^6} - \dots, \qquad s \gg 1. \tag{1.3.20}$$

Explicit forms for the general response functions $\Omega_{ll'}$ are given in Appendix B.

The possible modes of oscillation of the system are given by the poles, as a function of s, of the response functions ν_l/U, which depend only on the variable s. If a pole occurs at $s = s_0$, there exists a mode with dispersion relation $\omega = s_0 v_f q$. Thus the mode will have a sound-like dispersion relations with velocity $s_0 v_f$. The residue of the response function ν_l/U at s_0 measures the extent to which the lth harmonic of the distribution function is excited in this mode by the scalar field U.

If we take the $l = 0$ moment of the kinetic equation (1.3.8) directly, we derive the number conservation law

$$s\nu_0 = \frac{\nu_1}{3}\left(1 + \frac{F_1^s}{3}\right). \tag{1.3.21}$$

The $l = 1$ moment of the kinetic equation yields the equation of motion for the fluid momentum

$$s\nu_1 - (1 + F_0^s)\nu_0 - \frac{2}{5}\left(1 + \frac{F_2^s}{5}\right)\nu_2 = U. \tag{1.3.22}$$

Equations 1.3.21 and 1.3.22 are valid, of course, even in the presence of quasiparticle collisions. Eliminating ν_1 from these two equations we find

$$(s^2 - c_1^2/v_f^2)\nu_0 - \frac{2}{15}\left(1 + \frac{F_1^s}{3}\right)\left(1 + \frac{F_2^s}{5}\right)\nu_2 = \left(1 + \frac{F_1^s}{3}\right)\frac{U}{3}. \tag{1.3.23}$$

This equation is the generalization of (1.3.2), applying equally well for all ranges of $\omega\tau$. In order to use this equation, as we shall shortly, to determine the dispersion relation of zero sound, it is necessary to compute ν_2, the quadrupolar distortion of the Fermi surface, in terms of ν_0, which describes the local density variation. This may be done in general in the collisionless regime by solving (1.3.13).

To see clearly the nature of zero sound, we consider first the situation where only F_0^s differs appreciably from zero. This case is so simple that it does not require the full machinery of (1.3.22) and (1.3.23). From (1.3.13) we find directly

$$\nu_0 = \frac{-\Omega_{00}U}{1 + F_0^s\Omega_{00}} \tag{1.3.24}$$

and

$$\nu_l = (2l + 1)\frac{\Omega_{l0}}{\Omega_{00}}\nu_0, \qquad l \geqslant 1. \tag{1.3.25}$$

The zero sound mode is given by the poles of the response function ν_0/U. For $0 \leqslant s < 1$, Ω_{00} is complex, while as s increases from 1 to ∞, Ω_{00} is real and increases monotonically from $-\infty$ to 0. Thus, if $F_0^s > 0$, $[1 + F_0^s\Omega_{00}(s)]^{-1}$ will have a simple pole at $s = s_0 > 1$, where

$$s_0 = \frac{\omega}{q v_f} = \begin{cases} 1 + 2e^{-2(1+1/F_0^s)}, & F_0^s \ll 1 \qquad (1.3.26) \\ \sqrt{F_0^s/3}, & F_0^s \gg 1; \qquad (1.3.27) \end{cases}$$

s_0 increases monotonically with F_0^s between these limits.

The velocity of this zero sound mode is

$$c_0 = s_0 v_f. \tag{1.3.28}$$

For small positive F_0^s, c_0 is only slightly above v_f. In this mode the density oscillates locally, with $\delta n = N(0)\nu_0$, where $N(0)$ is the density of states at the Fermi surface; the local current is given by the continuity equation (1.3.21) (with $F_1^s = 0$), $\delta j = \hat{\mathbf{q}}N(0)v_f\nu_1/3 = c_0\hat{\mathbf{q}}\delta n$. In the first sound mode in the hydrodynamic regime, the Fermi surface, as it oscillates in momentum space, remains spherical. Here, however, higher order moments ν_l are excited and the Fermi surface becomes distorted from spherical shape according to

$$\delta n_{\mathbf{p}} \propto \frac{\mathbf{q}\cdot\mathbf{v}_{\mathbf{p}}}{\omega - \mathbf{q}\cdot\mathbf{v}_{\mathbf{p}}}. \tag{1.3.29}$$

Only for $F_0^s \gg 1$ (corresponding to $c_0 \gg v_f$) does the Fermi surface retain an approximately spherical shape, pulsating in radius, and with an oscillating center. In the transition from first sound to zero sound the sound velocity changes from $c_1 = v_f\sqrt{(1 + F_0^s)/3}$ to c_0. Note that for $F_0^s \gg 1$, $c_1 \simeq c_0$. We emphasize that in the zero sound case the restoring force is provided entirely by the effective field, through F_0^s, while in the case of first sound there is always a restoring force, due to quasiparticle collisions, that the effective field modifies.

When $F_0^s < 0$ the effective field is attractive and one finds that the equation $1 + F_0^s\Omega_{00}(s) = 0$ has no solutions for real s. There is no zero sound mode for $F_0^s < 0$ that is not strongly damped.

We turn now to consider the more general situation where more than one F_l^s is nonnegligible. We shall retain only F_0^s, F_1^s and F_2^s, as this is all that is necessary for the practical case of liquid ^{3}He. The zero sound mode is described most readily by (1.3.23). To compute ν_2 in terms of ν_0 we take (1.3.13) for $l = 0$ and $l = 2$ and eliminate U to find

$$\frac{\nu_2}{5} = \frac{\Omega_{20}\nu_0}{\Omega_{00} + F_2^s(\Omega_{00}\Omega_{22} - \Omega_{20}^2)}. \tag{1.3.30}$$

In deriving this equation, the generalization of (1.3.25) for $l = 2$, we have used (1.3.17). Substituting (1.3.30) into (1.3.23) we derive the equation for the density response

$$\left[(s^2 - c_1^2/v_f^2) - \frac{2}{3}\frac{(1 + F_1^s/3)(1 + F_2^s/5)\Omega_{20}}{\Omega_{00} + F_2^s(\Omega_{00}\Omega_{22} - \Omega_{20}^2)}\right]\nu_0 = \left(1 + \frac{F_1^s}{3}\right)\frac{U}{3}. \tag{1.3.31}$$

The zero sound frequency is determined by the value $s = s_0$ for which the square bracket in (1.3.31) vanishes; s_0 must generally be found numerically. However in ^{3}He because $F_0^s \gg 1$, s_0 is also large compared with unity [cf. (1.3.27)] and it suffices to expand the Ω's in inverse powers of s: $\Omega_{20} \to -2/15s^2 - 4/35s^4 + \ldots$, $\Omega_{22} \to -11/105s^2 + \ldots$. To leading order in s^{-2} the F_2^s term in the denominator in (1.3.31) can be neglected, and we find the expression for the velocity $c_0 = v_f s_0$ of zero sound

$$\frac{c_0^2 - c_1^2}{c_1^2} = \frac{4}{5}\frac{1 + F_2^s/5}{1 + F_0^s}. \tag{1.3.32}$$

More accurately the right side is multiplied by $[1 + (9v_f^2/35c_1^2)(1 + F_2^s/5) + \mathcal{O}((v_f/c_1)^4)]$.

When Landau parameters other than F_0^s and F_1^s are included, it is possible that more than one value of the velocity can satisfy the zero sound dispersion relation. Any two modes having different velocities will be orthogonal in the sense that

$$\sum_l \left(1 + \frac{F_l^s}{2l + 1}\right)\nu_l\nu_l' = 0, \tag{1.3.33}$$

where ν_l and ν_l' are the distortions of the Fermi surface associated with the two modes. In the case of liquid ^{3}He it seems unlikely that any longitudinal modes other than that resembling a pure density fluctuation, the zero sound mode described above, will have velocities very different from the Fermi velocity; as one sees from (1.3.38) in the next section, if such other modes have velocities near v_f they will be much more strongly damped by quasiparticle collisions than the high velocity zero sound mode [24]. As yet no direct way of determining experimentally if these other longitudinal modes exist in liquid ^{3}He has been suggested.

(c) Attenuation of Zero Sound. Quasiparticle collisions tend to damp zero sound at finite temperatures. Even in the absence of collisions there is a small Landau damping due to coupling of the zero sound to thermally excited quasiparticles travelling at the zero sound velocity. Since such quasiparticles are exponentially small in number, Landau damping is of no practical importance here. To compute the effect of collisions we write the zero sound amplitude attenuation coefficient as [25]

$$\alpha_0 = \frac{|\dot{E}_{\text{mech}}|}{2c_0\delta E}, \tag{1.3.34}$$

where δE is the mean energy density in a zero sound wave, $\dot{E}_{\text{mech}} = -T\dot{s}$ is the rate at which the mechanical energy of the wave is dissipated into heat, and $\dot{s}$ is the rate of at which entropy is produced by collisions.

The rate of entropy production can be computed from (1.2.71). When the wave is only weakly damped we may substitute for $\bar{\Phi}$ its value for a zero sound wave in the absence of collisions. Generally $\bar{\Phi}$ measures the deviation, as in (1.2.61), of the distribution function from a local equilibrium distribution evaluated at the true ε; however the deviation of this local equilibrium distribution from global equilibrium, $n^0(\varepsilon)$, evaluated at the true ε, does not contribute to the collision integral; we may therefore replace $\delta\bar{n}^{\text{l.e.}}$ by $n - n^0(\varepsilon)$, or equivalently replace $\bar{\Phi}_{\mathbf{p}}$ by $\delta n_{\mathbf{p}}/(-\partial n_p^0/\partial\varepsilon_p) + \delta\varepsilon_{\mathbf{p}}$. The first term is simply $\nu_{\mathbf{p}}$ while from (1.3.12) we see that the second is $\Sigma_l \nu_l F_l^s P_l/(2l+1)$. Thus in evaluating $\dot{s}$ we may let

$$\bar{\Phi}_{\mathbf{p}} \to \sum_l \nu_l\left(1 + \frac{F_l^s}{2l+1}\right)P_l(\cos\theta) \tag{1.3.35}$$

in (1.2.71).

The sums in (1.2.71) are done by the techniques shown in Section 1.2.4(b). The energy integral, which is decoupled from the angular integral, becomes simply $I_4(y \to 0) = 2\pi^2/3$; the spin sum is given by (1.2.85). There is no mixing of different l in the angular integrals. The result is

$$T\dot{s} = \frac{2\pi^2}{3}\frac{N(0)}{\tau}\sum_{l=2}^{\infty}\frac{(1-\lambda_{l1})}{2l+1}\nu_l^2\left(1 + \frac{F_l^s}{2l+1}\right)^2 \tag{1.3.36}$$

where $1 - \lambda_{l1} = 1 - w_{2l} - w_{3l} - w_{4l}$, and the w are given by (1.1.121). The lack of terms $l = 0$ and $l = 1$ in (1.3.36) results from collisions that conserve quasiparticle number ($l = 0$) and momentum ($l = 1$).

The average energy density contained in the wave may be found by the methods described in Section 1.1.3(e); it is given by (1.1.72) evaluated for $\nu_{l\uparrow} = \nu_{l\downarrow} = \nu_l$ and averaged over several wavelengths. Then the $\mu\delta n$ term vanishes, since on the average $\delta n = 0$, and we find

$$\delta E = \frac{N(0)}{2} \sum_{l=0}^{\infty} \frac{\nu_l^2}{2l+1}\left(1 + \frac{F_l^s}{2l+1}\right). \tag{1.3.37}$$

The resultant attenuation coefficient α_0 is then

$$\alpha_0 = \frac{2\pi^2}{3} \frac{1}{c_0\tau} \frac{\sum_{l=2}^{\infty} \frac{1-\lambda_{l1}}{2l+1} \nu_l^2 \left(1 + \frac{F_l^s}{2l+1}\right)^2}{\sum_{l=0}^{\infty} \frac{\nu_l^2}{2l+1}\left(1 + \frac{F_l^s}{2l+1}\right)} \sim T^2. \tag{1.3.38}$$

This result, first derived in Reference [26], is exact under the conditions that $\hbar\omega \ll \kappa T$ and that the damping is small, that is, $\alpha_0 \ll q$; note in particular that the full energy dependence of the collision integral has been taken into account. Results similar to (1.3.38) have also been derived under somewhat less general assumptions in References 27 and 28. In the collisionless regime and for $\hbar\omega \gtrsim \kappa T$, α_0 is given by (1.3.38) multiplied by a factor [2] $1 + (\hbar\omega/2\pi\kappa T)^2$.

In the collisionless regime the ν_l may be found by solving (1.3.13). The general results are rather complicated, and will not be quoted here. When the velocity of zero sound is very much greater than the Fermi velocity, then the quadrupolar distortion ν_2 dominates the entropy production, and only the $l = 2$ term in the numerator of (1.3.38) need be kept. Also, the average energy in the zero sound wave (the denominator of (1.3.38)) is then dominated by the $l = 0$ and $l = 1$ terms. Since in this limit $\nu_2 = 2\nu_0 = (2\nu_1/3s)(1 + F_1^s/3)$, (1.3.38) reduces to

$$\alpha_0 = \frac{2}{15} \frac{m^*}{m} \frac{v_f^2}{c_0^3(\tau_\eta)_{var}} \left(1 + \frac{1}{5}F_2^s\right)^2 \left[1 + \mathcal{O}\left(\frac{v_f^2}{c_0^2}\right)\right], \tag{1.3.39}$$

where the relaxation time $(\tau_\eta)_{var}$, given by

$$(\tau_\eta)_{var} = \frac{3}{2\pi^2} \frac{\tau}{1-\lambda_\eta}, \tag{1.3.40}$$

is the variational estimate of the relaxation time that enters the viscosity, since $\lambda_\eta \equiv \lambda_{21}$.

The calculation of the velocity and attenuation of sound when $\omega\tau \sim 1$ is complicated and to date it has been carried out in detail only in the relaxation time approximation [see Section 1.2.4.(a)] [24], although the machinery needed to carry out the full calculation has been described in detail by Brooker and Sykes [29], who have also studied the decay of localized disturbances in a Fermi liquid [30]. Upper and low bounds on the velocity and attenuation of sound in this regime have been calculated by Egilsson and Pethick [119].

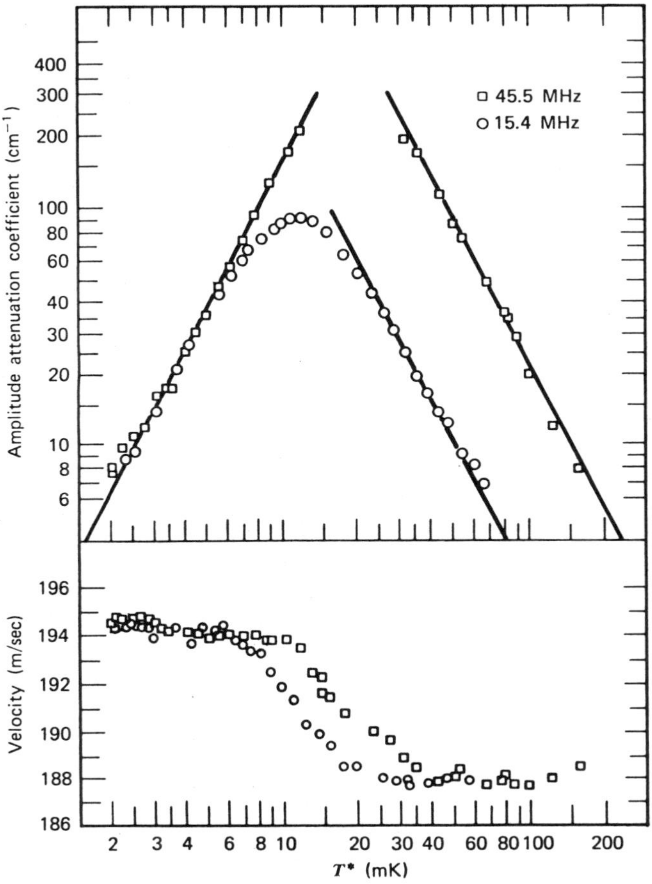

FIGURE 1.4. Amplitude attenuation coefficient and sound propagation velocity as a function of magnetic temperature in pure liquid ^3He at a pressure of 0.32 atm, for frequencies of 15.4 and 45.5 MHz, taken from the work of Abel, Anderson, and Wheatley [32]. The straight line at low temperatures has a T^2 behavior, and the two straight lines at higher temperatures have a ω^2/T^2 behavior.

(d) Experiments on Zero Sound. The transition between the first sound and zero sound regimes was first found experimentally by Keen, Matthews, and Wilks [31] in measurements of the acoustic impedance between liquid ^3He and a quartz crystal. The first direct measurements of the properties of zero sound were made by Abel, Anderson, and Wheatley [32]. In Figure 1.4 we show their data for the velocity and attenuation of sound as a function of temperature; this data illustrates very clearly the transition between the

first sound regime (at higher temperatures, where the quasiparticle collision time $\tau \sim T^{-2}$ is short) and the zero sound regime (at lower temperature). The attenuation in the first sound regime varies as $\omega^2\tau \sim \omega^2/T^2$, in accord with (1.3.5), and in the zero sound regime it varies as T^2, (1.3.38). The measurements of the zero sound velocity enable one to estimate F_2^s, if one assumes higher Landau parameters may be neglected; one finds [26] $F_2^s = 0.0 \pm 0.4$. This value is consistent with that estimated from the attenuation of zero sound and first sound [26] $F_2^s = 0.4 \pm 1.0$.

The acoustic impedance (in which the transition between the first and zero sound regimes was first observed) was calculated first by Bekarevich and Khalatnikov [33]; a simple approach to the problem, making use of the fact that in ^{3}He, as a consequence of the large value of F_0^s, a zero sound wave is almost a pure density fluctuation, was described by Gavoret [34], and Brooker [24] carried out detailed calculations in an attempt to explain the experimental measurements. The increase in the acoustic impedance observed experimentally [31, 35] is much larger than can be accounted for theoretically, and the reason for this is not clear. The theory of the acoustic impedance can also be used to estimate the Kapitza resistance between a solid and liquid ^{3}He; a review of the theories of Kapitza resistance, and the comparison of recent experiments with theory is given by Anderson and Johnson [36].

(e) Transverse Sound Modes. If one attempts to propagate a transverse oscillation in a fluid, say by oscillating a plane boundary in its own plane at frequency ω, then in the hydrodynamic regime one finds that a wave of complex wavenumber $\sqrt{i\omega mn/\eta}$ (where η is the viscosity) is set up [37]. Such a transverse wave is damped in a distance comparable to its wavelength. The analogue of this mode in the collisionless regime in a Fermi liquid is called transverse zero sound; such modes have the possibility of being only weakly attenuated.

In a transverse zero sound mode the quasiparticle distribution function at low T has the form (1.3.9) where now $\nu_{\mathbf{p}}$ is no longer symmetric about the azimuthal axis. In general we may expand $\nu_{\mathbf{p}}$ as

$$\nu_{\mathbf{p}} = \sum_{lm} Y_{lm}(\theta, \phi)\nu_{lm}. \tag{1.3.41}$$

Let us assume that the system is being driven by an external potential $U(\mathbf{r}, t; \hat{p})$ dependent on the direction of $\mathbf{p}$, and write

$$U(\hat{p}) = \sum_{lm} Y_{lm}(\theta, \phi)U_{lm}. \tag{1.3.42}$$

Then from the kinetic equation (1.3.10), we may derive the equation for the ν_{lm} analogous to (1.3.13):

$$\nu_{lm} + \sum_{l' \geqslant |m|} B_{ll'}^{(m)}(s)\left[\frac{F_{l'}^s}{2l'+1}\nu_{l'm} + U_{l'm}\right] = 0 \qquad (1.3.43)$$

where

$$B_{ll'}^{(m)}(s) = \int d\Omega_{\mathbf{p}} Y_{lm}^* \frac{\mathbf{q}\cdot\mathbf{v_p}}{\mathbf{q}\cdot\mathbf{v_p} - \omega} Y_{l'm} \qquad (1.3.44)$$

Note that for a mode with a given m, only Landau parameters $F_{l'}^s$, with $l' \geqslant m$ enter.

The transverse zero sound mode is well illustrated by the case in which only F_0^s and F_1^s are appreciably different from zero. Then the dispersion relation for the $m = 1$ mode is simply

$$1 + \frac{F_1^s}{3} B_{11}^{(1)}(s) = 0. \qquad (1.3.45)$$

Using the relation $B_{11}^{(1)} = -\frac{1}{2}(1 + 3(s^2 - 1)\Omega_{00})$ we have the equation

$$\Omega_{00}(s) = \frac{6 - F_1^s}{3F_1^s(s^2 - 1)}, \qquad (1.3.46)$$

which has a real solution only for $F_1^s \geqslant 6$; in this case there is one real solution. Note that the value of F_0^s does not affect the velocity of the mode since the mode, being transverse, does not have a density fluctuation associated with it. For liquid ^3He the velocity of this mode is expected to be slightly above v_f. Even for F_1^s as large as 14, the value near the melting pressure, the velocity would be only 1.2 v_f.

In general the attenuation $\alpha_0^{(m)}$ of transverse zero sound with azimuthal quantum number $m \geqslant 1$ is given by [cf. (1.3.38)]

$$\alpha_0^{(m)} = \frac{2\pi^2}{3c_0^{(m)}\tau} \frac{\sum_{l=m}^{\infty}(1 - \lambda_{l1})|\nu_{lm}|^2\left(1 + \frac{F_l^s}{2l+1}\right)^2}{\sum_{l=m}^{\infty}|\nu_{lm}|^2\left(1 + \frac{F_l^s}{2l+1}\right)} \qquad (1.3.47)$$

where $c_0^{(m)}$ is the velocity of the mode. The transverse modes with $m \geqslant 1$ are more strongly damped than the $m = 0$ longitudinal zero sound mode, due to the absence in (1.3.47) of terms with $l < m$ in the denominator, while all terms with $l \geqslant 0$ are present in the denominator of (1.3.38). The reason that there are fewer terms in the numerator than the denominator for the $m = 0$ zero sound is the existence of the particle and momentum conservation laws, which always cause the terms with $l = 0$ and $l = 1$ in the numerator to vanish identically. (Note also that $\lambda_{11} = 1$.) $\alpha_0^{(m)}$ is of order $1/(c_0^{(m)}\tau)$ for $m \geqslant 1$, whereas for longitudinal zero sound the attenuation α_0 is of order $v_f^2/(c_0^3\tau)$ [cf. (1.3.39)]. For liquid ^3He, the factor $(v_f/c_0)^2$ is less than 0.1 at low pressures, decreasing with increasing pressure. Detailed computations of the velocity and attenuation of the transverse zero sound modes using the relaxation time approximation are given by Corruccini, Clarke, Mermin, and Wilkins [27]. Brooker and Sykes [29] also discuss these modes. Calculations of the acoustic impedance and the damping of transverse zero sound have been carried out using the relaxation time approximation [38]. Measurements of transverse zero sound are reported in Reference [116].

(f) Spin Zero-Sound Modes. Up to this point we have considered only collective modes in which the spin-up and spin-down quasiparticle distributions undergo identical oscillations. We want now to describe briefly, for the case of no static magnetic field, the modes that correspond to the spin-up and spin-down distributions oscillating 180° out of phase, thus creating a local spin polarization of quasiparticles of a given momentum. We shall neglect forces that rotate spins until the following section, where we derive the transport equation that includes spin precession.

Let us consider then the case that

$$\delta n_{\mathbf{p}\uparrow}(\mathbf{r}, t) = -\delta n_{\mathbf{p}\downarrow}(\mathbf{r}, t) = -\frac{\partial n_p^0}{\partial \varepsilon_p}\nu_{\mathbf{p}}. \tag{1.3.48}$$

If we in general decompose ν_p into spherical harmonics, as in (1.3.41), then the dispersion relation of these "spin" modes is given by equations analogous to (1.3.43), only with F_l^s replaced by F_l^a.

For $m = 0$, the equations are completely analogous to those for longitudinal zero sound, and the possible spin modes are simply longitudinal zero sound oscillations of the spin-up and spin-down quasiparticle distributions, 180° out of phase, in which there is an oscillating local spin density, but constant quasiparticle density. For example, if only F_0^a is nonzero, the dispersion relation is given by the solution of

$$1 + F_0^a\,\Omega_{00}(s) = 0; \tag{1.3.49}$$

this has a real solution if and only if $F_0^a > 0$, which is not the case in liquid ^{3}He. In general, it does not appear that the Landau parameters in liquid ^{3}He permit the existence of these spin modes for $m = 0$ and $m = 1$.

An interesting result is Mermin's theorem that states that at least one of the $m = 0$ modes, zero sound or spin zero sound, must exist in a Fermi liquid at sufficiently low temperatures. The proof, which may be found in Reference [39], hinges upon the forward scattering sum rule (1.4.36), which is discussed later.

1.3.2. Spin Waves and Related Phenomena

(a) Kinetic Equations. We have, until now, considered only processes in which the quasiparticle distribution can be specified by giving the numbers

of up-spin and down-spin quasiparticles. We have not yet developed the machinery to describe phenomena, such as the precession of the magnetization in a uniform external magnetic field, in which the axis of the net spin polarization is locally in an arbitrary direction; it is to this task that we now turn.

In general the quasiparticle distribution function is given by a 2×2 density matrix in spin space $[n_{\mathbf{p}}(\mathbf{r}, t)]_{\alpha\alpha'}$ [cf. (1.2.3)]. Equivalently, as in (1.1.19) and (1.1.20), the distribution function can be specified by the four quantities:

$$n_{\mathbf{p}}(\mathbf{r}, t) = \frac{1}{2} \sum_{\alpha} [n_{\mathbf{p}}(\mathbf{r}, t)]_{\alpha\alpha}$$

$$\boldsymbol{\sigma}_{\mathbf{p}}(\mathbf{r}, t) = \frac{1}{2} \sum_{\alpha\alpha'} \boldsymbol{\tau}_{\alpha\alpha'}[n_{\mathbf{p}}(\mathbf{r}, t)]_{\alpha'\alpha}; \tag{1.3.50}$$

$2n_{\mathbf{p}}(\mathbf{r}, t)$ is the total density of quasiparticles at $\mathbf{r}$, t, while $2\boldsymbol{\sigma}_{\mathbf{p}}(\mathbf{r}, t)$ is the local spin polarization. In terms of these quantities the density matrix can be written as

$$[n_{\mathbf{p}}(\mathbf{r}, t)]_{\alpha\alpha'} = n_{\mathbf{p}}(\mathbf{r}, t)\delta_{\alpha\alpha'} + \boldsymbol{\sigma}_{\mathbf{p}}(\mathbf{r}, t)\cdot\boldsymbol{\tau}_{\alpha\alpha'}. \tag{1.3.51}$$

Similarly, the quasiparticle energy matrix $[\epsilon_{\mathbf{p}}(\mathbf{r}, t)]_{\alpha\alpha'}$ (1.2.15), can be written as in (1.1.23) as

$$[\epsilon_{\mathbf{p}}(\mathbf{r}, t)]_{\alpha\alpha'} = \varepsilon_{\mathbf{p}}(\mathbf{r}, t)\delta_{\alpha\alpha'} + \mathbf{h}_{\mathbf{p}}(\mathbf{r}, t)\cdot\boldsymbol{\tau}_{\alpha\alpha'}. \tag{1.3.52}$$

The total quasiparticle current density in direction i for quasiparticles of a given momentum $\mathbf{p}$ is

$$\begin{aligned} j_{\mathbf{p}i}(\mathbf{r}, t) &= \sum_{\alpha\alpha'} \frac{\partial}{\partial p_i} (\epsilon_{\mathbf{p}})_{\alpha\alpha'}(n_{\mathbf{p}})_{\alpha'\alpha} \\ &= 2\frac{\partial \epsilon_{\mathbf{p}}}{\partial p_i} n_{\mathbf{p}} + 2\frac{\partial \mathbf{h}_{\mathbf{p}}}{\partial p_i}\cdot\boldsymbol{\sigma}_{\mathbf{p}}. \end{aligned} \tag{1.3.53}$$

This expression may be interpreted as follows: when there is a net spin polarization the unit vector $\hat{\boldsymbol{\sigma}}_{\mathbf{p}}(\mathbf{r}, t)$ can be chosen as a local spin quantization axis for quasiparticles of momentum $\mathbf{p}$. Then the current is the sum of terms $\mathbf{v}_{\mathbf{p}\uparrow}n_{\mathbf{p}\uparrow} + \mathbf{v}_{\mathbf{p}\downarrow}n_{\mathbf{p}\downarrow}$ where

$$\begin{aligned} (v_{\mathbf{p}\uparrow})_i &= \frac{\partial \varepsilon_{\mathbf{p}}}{\partial p_i} + \left(\frac{\partial \mathbf{h}_{\mathbf{p}}}{\partial p_i}\cdot\hat{\boldsymbol{\sigma}}_{\mathbf{p}}\right) \\ (v_{\mathbf{p}\downarrow})_i &= \frac{\partial \varepsilon_{\mathbf{p}}}{\partial p_i} - \left(\frac{\partial \mathbf{h}_{\mathbf{p}}}{\partial p_i}\cdot\hat{\boldsymbol{\sigma}}_{\mathbf{p}}\right), \end{aligned} \tag{1.3.54}$$

and

$$n_{\mathbf{p}\uparrow} = n_{\mathbf{p}} + \boldsymbol{\sigma}_{\mathbf{p}}\cdot\hat{\boldsymbol{\sigma}}_{\mathbf{p}}, \qquad n_{\mathbf{p}\downarrow} = n_{\mathbf{p}} - \boldsymbol{\sigma}_{\mathbf{p}}\cdot\hat{\boldsymbol{\sigma}}_{\mathbf{p}}. \tag{1.3.55}$$

The local spin current $j_{\sigma_i, i}(\mathbf{p}, \mathbf{r}, t)$, representing the current in the ith spatial direction of the jth component of the spin polarization carried by quasiparticles of momentum **p**, can similarly be written as

$$j_{\sigma,i}(\mathbf{p},\mathbf{r},t) = \sum_{\alpha\alpha'\alpha''} [n_{\mathbf{p}}(\mathbf{r}, t)]_{\alpha\alpha'} \frac{\partial}{\partial p_i} [\epsilon_{\mathbf{p}}(\mathbf{r}, t)]_{\alpha'\alpha''} \tau_{\alpha''\alpha}. \tag{1.3.56}$$

The only terms that survive the trace operation are those with two Pauli matrices (and so the result is independent of the order of the matrices in the trace); the result is

$$j_{\sigma,i}(\mathbf{p},\mathbf{r},t) = 2\left[\frac{\partial \varepsilon_{\mathbf{p}}}{\partial p_i} \boldsymbol{\sigma}_{\mathbf{p}}(\mathbf{r}, t) + \frac{\partial \mathbf{h}_{\mathbf{p}}}{\partial p_i} n_{\mathbf{p}}(\mathbf{r}, t)\right]. \tag{1.3.57}$$

The first term represents the drift of the net spin polarization at the average quasiparticle velocity, while the second term takes into account the dependence of the quasiparticle velocity on the spin direction; this dependence gives rise to a spin current even in the absence of a local spin polarization.

The full kinetic equations obeyed by $n_{\mathbf{p}}$ and $\boldsymbol{\sigma}_{\mathbf{p}}$ can be written in forms analogous to the simple case previously considered, (1.2.9). The equation for $n_{\mathbf{p}}$ takes the form of a differential conservation law in position and momentum space:

$$\frac{\partial n_{\mathbf{p}}}{\partial t} + \frac{\partial}{\partial r_i}\left(\frac{\partial \varepsilon_{\mathbf{p}}}{\partial p_i} n_{\mathbf{p}} + \frac{\partial \mathbf{h}_{\mathbf{p}}}{\partial p_i} \cdot \boldsymbol{\sigma}_{\mathbf{p}}\right) + \frac{\partial}{\partial p_i}\left(-\frac{\partial \varepsilon_{\mathbf{p}}}{\partial r_i} n_{\mathbf{p}} - \frac{\partial \mathbf{h}_{\mathbf{p}}}{\partial r_i} \cdot \boldsymbol{\sigma}_{\mathbf{p}}\right) = I[n_{\mathbf{p}'}, \boldsymbol{\sigma}_{\mathbf{p}'}]. \tag{1.3.58}$$

The term representing drift in position space is $\nabla_r . \mathbf{j}_{\mathbf{p}}(\mathbf{r}, t)$, and the term representing drift in momentum space has an analogous structure; I is the collision integral. If either $\boldsymbol{\sigma}_{\mathbf{p}}$ or $\mathbf{h}_{\mathbf{p}}$ is identically zero, (1.3.58) reduces to our previous kinetic equation (1.2.12a).

The kinetic equation for the spin density is also in the form of a continuity equation. However, here one must take into account the fact that the local spin polarization can also change as a result of precession of the local spin polarization about the local effective magnetic field (proportional to $-\mathbf{h}_{\mathbf{p}}$). The rate of change of the spin due to precession is

$$\left(\frac{\partial \boldsymbol{\sigma}_{\mathbf{p}}}{\partial t}\right)_{\text{precession}} = -\frac{2}{\hbar} \boldsymbol{\sigma}_{\mathbf{p}} \times \mathbf{h}_{\mathbf{p}} \tag{1.3.59}$$

and thus the kinetic equation for $\boldsymbol{\sigma}$ becomes

$$\frac{\partial \boldsymbol{\sigma}_{\mathbf{p}}}{\partial t} + \frac{\partial}{\partial r_i}\left(\frac{\partial \varepsilon_{\mathbf{p}}}{\partial p_i} \boldsymbol{\sigma}_{\mathbf{p}} + \frac{\partial \mathbf{h}_{\mathbf{p}}}{\partial p_i} n_{\mathbf{p}}\right) + \frac{\partial}{\partial p_i}\left(-\frac{\partial \varepsilon_{\mathbf{p}}}{\partial r_i} \boldsymbol{\sigma}_{\mathbf{p}} - \frac{\partial \mathbf{h}_{\mathbf{p}}}{\partial r_i} n_{\mathbf{p}}\right) = -\frac{2}{\hbar} \boldsymbol{\sigma}_{\mathbf{p}} \times \mathbf{h}_{\mathbf{p}} + \left(\frac{\partial \boldsymbol{\sigma}_{\mathbf{p}}}{\partial t}\right)_{\text{collision}}. \tag{1.3.60}$$

The drift terms in position and momentum space are written analogously to those in (1.3.58). Equations (1.3.58) and (1.3.60) are valid provided n and $\boldsymbol{\sigma}$ may be treated as classical distribution functions. When the characteristic times and distances over which the distribution function varies are small compared with $\hbar/\kappa T$ and $\hbar v_f/\kappa T$ respectively, the distribution function must be regarded as a Wigner distribution function and its equation of motion is given by the quantum kinetic equation (1.2.16). In the classical limit the quantum-mechanical equation reduces to (1.3.58) and (1.3.60); the precession term in (1.3.60) arises from the lack of commutativity of the ϵ and n matrices at the same point in space when ϵ and n are not diagonal in the same representation.

(b) Homogeneous Spin Waves in the Presence of a Uniform Magnetic Field. Let us consider a noninteracting uniform system containing a spatially uniform spin polarization $\boldsymbol{\sigma}_{\mathbf{p}}$. In the presence of a uniform magnetic field the spin polarization will precess about the field according to

$$\frac{\partial \boldsymbol{\sigma}_{\mathbf{p}}}{\partial t} = \boldsymbol{\sigma}_{\mathbf{p}} \times \gamma \boldsymbol{\mathscr{H}}; \tag{1.3.61}$$

each spin component $\boldsymbol{\sigma}_{\mathbf{p}}$ thus precesses independently about $\boldsymbol{\mathscr{H}}$ at the Larmor frequency

$$\omega_0 = \gamma \mathscr{H}. \tag{1.3.62}$$

Fermi liquid interactions between the spins qualitatively alter this behavior. To see this we consider the same situation in an interacting Fermi liquid, and neglect collisions. Then $\boldsymbol{\sigma}_{\mathbf{p}}(t)$ obeys (1.3.59). The term $\mathbf{h}_{\mathbf{p}}$ is a sum of

$$\mathbf{h}_{\mathbf{p}}^0 = -\gamma\hbar\boldsymbol{\mathscr{H}}/2, \tag{1.3.63}$$

which is the coupling to the external field, plus an internal field term. In evaluating the internal field term we may, when the density of polarized spins is a small fraction of the total density of particles, neglect the effect of the spin polarization on the quasiparticle interaction $f_{\mathbf{pp}'}$. Then $\mathbf{h}_{\mathbf{p}}$ is given by [cf. (1.1.28)]

$$\mathbf{h}_{\mathbf{p}} = -\gamma\frac{\hbar}{2}\boldsymbol{\mathscr{H}} + 2\int\frac{d^3p'}{(2\pi\hbar)^3} f^a_{\mathbf{pp}'}\boldsymbol{\sigma}_{\mathbf{p}'}. \tag{1.3.64}$$

The quantity $-4/(\hbar\gamma)\int d^3p'/(2\pi\hbar)^3 f^a_{\mathbf{pp}'}\boldsymbol{\sigma}_{\mathbf{p}'}$ is the effective magnetic field produced by quasiparticle interactions. The spin polarization thus obeys

$$\frac{\partial \boldsymbol{\sigma}_{\mathbf{p}}}{\partial t} = \gamma\boldsymbol{\sigma}_{\mathbf{p}} \times \boldsymbol{\mathscr{H}} - \frac{4}{\hbar}\int\frac{d^3p'}{(2\pi\hbar)^3} f^a_{\mathbf{pp}'}(\boldsymbol{\sigma}_{\mathbf{p}} \times \boldsymbol{\sigma}_{\mathbf{p}'}). \tag{1.3.65}$$

Note that if $\boldsymbol{\sigma}_{\mathbf{p}}$ has the same direction for all values of $\mathbf{p}$, the quasiparticle interaction term in (1.3.65) vanishes, and $\boldsymbol{\sigma}_{\mathbf{p}}$ again precesses about the external field at the Larmor frequency.

In cases of interest the spin polarization $\boldsymbol{\sigma}_{\mathbf{p}}$ is nonzero only for $\mathbf{p}$ in the neighborhood of the Fermi surface; it is then convenient to introduce the variable

$$\boldsymbol{\sigma}(\hat{\mathbf{p}}) = 2\int \frac{d^3p'}{(2\pi\hbar)^3}\, \boldsymbol{\sigma}_{\mathbf{p}'}\delta^{(2)}(\hat{\mathbf{p}} - \hat{\mathbf{p}}'). \tag{1.3.66}$$

obtained by summing $\boldsymbol{\sigma}_{\mathbf{p}'}$, over all values of $\mathbf{p}'$ in the direction of the unit vector $\hat{\mathbf{p}}'$; note that $\int (d\Omega'/4\pi)\delta^{(2)}(\hat{\mathbf{p}} - \hat{\mathbf{p}}') = 1$. When we write (1.3.65) in terms of $\boldsymbol{\sigma}(\hat{\mathbf{p}})$, only $f^a_{\mathbf{pp}'}$ at the Fermi surface enters and we have

$$\frac{\partial \boldsymbol{\sigma}(\hat{\mathbf{p}})}{\partial t} = \gamma \boldsymbol{\sigma}(\hat{\mathbf{p}}) \times \mathscr{H} - \frac{2}{\hbar}\int \frac{d\Omega'}{4\pi} f^a_{\mathbf{pp}'}\, \boldsymbol{\sigma}(\hat{\mathbf{p}}) \times \boldsymbol{\sigma}(\hat{\mathbf{p}}'). \tag{1.3.67}$$

Here $f^a_{\mathbf{pp}'}$ denotes the quasiparticle interaction evaluated at the Fermi surface.

The internal field exerts an influence when the direction of $\boldsymbol{\sigma}_{\mathbf{p}}$ is not the same for all quasiparticle states. To illustrate these effects we consider small oscillations of a Fermi liquid about a state of polarization $\boldsymbol{\sigma}^0$ in the direction of the external field. We write

$$\boldsymbol{\sigma}(\hat{\mathbf{p}}) = \boldsymbol{\sigma}^0 + \delta\boldsymbol{\sigma}(\hat{\mathbf{p}}), \tag{1.3.68}$$

where $\delta\boldsymbol{\sigma}(\hat{\mathbf{p}})$ is a small perturbation. Substituting (1.3.68) into (1.3.67) and assuming $\delta\boldsymbol{\sigma}(\hat{\mathbf{p}}) \ll \sigma^0$ one finds the equation of motion for $\delta\boldsymbol{\sigma}(\hat{\mathbf{p}})$:

$$\begin{aligned}\frac{\partial}{\partial t}\, \delta\boldsymbol{\sigma}(\hat{\mathbf{p}}) &= \gamma \delta\boldsymbol{\sigma}(\hat{\mathbf{p}}) \times \mathscr{H} \\ &\quad - \frac{2}{\hbar}\int \frac{d\Omega'}{4\pi} f^a_{\mathbf{pp}'}[\delta\boldsymbol{\sigma}(\hat{\mathbf{p}}) \times \boldsymbol{\sigma}^0 + \boldsymbol{\sigma}^0 \times \delta\boldsymbol{\sigma}(\hat{\mathbf{p}}')]. \end{aligned} \tag{1.3.69}$$

In the latter integral the first term is the precession of $\delta\boldsymbol{\sigma}$ due to the internal field while the second term is the precession of the initial spin polarization $\boldsymbol{\sigma}_0$ because of modification of the internal field. As we remarked above, the component of $\boldsymbol{\sigma}$ in the direction of $\mathscr{H}$ (which we take to be the z direction) does not vary. The perpendicular components

$$\delta\sigma^{\pm}(\hat{\mathbf{p}}) \equiv \delta\sigma_x(\hat{\mathbf{p}}) \pm i\delta\sigma_y(\hat{\mathbf{p}}) \tag{1.3.70}$$

obey

$$\begin{aligned}\frac{\partial}{\partial t}\, \delta\sigma^{\pm}(\hat{\mathbf{p}}) &= \mp\, i\left[\omega_0\delta\sigma^{\pm}(\hat{\mathbf{p}}) - \frac{2\sigma^0}{\hbar}\int \frac{d\Omega'}{4\pi} f^a_{\mathbf{pp}'}(\delta\sigma^{\pm}(\hat{\mathbf{p}}) - \delta\sigma^{\pm}(\hat{\mathbf{p}}'))\right] \\ &= \mp\, i[(\omega_0 - \frac{2}{\hbar} f^a_0\sigma^0)\delta\sigma^{\pm}(\hat{\mathbf{p}}) \\ &\qquad + \frac{2}{\hbar}\sigma^0 \int \frac{d\Omega'}{4\pi} f^a_{\mathbf{pp}'}\, \delta\sigma^{\pm}(\hat{\mathbf{p}}')]. \end{aligned} \tag{1.3.71}$$

The normal modes of the system are the eigenfunctions of (1.3.71), which are proportional to $Y_{lm}(\hat{\mathbf{p}})$. The corresponding eigenfrequencies $\omega_l^{\pm}$ of the modes are independent of m and are given by

$$\omega_l^{\pm} = \pm\left[\omega_0 - \frac{2}{\hbar}\sigma^0\left(f_0^a - \frac{f_l^a}{2l+1}\right)\right]. \qquad (1.3.72)$$

The term $\omega_0 - 2\sigma^0 f_0^a/\hbar$ represents the precession of the spin deviation in the effective field, while the f_l^a term represents the effect of the modification of the field due to the distortion of the Fermi surface. For an $l = 0$ mode the Fermi liquid effects exactly cancel as a consequence of the conservation of spin; an $l = 0$ mode is a precession of the net spin density, which is driven only by the external magnetic field. When σ^0 is given by its equilibrium value in the magnetic field

$$\sigma^0 = \frac{\gamma\hbar}{2}\frac{N(0)}{1 + F_0^a}\mathscr{H} \qquad (1.1.53)$$

the eigenfrequencies of the modes become

$$\omega_l^{\pm} = \pm\,\omega_0\frac{1 + F_l^a/(2l+1)}{1 + F_0^a}. \qquad (1.3.73)$$

Many of the properties of spin waves in polarized Fermi liquids were first investigated by Silin [11].

(c) Spin Hydrodynamic Equations. In Section 1.2.4(d) we discussed spin diffusion in the situation in which the system could be described by a single universal quantization axis. To discuss the more general situation, in which for example we consider the transport of spin components perpendicular to an external magnetic field, we must first derive the hydrodynamic equations obeyed by the local spin polarization density

$$\boldsymbol{\sigma}(\mathbf{r}, t) = 2\int\frac{d^3p}{(2\pi\hbar)^3}\,\boldsymbol{\sigma}_{\mathbf{p}}(\mathbf{r}, t). \qquad (1.3.74)$$

To derive the spin conservation law we sum both sides of (1.3.60) over momenta. In the absence of spin-orbit interactions, spin angular momentum is conserved in collisions, and thus the contribution of the collision term vanishes. The precession term (1.3.59), when summed over momenta, reduces simply to $\gamma\boldsymbol{\sigma}(\mathbf{r}, t) \times \boldsymbol{\mathscr{H}}(\mathbf{r}, t)$, where $\boldsymbol{\mathscr{H}}(\mathbf{r}, t)$ is the external magnetic field. Physically, this is because the internally produced fields cannot make the net magnetization (which produces the fields) precess. One can derive this result in general by using rotational invariance arguments, and for the simple case in which only terms to first order in the deviation from the unpolarized ground state are kept in $\mathbf{h}_{\mathbf{p}}$, as in (1.3.64), then one sees directly from the symmetry of $f_{\mathbf{pp}'}^a$ in $\mathbf{p}$ and $\mathbf{p}'$ that the internal field term

$$-\frac{4}{\hbar}\sum_{\mathbf{pp}'} f^a_{\mathbf{pp}'}\,\boldsymbol{\sigma}_{\mathbf{p}} \times \boldsymbol{\sigma}_{\mathbf{p}'} \tag{1.3.75}$$

vanishes. The net spin conservation law is thus

$$\frac{\partial \boldsymbol{\sigma}(\mathbf{r}, t)}{\partial t} + \frac{\partial}{\partial r_i} j_{\sigma, i}(\mathbf{r}, t) = \gamma \boldsymbol{\sigma}(\mathbf{r}, t) \times \mathscr{H}(\mathbf{r}, t) \tag{1.3.76}$$

where the spin current $\mathbf{j}_\sigma$ is given by

$$j_{\sigma, i}(\mathbf{r}, t) = 2\int \frac{d^3p}{(2\pi\hbar)^3}\left[\frac{\partial \varepsilon_{\mathbf{p}}}{\partial p_i}\,\boldsymbol{\sigma}_{\mathbf{p}} + \frac{\partial \mathbf{h}_{\mathbf{p}}}{\partial p_i}\,n_{\mathbf{p}}\right]. \tag{1.3.77}$$

Here i refers to the spatial direction in which spin $\boldsymbol{\sigma}$ is being transported. Compare (1.3.76) with the earlier equation (1.2.139).

The next result we need is the equation of motion for $\mathbf{j}_\sigma$; this equation is essentially the generalization of (1.2.144). We shall assume, for simplicity in the derivation, that we can treat the spin polarization $\boldsymbol{\sigma}_{\mathbf{p}}$ as well as $\delta n_{\mathbf{p}}$ as small quantities, and also that the spatial and temporal variations are slow. We first note, from integrating the second term in (1.3.77) by parts, that the linearized spin current is given by

$$j_{\sigma, i}(\mathbf{r}, t) = 2\int \frac{d^3p}{(2\pi\hbar)^3}\, v_{\mathbf{p}_i}\left(\boldsymbol{\sigma}_{\mathbf{p}} - \frac{\partial n^0_p}{\partial \varepsilon_p}\,\mathbf{h}_{\mathbf{p}}\right) \tag{1.3.78}$$

where $\mathbf{v}_{\mathbf{p}} = \nabla_p \varepsilon^0_p$ is the equilibrium quasiparticle velocity. [Compare with (1.2.26) for the linearized particle current.] If we then use (1.3.64) for $\mathbf{h}_{\mathbf{p}}$ and assume that the spin polarization $\boldsymbol{\sigma}_{\mathbf{p}}$ is nonzero only for $\mathbf{p}$ near the Fermi surface we find [cf. (1.2.143)]

$$j_{\sigma, i}(\mathbf{r}, t) = 2\int \frac{d^3p}{(2\pi\hbar)^3}\, v_{\mathbf{p}_i}\boldsymbol{\sigma}_{\mathbf{p}}(\mathbf{r}, t)\left(1 + \frac{F^a_1}{3}\right). \tag{1.3.79}$$

To derive the equation of motion for $\mathbf{j}_\sigma$ we turn to the kinetic equation (1.3.60) and drop terms there that are of third or higher order in ∇, $\delta n_{\mathbf{p}}$ or $\boldsymbol{\sigma}_{\mathbf{p}}$. Since $\mathbf{h}_{\mathbf{p}} - \mathbf{h}^0_{\mathbf{p}}$ is a first order term, and $\partial \varepsilon_{\mathbf{p}}/\partial r_i$ is itself second order (1.3.60) becomes, to second order,

$$\frac{\partial \boldsymbol{\sigma}_{\mathbf{p}}}{\partial t} + v_{\mathbf{p}_i}\frac{\partial}{\partial r_i}\left(\boldsymbol{\sigma}_{\mathbf{p}} - \frac{\partial n^0_p}{\partial \varepsilon_p}\,\mathbf{h}_{\mathbf{p}}\right) = -\frac{2}{\hbar}\,\boldsymbol{\sigma}_{\mathbf{p}} \times \mathbf{h}_{\mathbf{p}} + \left(\frac{\partial \boldsymbol{\sigma}_{\mathbf{p}}}{\partial t}\right)_{\text{collision}}. \tag{1.3.80}$$

Combining this result with (1.3.79) we find the equation of motion for $\mathbf{j}_\sigma$:

$$\frac{\partial}{\partial t} j_{\sigma, i}(\mathbf{r}, t) + \frac{\partial}{\partial r_k}(\Pi_\sigma)_{ik} = \left(\frac{\partial j_{\sigma, i}}{\partial t}\right)_{\text{precession}} + \left(\frac{\partial j_{\sigma, i}}{\partial t}\right)_{\text{collision}} \tag{1.3.81}$$

where

$$(\Pi_\sigma)_{ik} = \left(1 + \frac{F^a_1}{3}\right)2\int \frac{d^3p}{(2\pi\hbar)^3}\, v_{\mathbf{p}_i}v_{\mathbf{p}_k}\left(\boldsymbol{\sigma}_{\mathbf{p}} - \frac{\partial n^0_{\mathbf{p}}}{\partial \varepsilon_p}\,\mathbf{h}_{\mathbf{p}}\right) \tag{1.3.82}$$

is the "spin stress-tensor,"

$$\left(\frac{\partial j_{\sigma,i}}{\partial t}\right)_{\text{precession}} = -\frac{2}{\hbar}\left(1+\frac{F_1^a}{3}\right)2\int\frac{d^3p}{(2\pi\hbar)^3}\, v_{\mathbf{p}_i}\boldsymbol{\sigma}_{\mathbf{p}} \times \mathbf{h}_{\mathbf{p}} \qquad (1.3.83)$$

and

$$\left(\frac{\partial j_{\sigma,i}}{\partial t}\right)_{\text{collision}} = \left(1+\frac{F_1^a}{3}\right)2\int\frac{d^3p}{(2\pi\hbar)^3}\, v_{\mathbf{p}_i}\left(\frac{\partial \boldsymbol{\sigma}_{\mathbf{p}}}{\partial t}\right)_{\text{collision}}. \qquad (1.3.84)$$

Unlike in the case of particle current, the collision term does not vanish, since spin current is not conserved in collisions. Equations 1.3.81 and 1.3.76 constitute the hydrodynamic equations for the spin.

(d) Spin Diffusion. We now apply the spin hydrodynamic equations to the general problem of spin diffusion. There are three characteristic lengths in the transport equation: L, the mean distance over which the quasiparticle distribution function varies in space; $\lambda = v_f\tau$, the quasiparticle mean free path, where τ is a characteristic collision time; and r_L, the mean distance a quasiparticle travels while its spin precesses by an appreciable angle ($\sim\pi/2$) due to the *internal* magnetic field. This latter length measures the distance over which different harmonics of the spin distribution get out of phase with respect to one another; note that all harmonics precess by the same amount in the external field. From (1.3.72) we see that r_L is essentially of order

$$r_L \sim v_f/(\omega_l^+ - \omega_0) \sim \hbar v_f \sigma^0 f^a \qquad (1.3.85)$$

where σ^0 is the net spin polarization and f^a is a typical spin antisymmetric Landau parameter.

When λ is the shortest of these three lengths, that is, collisions dominate, then spin precession effects are unimportant and spin diffusion is described by the simple theory in Section 1.2.4(d). On the other hand, when the precession length r_L is the shortest of the three, then precession of the spins is the dominant physical effect and the equations of spin diffusion are radically altered from those of the collision-dominated regime. From (1.3.85) we see that the condition $r_L \ll \lambda$ is roughly the condition that $(\omega_l^+ - \omega_0)\tau \gg 1$. Typically, in He^3 at low pressure $\tau \sim (10^{-12}/T^2)$ sec, where T is in Kelvin; also $(\omega_l^+ - \omega_0) \sim 10^4$ (rad/sec)/gauss, so that, for example, at fields $\geq 10^4$G one enters the precession regime at $T \leq 10^{-2}K$.

The two situations $\lambda \ll r_L$, L and $r_L \ll \lambda$, L are those that are realized in practice and are those in which the theory is straightforward. If one examines the kinetic equation (1.3.80) in the spatially inhomogeneous case, one finds that an external time varying magnetic field will excite spherical harmonics of the spin distribution function with all l. However, the relative size of the $l + 1$ harmonic component compared with the lth harmonic component is of order

$$\frac{v_f/L}{\frac{1}{\tau} + (\omega_{l+1}^+ - \omega_l^+)} \sim \frac{L^{-1}}{\lambda^{-1} + r_L^{-1}}, \tag{1.3.86}$$

[cf. (1.2.78)] and is thus small in the long wavelength limit. To compute the terms (1.3.82) – (1.3.84) in the spin hydrodynamic equations, it is sufficient to use only the $l = 0$ and $l = 1$ harmonics of the spin distribution function. The situation is analogous to that of simple spin diffusion ($\lambda \ll r_L, L$) where the $l = 1$ component of the spin distribution that produces the spin current is of order λ/L times a typical scale of spin polarization. It is also analogous to the case of first sound where the $l = 2$ component of the distribution function produces viscous damping terms in the stress tensor that are of relative order λ/L compared with the contribution of the $l = 0$ component.

We now turn to the evaluation of the terms in (1.3.81). The leading contributions to Π_σ come from the $l = 0$ component of $\boldsymbol{\sigma}_\mathbf{p}$ and $f^a_{\mathbf{pp}'}\boldsymbol{\sigma}_{\mathbf{p}'}$, and from the external field term in $\mathbf{h}_\mathbf{p}$. Then in (1.3.82), $v_{\mathbf{p}_i}v_{\mathbf{p}_j} \to \delta_{ij}\, v_f^2/3$; the integral of $2\boldsymbol{\sigma}_\mathbf{p}$ yields the net spin polarization, while twice the integral of the $\mathbf{h}_\mathbf{p}$ term becomes $N(0)\,[f_0^a\boldsymbol{\sigma} - (\gamma\hbar/2)\boldsymbol{\mathscr{H}}]$. Thus in the long wavelength limit

$$(\Pi_\sigma)_{ik} = \delta_{ik}\left(1 + \frac{F_1^a}{3}\right)(1 + F_0^a)\,\frac{v_f^2}{3}\,[\boldsymbol{\sigma}(\mathbf{r}, t) - \boldsymbol{\sigma}^0(\mathbf{r}, t)], \tag{1.3.87}$$

where

$$\boldsymbol{\sigma}^0(\mathbf{r}, t) \equiv \frac{\gamma\hbar}{2}\,\frac{N(0)\boldsymbol{\mathscr{H}}(\mathbf{r}, t)}{1 + F_0^a} \tag{1.3.88}$$

is the local equilibrium magnetization that would be produced by a static field of the same value as the instantaneous external field.

To evaluate the precession term (1.3.83) we use (1.3.64) for $\mathbf{h}_\mathbf{p}$. The external field contribution to (1.3.83) is, by use of (1.3.79), simply $\gamma\mathbf{j}_\sigma \times \boldsymbol{\mathscr{H}}$. In the $\boldsymbol{\sigma}_\mathbf{p} \times \boldsymbol{\sigma}_{\mathbf{p}'}$ term the leading contribution is from the $l = 0$ component of $\boldsymbol{\sigma}_\mathbf{p}$ cross the $l = 1$ component of $\boldsymbol{\sigma}_{\mathbf{p}'}$ and vice versa. The net result for the precession term is then

$$\left(\frac{\partial j_{\sigma,i}}{\partial t}\right)_{\text{precession}} = \gamma j_{\sigma,i} \times \boldsymbol{\mathscr{H}} - \frac{2}{\hbar}\left(f_0^a - \frac{f_1^a}{3}\right)(j_{\sigma,i} \times \boldsymbol{\sigma}) \tag{1.3.89}$$

(the cross products involve the *spin* components of $\mathbf{j}_\sigma$).

We will make a simple relaxation time approximation to the collision term by writing for the $l = 1$ component of $\boldsymbol{\sigma}_\mathbf{p}$:

$$\left(\frac{\partial \boldsymbol{\sigma}_\mathbf{p}^{l=1}}{\partial t}\right)_{\text{collision}} = -\frac{\left(\boldsymbol{\sigma}_\mathbf{p} - \dfrac{\partial n_p^0}{\partial \varepsilon_p}\,\mathbf{h}_\mathbf{p}\right)^{l=1}}{\tau_D} = -\left(1 + \frac{F_1^a}{3}\right)\frac{\boldsymbol{\sigma}_\mathbf{p}^{l=1}}{\tau_D}. \tag{1.3.90}$$

The collision integral depends on the combination $\boldsymbol{\sigma}_\mathbf{p} - \partial n_p^0/\partial\varepsilon_p\,\mathbf{h}_\mathbf{p}$ as a consequence of the conservation of the true quasiparticle energies in collisions. [Cf. (1.2.60), as well as (1.2.72)]. Then

$$\left(\frac{\partial j_{\sigma,i}}{\partial t}\right)_{\text{collision}} = -\left(1 + \frac{F_1^a}{3}\right)\frac{j_{\sigma,i}}{\tau_D}. \tag{1.3.91}$$

Using the results (1.3.87), (1.3.89), and (1.3.91) in (1.3.81), we find the equation for the spin current:

$$\begin{aligned}\frac{\partial}{\partial t} j_{\sigma,i}(\mathbf{r}, t) &+ \left(1 + \frac{F_1^a}{3}\right)(1 + F_0^a)\frac{v_f^2}{3}\frac{\partial}{\partial r_i}[\sigma(\mathbf{r}, t) - \boldsymbol{\sigma}^0(\mathbf{r}, t)] \\ &= \gamma j_{\sigma,i}(\mathbf{r}, t) \times \mathscr{H}(\mathbf{r}, t) - \frac{2}{\hbar}\left(f_0^a - \frac{f_1^a}{3}\right) j_{\sigma,i}(\mathbf{r}, t) \times \boldsymbol{\sigma}(\mathbf{r}, t) \\ &\quad - \left(1 + \frac{F_1^a}{3}\right)\frac{j_{\sigma,i}(\mathbf{r}, t)}{\tau_D}.\end{aligned} \tag{1.3.92}$$

This equation, together with the spin conservation law (1.3.76), describes spin diffusion in the long wavelength limit. The gradient of the deviation $\delta\boldsymbol{\sigma} = \boldsymbol{\sigma} - \boldsymbol{\sigma}^0$ of the local spin density acts as a source of the spin current. Note that the equation for the spin current contains a precession term from the internal field, whereas the equation for the spin polarization contains only the precession in the external field.

In the weak field and static limit where precession and $\partial j/\partial t$ can be neglected (1.3.92) reduces to the earlier equation (1.2.144) for spin diffusion, with τ_D related to the spin diffusion coefficient by

$$D_\sigma = \frac{v_f^2}{3}(1 + F_0^a)\tau_D. \tag{1.3.93}$$

Comparing this equation with (1.2.159) one finds the exact expression for τ_D in this limit. More generally, when precession is important, this result for τ_D is somewhat modified.*

In the spatially uniform case (1.3.92) describes the homogeneous spin modes for $l = 1$ considered in the previous section, only now with the inclusion of damping (in the relaxation time approximation). The complex frequencies of the $l = 1$ modes are

$$\omega_1^{\pm} = \frac{1 + F_1^a/3}{i\tau_D} \pm \left[\omega_0 - \frac{2}{\hbar}\sigma^0\left(f_0^a - \frac{f_1^a}{3}\right)\right] \tag{1.3.94}$$

which reduces to (1.3.72) for $l = 1$ in the absence of quasiparticle collisions.

In a typical spin echo experiment one first polarizes the spins in a slightly inhomogeneous external field $\mathscr{H}(\mathbf{r})$, and then applies a radio frequency pulse that rotates the spin directions with respect to the applied field. The spin

*In particular, when the molecular field precession time is small compared with the collision time, the appropriate collision time to use is the simplest variational estimate of the collision time for spin diffusion, $(\tau_D)_{\text{var}} = (3/2\pi^2)\tau(1 - \lambda_D)^{-1}$, described in Section 1.2.4 (d). [cf. the discussion of zero-sound attenuation in Section 1.3.1 (c)].

polarizations at different points precess at slightly different rates due to the inhomogeneity of the external field. This produces a gradient of the spin polarization, and, hence, from (1.3.92), a spin current. After a few relaxation times τ_D the spin current reaches a quasi-steady state value. The spin polarization precesses with the local Larmor frequency, and therefore in the quasi-steady state the spin current also precesses about $\mathscr{H}(\mathbf{r})$ with the local Larmor frequency, that is,

$$\frac{\partial}{\partial t} j_{\sigma,i}(\mathbf{r}, t) = \gamma j_{\sigma,i}(\mathbf{r}, t) \times \mathscr{H}(\mathbf{r}). \tag{1.3.95}$$

Using this result in (1.3.92) we see that $j_{\sigma,i}$ is given as the solution of

$$j_{\sigma,i}(\mathbf{r}, t) + \Upsilon j_{\sigma,i}(\mathbf{r}, t) \times \boldsymbol{\sigma}(\mathbf{r}, t) = -D_\sigma \frac{\partial}{\partial r_i}[\boldsymbol{\sigma}(\mathbf{r}, t) - \boldsymbol{\sigma}^0(\mathbf{r}, t)], \tag{1.3.96}$$

where

$$\Upsilon = \frac{2}{\hbar} \tau_D \frac{(f_0^a - f_1^a/3)}{1 + F_1^a/3}. \tag{1.3.97}$$

Equation 1.3.96 is a simple linear equation for $j_{\sigma,i}$. If we take the scalar product (in spin space) of both sides of (1.3.96) with the unit vector $\hat{\sigma}$ in the direction of the local spin polarization, the Υ term drops out and we see that the component of the right side that is parallel to $\boldsymbol{\sigma}$ induces a current of the parallel component of spin, the current being given by a simple Fick equation [as in (1.2.144)]. On the other hand the component of the right side of (1.3.96) that is perpendicular to $\boldsymbol{\sigma}$ drives a current of the spin component that is also perpendicular to $\boldsymbol{\sigma}$ but at an *angle* $\theta = \tan^{-1}\Upsilon\sigma$ to the driving term. The general solution for $j_{\sigma,i}$ is

$$j_{\sigma,i} = -\frac{D_\sigma}{1 + \Upsilon^2\sigma^2}\left[\frac{\partial\delta\boldsymbol{\sigma}}{\partial r_i} + \Upsilon\boldsymbol{\sigma} \times \frac{\partial\delta\boldsymbol{\sigma}}{\partial r_i} + \Upsilon^2\boldsymbol{\sigma}\left(\boldsymbol{\sigma}\cdot\frac{\partial\delta\boldsymbol{\sigma}}{\partial r_i}\right)\right], \tag{1.3.98}$$

where

$$\delta\boldsymbol{\sigma}(\mathbf{r}, t) = \boldsymbol{\sigma}(\mathbf{r}, t) - \boldsymbol{\sigma}^0(\mathbf{r}, t). \tag{1.3.99}$$

That θ is nonzero, a phenomenon known as the Leggett–Rice effect, is entirely a consequence of the interactions between quasiparticles; it has no analogue in a free Fermi gas. If $f_0^a = 0 = f_1^a$ then Υ vanishes and the spin current obeys $j_{\sigma,i} = -D_\sigma \partial\delta\boldsymbol{\sigma}/\partial r_i$.

Finally we substitute (1.3.98) into (1.3.76) to find the Leggett–Rice equations for spin diffusion:*

*Note that this equation correctly involves $\delta\boldsymbol{\sigma}$ rather than $\boldsymbol{\sigma}$ itself as in equation (25) of Reference [41]; this modification, however, does not affect the principal results on spin echoes in that paper.

$$\frac{\partial \boldsymbol{\sigma}(\mathbf{r}, t)}{\partial t} - \gamma \boldsymbol{\sigma}(\mathbf{r}, t) \times \mathscr{H}(\mathbf{r})$$
$$= D_\sigma \sum_{i=1}^{3} \frac{\partial}{\partial r_i} \left\{ \frac{1}{1 + \Upsilon^2 \sigma^2} \left[\frac{\partial \delta \boldsymbol{\sigma}}{\partial r_i} + \Upsilon \boldsymbol{\sigma} \times \frac{\partial \delta \boldsymbol{\sigma}}{\partial r_i} + \Upsilon^2 \boldsymbol{\sigma} \left(\boldsymbol{\sigma} \cdot \frac{\partial \delta \boldsymbol{\sigma}}{\partial r_i} \right) \right] \right\}. \quad (1.3.100)$$

The theory of the above effects was first discussed by Leggett and Rice [40], and the application of (1.3.100) to the theory of spin echo experiments was given by Leggett [41]. Doniach [42] has given a discussion of how one may understand a number of the results for spin echo experiments in terms of the spin waves discussed in Section 1.3.2(b).

One of the principal theoretical results of the Leggett–Rice equations is that in an experiment in which the spins are rotated by an angle ϕ from the z axis by the r.f. field and then rotated by 180° about an axis in the x–y plane at times $t_0/2$, $3t_0/2$, $5t_0/2$ . . . $(n - \frac{1}{2})t_0$ . . ., the relative height h_n of the nth spin echo (at time nt_0) is given by the solution of

$$\ln h_n - \frac{\frac{1}{2}(\Upsilon \sigma_0 \sin \phi)^2 (1 - h_n^2)}{1 + (\Upsilon \sigma_0 \cos \phi)^2} = - \frac{1}{12} \frac{n D_\sigma \gamma^2 G^2 t_0^3}{1 + (\Upsilon \sigma_0 \cos \phi)^2}, \quad (1.3.101)$$

where G is the gradient of the applied magnetic field and σ_0 is the initial spin polarization. When $\Upsilon \sigma_0$ is $\ll 1$, (1.3.101) reduces to the familiar exponential envelope function for h_n. Experiments measure the effective diffusion constant

$$D_{\text{eff}} \equiv \frac{D_\sigma}{1 + (\Upsilon \sigma_0 \cos \phi)^2}; \quad (1.3.102)$$

if we write

$$\Upsilon \sigma_0 \cos \phi \equiv \Omega_m \tau_D, \quad (1.3.103)$$

where

$$\Omega_m = \frac{2}{\hbar} \frac{f_0^a - f_1^a/3}{1 + F_1^a/3} \sigma_0 \cos \phi \quad (1.3.104)$$

is an effective precession frequency due to the internal molecular field, then

$$D_{\text{eff}} \sim \frac{\tau_D}{1 + \Omega_m^2 \tau_D^2}. \quad (1.3.105)$$

In the collision dominated regime at higher T, $\Omega_m \tau_D \ll 1$ and $D_{\text{eff}} \propto T^{-2}$; however, at low T where precession effects dominate, $\Omega_m \tau_D \gg 1$ and $D_{\text{eff}} \propto T^2$.

Experimental evidence for the existence of the Leggett–Rice effect in liquid ^{3}He was first obtained by Abel and Wheatley [43], and detailed experiments

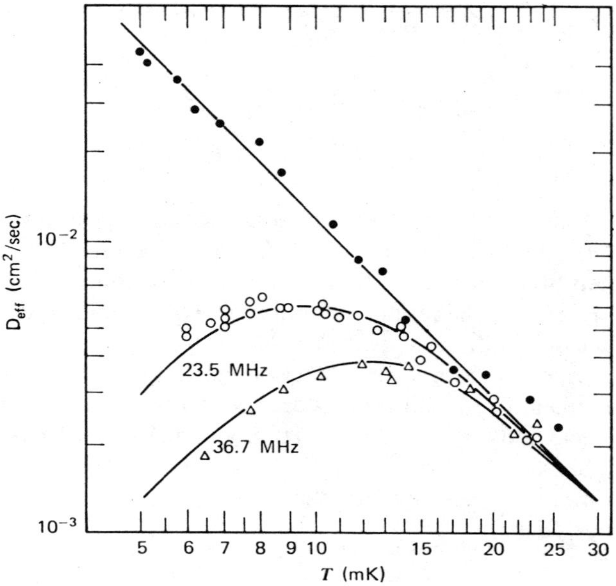

FIGURE 1.5. Values of effective spin diffusion coefficient $D_{\rm eff}$ measured in spin echo experiment in ^{3}He at $P = 0$ by Corruccini, Osheroff, Lee, and Richardson [44]. The solid points are data for $\phi = 90°$, and give directly the spin diffusion coefficient D_σ, as one can see from (1.3.102). The open circles and triangles are data for $\phi = 18°$. The magnetization σ_0 is the equilibrium polarization in the magnetic fields that give rise to the Larmor frequencies ν_L shown in the graph ($\sigma_0 = h\nu_L N(0)/(1 + F_0^a)$). The solid lines are fits to the data. These are given by (1.3.102), with τ_D assumed to vary as T^{-2}. From the fits one finds $|A_1^a/3 - A_0^a| = |1/(1 + F_0^a) - 1/(1 + F_1^a/3)| = 1.95 \pm 0.1$ or $F_1^a = -0.15 \pm 0.3$. The quoted uncertainty indicates only the scatter in the data and does not take into account possible errors in the determination of τ_D from D_σ.

to verify (1.3.100) were carried out by Corruccini et al. [44]. In Figure 1.5 we show some of the data of Corruccini et al.; this exhibits well the transition between the collision-dominated and collisionless regimes. The experiments agree well with Leggett's predictions, and provide conclusive evidence for the existence of spin waves in liquid ^{3}He (and solutions of ^{3}He in ^{4}He); this is very satisfying, since the spin waves are a very direct consequence of the molecular field produced by the quasiparticle interactions.

The advantage of low temperature spin echo measurements is that they provide in principle a way to measure F_1^a directly. Corruccini et al. [44] deduce from their spin echo experiments in low pressure ^{3}He that $F_1^a = 0.15 \pm 0.3$;

but finite temperature effects, which were not included in their analysis, increase the uncertainty in this determination of F_1^a.

1.3.3. Response Functions, Inequalities, and Form Factors

In the previous sections (1.3.1, 1.3.2) we investigated the collective modes of a Fermi liquid by considering the response of the system to an external potential. In this section we shall consider the response in more general terms, using the language of response functions, to discuss properties of the dynamic and static structure factors, as well as derive some inequalities for the Landau parameters.

According to standard linear response theory [6, 45] the change in the density, $\delta n(\mathbf{r}, t) = e^{i(\mathbf{q}\cdot\mathbf{r}-(\omega+i\eta)t)}\,\delta n(\mathbf{q}, \omega)$, produced by a space- and time-dependent external scalar potential, $U(\mathbf{r}, t) = e^{i(\mathbf{q}\cdot\mathbf{r}-(\omega+i\eta)t)}\,U(\mathbf{q}, \omega)$, acting on a system initially in its ground state, via the interaction Hamiltonian $\int d^3r n(\mathbf{r})U(\mathbf{r}, t)$, is given by

$$\delta n(\mathbf{q}, \omega) = \chi(\mathbf{q}, \omega)U(\mathbf{q}, \omega), \tag{1.3.106}$$

where the response function χ has the form*

$$\chi(\mathbf{q}, \omega) = \sum_{n\neq 0} |(\rho_{\mathbf{q}}^{\dagger})_{n0}|^2 \frac{2\omega_{n0}}{(\omega + i\eta)^2 - \omega_{n0}^2}\ ; \tag{1.3.107}$$

$\rho_{\mathbf{q}}$ is the spatial Fourier transform of the (Schrödinger representation) particle density operator, and $(\rho_{\mathbf{q}})_{n0}$ denotes its matrix element between an excited state, labelled by the index n, and the ground state 0. The excitation energy ω_{n0} is given by

$$\omega_{n0} = E_n - E_0, \tag{1.3.108}$$

where E_i is the energy of the state $|i\rangle$.

When the density operator acts on the ground state of a Fermi liquid, the simplest thing that can happen is that a quasiparticle initially below the Fermi surface becomes promoted to a state above the Fermi surface, leaving behind a quasihole below the Fermi surface; in other words a quasiparticle–quasihole pair is created. Another possibility is that a zero sound quantum is created. By comparing (1.3.107), which describes the response in microscopic terms, with the macroscopic results of Section 1.3.1 we can gain much information on the nature of the low-lying states of the system and the magnitude of the corresponding matrix elements. The Landau kinetic equation used to describe the response in Section 1.3.1 takes into account, through the $f_{\mathbf{pp}'}$, repeated interactions of a single quasiparticle–quasihole pair, and it may therefore be used to investigate the single pair states and the collective

*For simplicity we shall suppress factors of $\hbar$ and take the volume of the system to be unity in the calculations described in the rest of the Chapter.

modes, which come about as a result of repeated interactions of a single pair. At long wavelengths the quasiparticle contribution to $\delta n(\mathbf{q}, \omega)$ is, according to (1.3.08), $N(0)\nu_0(\mathbf{q}, \omega)$, where ν_0, given by (1.3.11), is the angular average of the spin-symmetric distortion of the Fermi surface. We show now that the matrix elements $(\rho_{\mathbf{q}}^{\dagger})_{n0}$ for creating a single pair are independent of q for small q, as they are in the free Fermi gas, and also that the matrix element for creating a zero sound quantum behaves as $\sqrt{q}$, exactly as does the matrix element $(\rho_{\mathbf{q}}^{\dagger})_{n0}$ for creating an acoustic phonon in a solid. To see these results we use the fact that the quasiparticle part of the density response function

$$\chi_{\text{Landau}}(\mathbf{q}, \omega) = \frac{N(0)\nu_0(\mathbf{q}, \omega)}{U(\mathbf{q}, \omega)}$$

depends only on the variable $s = \omega/v_f q$. [Cf. (1.3.13).] Thus, both the contribution from the quasiparticle–quasihole states and the collective modes are separately functions of s. The contribution to χ, (1.3.107), coming from an intermediate state $|c\rangle$ containing a single collective mode quantum is $|(\rho_{\mathbf{q}}^{\dagger})_{c0}|^2 2\omega_{c0}/[(\omega + i\eta)^2 - \omega_{c0}^2]$. Since $\omega_{c0} \propto q$, one must have $(\rho_{\mathbf{q}}^{\dagger})_{c0} \propto q^{1/2}$ if this contribution is to depend only on $s = \omega/v_f q$.

The matrix element $(\rho_{\mathbf{q}}^{\dagger})_{n0}$ between the ground state and a single pair state consisting of a quasiparticle $\mathbf{p} + \mathbf{q}, \sigma$ and a quasihole $\mathbf{p}, \sigma$, can depend, in the small $\mathbf{q}$ limit and for p in the vicinity of the Fermi surface, only on $\omega_{n0} = \varepsilon_{\mathbf{p}+\mathbf{q}} - \varepsilon_{\mathbf{p}} = \mathbf{q} \cdot \mathbf{v}_{\mathbf{p}}$ and on q. Let us write $(\rho_{\mathbf{q}}^{\dagger})_{n0} = M(\omega_{n0}, q)$; as we shall now see, M depends only on ω_{n0}/q—that is, only on the angle between $\mathbf{p}$ and $\mathbf{q}$. To show this we look at the imaginary part of χ, given for $\omega > 0$ by

$$\operatorname{Im}\chi(\mathbf{q}, \omega) = -\pi \sum_{n \neq 0} |(\rho_{\mathbf{q}}^{\dagger})_{n0}|^2 \delta(\omega - \omega_{n0}). \tag{1.3.109}$$

Then the single pair contribution can be written as

$$\begin{aligned} \operatorname{Im}\chi_{\text{Pair}}(\mathbf{q}, \omega) &= -\pi |M(\omega, q)|^2 \sum_{\mathbf{p}, \sigma} n_{p\sigma}^0 (1 - n_{\mathbf{p}+\mathbf{q}, \sigma}^0)\delta(\omega - \varepsilon_{\mathbf{p}+\mathbf{q}} + \varepsilon_{\mathbf{p}}) \\ &= -|M(\omega, q)|^2 \frac{\pi N(0)}{2} \frac{\omega}{v_f q} \Theta(v_f q - \omega). \end{aligned} \tag{1.3.110}$$

The distribution functions ensure that in the ground state the quasiparticle state $\mathbf{p}, \sigma$ is occupied and the quasiparticle state $\mathbf{p} + \mathbf{q}, \sigma$ unoccupied, and therefore creation of a pair with a quasiparticle $\mathbf{p} + \mathbf{q}, \sigma$ and a quasihole $\mathbf{p}, \sigma$ is not forbidden by the Pauli principle. Since the single pair contribution $\operatorname{Im}\chi_{\text{Pair}}$ and the density of single pair states depend only on s, from (1.3.109) it follows that the matrix element $(\rho_{\mathbf{q}}^{\dagger})_{n0}$ for small q can depend only on $\omega_{n0}/v_f q$, and hence is independent of q.

The density operator can also couple the ground state to states that contain more than just a single pair or zero sound quantum; we refer to these more complicated states as *multipair* excited states. The behavior of the matrix elements for multipair excitations may be found from the equation of continuity, which expresses conservation of particles. In operator form the equation is $-i[\rho(\mathbf{r}), H] + \nabla\cdot\mathbf{j}(\mathbf{r}) = 0$, where H is the Hamiltonian of the system. Taking matrix elements of this equation one finds

$$\omega_{n0}(\rho_{\mathbf{q}}^{\dagger})_{n0} = -\mathbf{q}\cdot(\mathbf{j}_{\mathbf{q}}^{\dagger})_{n0}. \tag{1.3.111}$$

Here $\mathbf{j}(\mathbf{r})$ is the operator for the particle current and $\mathbf{j}_{\mathbf{q}}$ its Fourier transform. In contrast to the excitation energy of the single pair and single zero-sound quantum states, which vary as $q \to 0$, the excitation energy of the multipair states tends in general to a finite value. Also the matrix elements of the current operator generally tend to a constant in this limit. Thus from (1.3.111) it follows immediately that $(\rho_{\mathbf{q}}^{\dagger})_{n0} \sim q$ as $q \to 0$. For a single component Galilean invariant system $(\rho_{\mathbf{q}}^{\dagger})_{n0}$ must fall off even more rapidly. In the limit $q \to 0$, $\mathbf{j}_{\mathbf{q}}$ tends to the total current $\mathbf{J}$. If the total current is conserved, this implies that for $q \to 0$, $\mathbf{j}_{\mathbf{q}}$ can have nonvanishing matrix elements only to states whose excitation energy vanishes. Thus as $q \to 0$, $\omega_{n0}(\mathbf{j}_{\mathbf{q}}^{\dagger})_{n0} \to 0$ and for multipair states must behave as q^{α}, where $\alpha > 0$; therefore, from the continuity equation $(\rho_{\mathbf{q}}^{\dagger})_{n0}$ must behave as $q^{1+\alpha}$.

From these results we can determine the conditions under which the multipair contribution to χ is negligible. We first write χ as

$$\chi = \chi_{\text{Landau}} + \chi_{\text{Multipair}} \tag{1.3.112}$$

where χ_{Landau} is the contribution to χ (1.3.107) coming from intermediate states containing a single quasiparticle–quasihole pair or a zero sound quantum, and $\chi_{\text{Multipair}}$ is the contribution to χ coming from multipair intermediate states. χ_{Landau} may be calculated using Landau theory (Section 1.3.1) and is equal to $N(0)\nu_0/U$; it depends on ω and q only through the variable $s = \omega/v_f q$, and for $s \gg 1$ it behaves as $1/s^2$. At small wavenumbers $\chi_{\text{Multipair}}$ falls off at least as rapidly as q^2, since the matrix elements fall off at least as rapidly as q. From these properties of χ_{Landau} and $\chi_{\text{Multipair}}$, it follows that in the long-wavelength low-frequency limit $\chi_{\text{Multipair}}$ is negligible, compared with χ_{Landau}, independent of how the limit is taken.

With the help of the f sum rule we can derive an inequality for the Landau parameters F_1^s and F_1^a. Consider the quantity $\text{Lim}_{\omega\to 0}\,\text{Lim}_{q\to 0}(\omega^2/q^2)\chi(q, \omega)$; since $\chi_{\text{Multipair}}$ behaves as q^2, or $q^{2(1+\alpha)}$, and is independent of frequency for $\omega \to 0$, this quantity contains only contributions from χ_{Landau}. As far as the quasiparticles are concerned, the limit considered is a high frequency one, since the excitation frequencies of the single pairs and collective modes ($\sim q$) are made to go to zero before ω. From (1.3.107) we see that in this limit

$$\chi_{\text{Landau}} \to \frac{1}{\omega^2} \sum_n{}' \left|(\rho_{\mathbf{q}}^{\dagger})_{n0}\right|^2 2\omega_{n0}, \tag{1.3.113}$$

where the prime on the sum denotes that multipair states are not to be included in the sum; thus we find

$$\lim_{\omega\to 0}\lim_{q\to 0}\frac{\omega^2}{q^2}\chi(q,\omega) = \lim_{\omega\to 0}\lim_{q\to 0}\frac{\omega^2}{q^2}\chi_{\text{Landau}}(q,\omega)$$
$$= \lim_{q\to 0}\frac{1}{q^2}\sum_n{}' |(\rho_{\mathbf{q}}^\dagger)_{n0}|^2 2\omega_{n0}. \tag{1.3.114}$$

To derive the inequality we first observe that since $|(\rho_{\mathbf{q}}^\dagger)_{n0}|^2\omega_{n0}$ is positive for each n,

$$\sum_n{}' |(\rho_{\mathbf{q}}^\dagger)_{n0}|^2 2\omega_{n0} \leq \sum_n |(\rho_{\mathbf{q}}^\dagger)_{n0}|^2 2\omega_{n0}, \tag{1.3.115}$$

where the sum on the right is over all states, including the multipair ones. The right side is given by the f-sum rule

$$\sum_n |(\rho_{\mathbf{q}}^\dagger)_{n0}|^2 2\omega_{n0} = \langle 0|[[\rho_{\mathbf{q}}, H], \rho_{\mathbf{q}}^\dagger]|0\rangle = n\frac{q^2}{m}, \tag{1.3.116}$$

which follows simply from particle conservation, and the assumption that the interaction between particles is independent of velocity; m is the bare mass. On the other hand $\chi_{\text{Landau}} = N(0)\nu_0/U$ may be calculated directly from (1.3.10). All we need is the $1/\omega^2$ term in χ, which is most easily determined by expanding $\nu_{\mathbf{p}}$ in powers of $1/\omega$:

$$\frac{\nu_{\mathbf{p}}}{U} = \frac{\mathbf{q}\cdot\mathbf{v}_{\mathbf{p}}}{\omega - \mathbf{q}\cdot\mathbf{v}_{\mathbf{p}}}\left[1 - \sum_{\mathbf{p}'} f^s_{\mathbf{p}\mathbf{p}'}\frac{\partial n^0_{p'}}{\partial\varepsilon_{p'}}\frac{\mathbf{q}\cdot\mathbf{v}_{\mathbf{p}'}}{\omega - \mathbf{q}\cdot\mathbf{v}_{\mathbf{p}'}} + \cdots\right] \tag{1.3.117}$$

$$= \frac{\mathbf{q}\cdot\mathbf{v}_{\mathbf{p}}}{\omega} + \frac{(\mathbf{q}\cdot\mathbf{v}_{\mathbf{p}})^2}{\omega^2} - \sum_{\mathbf{p}'}\frac{\mathbf{q}\cdot\mathbf{v}_{\mathbf{p}}}{\omega} f^s_{\mathbf{p}\mathbf{p}'}\frac{\partial n^0_{p'}}{\partial\varepsilon_{p'}}\frac{\mathbf{q}\cdot\mathbf{v}_{\mathbf{p}'}}{\omega} + \mathcal{O}\left(\frac{1}{\omega^3}\right). \tag{1.3.118}$$

Summing over $\mathbf{p}$ we find

$$\frac{N(0)\nu_0}{U} = \frac{N(0)}{3}\frac{v_f^2 q^2}{\omega^2}\left(1 + \frac{1}{3}F_1^s\right) + \mathcal{O}\left(\frac{1}{\omega^4}\right), \tag{1.3.119}$$

or

$$\lim_{\omega\to 0}\lim_{q\to 0}\frac{\omega^2}{q^2}\chi(q,\omega) = \frac{n}{m^*}\left(1 + \frac{1}{3}F_1^s\right). \tag{1.3.120}$$

From (1.3.114), (1.3.115), (1.3.116), and (1.3.120) it therefore follows that

$$\frac{m}{m^*}\left(1 + \frac{1}{3}F_1^s\right) \leq 1. \tag{1.3.121}$$

This result was first derived by Leggett [46].

For the case of a single component Galilean invariant system the multipair contribution to the f-sum rule behaves as $q^{2(1+\alpha)}$, and is therefore negligible in the long-wavelength limit. Thus for $q \to 0$ the equality holds in (1.3.115) and therefore also in (1.3.121), and one immediately recovers the well-known relation (1.1.56) for the effective mass.

By considering the response of a Fermi liquid to a space- and time-dependent magnetic field one can show by precisely similar methods that

$$\frac{m}{m^*}\left(1 + \frac{1}{3}F_1^a\right) \leq 1. \tag{1.3.122}$$

In this case the equality holds only if the interaction between fermions conserves spin current; this is true only for pathological forms of the interaction, for example, if there were no interaction between opposite spins. For systems with interactions that conserve fermion particle current one can see immediately from the effective mass relation (1.1.56) and (1.3.122) that

$$F_1^a \leq F_1^s. \tag{1.3.123}$$

This inequality seems to be well satisfied for liquid ^{3}He.

We have seen above that in the long-wavelength low-frequency limit the density response function may be calculated within the framework of Landau theory. Generally, the linear response $\delta\langle A\rangle(\mathbf{q}, \omega)$ of the physical quantity corresponding to the operator A caused by a probe, of wavenumber $\mathbf{q}$ and frequency ω, coupled to some other operator B, is given, analogously to (1.3.106), in terms of the response function:

$$\chi^{AB}(\mathbf{q}, \omega) = \sum_n \left\{\frac{(A_\mathbf{q})_{0n}(B_{-\mathbf{q}})_{n0}}{\omega - \omega_{n0} + i\eta} - \frac{(B_{-\mathbf{q}})_{0n}(A_\mathbf{q})_{n0}}{\omega + \omega_{n0} + i\eta}\right\}. \tag{1.3.124}$$

The response function may again be divided into a quasiparticle part, and a multipair part. If at least one of the operators $A_{\mathbf{q}=0}$ and $B_{\mathbf{q}=0}$ is a conserved quantity, the multipair contribution to χ^{AB} is negligible compared with the quasiparticle part, at long wavelengths and low frequencies, since then at least one of the multipair matrix elements is of order q; one can see this by inspecting the equation of motion for $A_\mathbf{q}$ or $B_\mathbf{q}$, using the same argument that we used to show that $(\mathbf{j}_\mathbf{q})_{n0} \sim q^\alpha$ in a single component Galilean invariant system. However, if neither of the operators $A_{\mathbf{q}=0}$ or $B_{\mathbf{q}=0}$ are conserved quantities (as in the case of the spin current–spin current response function) there will be multipair contributions to $\chi^{AB}(\mathbf{q}, \omega)$ at long wavelengths and low frequencies. The existence of such multipair contributions in sum rules and reponse functions does not indicate a breakdown [47] of the applicability of the Landau theory to long wavelength, low frequency phenomena.

The operators of physical interest are generally either quantities conserved in the $\mathbf{q} \to 0$ limit (particle number, total spin, and the particle current in a Galilean invariant system) or currents of conserved quantities (the spin current, the stress tensor in a Galilean invariant system, and the particle current in a non-Galilean-invariant system), since these operators can be directly coupled to using the long wavelength probes available in the laboratory. One has explicit expressions for such physical quantities in terms of the quasiparticle distribution function [cf. (1.2.24), (1.2.25), (1.2.36), (1.2.140), and (1.2.141)]. The Landau contribution to χ^{AB} for such operators is proportional to the correlation function of the distortions of the Fermi surface having the same rotational and spin symmetries as A and B. Leggett [48] discusses in detail the general determination of the Landau contributions.

A quantity closely related to the density response function is the dynamic form factor, which at zero temperature is given by

$$S(\mathbf{q}, \omega) = \sum_n |(\rho_{\mathbf{q}}^\dagger)_{n0}|^2 \delta(\omega - \omega_{n0}), \tag{1.3.125}$$

$$= -\frac{1}{\pi} \operatorname{Im}\chi(\mathbf{q}, \omega)\Theta(\omega). \tag{1.3.126}$$

A number of properties of this function can be measured by neutron, X-ray, and light scattering experiments. For small frequencies and wavenumbers χ may, as described above, be computed by the Landau theory, and $S(\mathbf{q}, \omega)$ then evaluated using (1.3.126); such calculations are described in detail by Tan [49] for the collisionless regime and by Safier and Widom [50], who allowed for a finite collision time.*

In the hydrodynamic regime $S(\mathbf{q}, \omega)$ is given by the well-known results obtained from hydrodynamics [52]; for the case of a Fermi liquid at low temperature the central diffusive peak due to entropy fluctuations, which is of strength $1 - c_V/c_P$ relative to the strength of the doublet due to scattering by sound waves, is absent since as $T \to 0$ the specific heat at constant volume c_V becomes equal to the specific heat at constant pressure c_P.

In the collisionless regime and for $\mathbf{q} \to 0$, there are two types of contribution to $S(\mathbf{q}, \omega)$, corresponding to the excitation of longitudinal zero sound waves, and quasiparticle–quasihole pairs. For liquid ^{3}He the contribution of the zero sound mode is far more important than that of the quasiparticle–quasihole pairs because, due to the large positive value of F_0^s, the density operator couples to the single pair states rather weakly. We refer the reader to the original papers for further details.

X-ray scattering measurements provide a direct measurement of the static structure factor

$$S(q) = \frac{1}{n}\int_0^\infty S(\mathbf{q}, \omega)d\omega. \tag{1.3.127}$$

*The first calculations of $S(\mathbf{q}, \omega)$ were carried out by Abrikosov and Khalatnikov [51], using a somewhat different approach. It is illuminating to compare their calculations with those of References [49] and [50].

[This may be calculated directly from $S(\mathbf{q}, \omega)$; however, the results are not particularly illuminating so we shall not describe them here.] Two useful bounds on $S(q)$ may be obtained from sum rules. The f-sum rule (1.3.116), when written in terms of $S(\mathbf{q}, \omega)$, is

$$\int_0^\infty d\omega\, \omega\, S(\mathbf{q}, \omega) = \frac{nq^2}{2m}. \tag{1.3.128}$$

The compressibility sum rule [6], states that $-\chi(\mathbf{q}, 0)$ in the long-wavelength limit tends to the thermodynamic derivative $\partial n/\partial\mu = n/mc_1^2$, where c_1 is the (isothermal) first sound velocity. Combining (1.3.125) and (1.3.107) we find

$$-\lim_{q\to 0} \frac{1}{2}\chi(\mathbf{q}, 0) = \lim_{q\to 0} \int_0^\infty d\omega \frac{S(\mathbf{q}, \omega)}{\omega} = \frac{n}{2mc_1^2}. \tag{1.3.129}$$

The first bound is obtained by exploiting the fact that

$$\overline{(\omega - \bar{\omega})^2} \geq 0, \tag{1.3.130}$$

where

$$\overline{g(\omega)} \equiv \int_0^\infty d\omega\, g(\omega) \frac{S(\mathbf{q}, \omega)}{\omega} \Big/ \int_0^\infty d\omega \frac{S(\mathbf{q}, \omega)}{\omega}; \tag{1.3.131}$$

this is a consequence of the fact that $S(\mathbf{q}, \omega)/\omega$ is positive for $\omega > 0$. Using this inequality, the sum rules (1.3.128) and (1.3.129), and the definition of $S(q)$ (1.3.127) one finds [53].

$$S(q) \leq \frac{q}{2mc_1}, \qquad q \to 0. \tag{1.3.132}$$

A second inequality may be obtained from the f-sum rule; we shall assume that the temperature is so low that quasiparticle collisions can be neglected. We showed earlier in this section that multipair excitations do not contribute to the f-sum rule in the long-wavelength limit if the system is Galilean invariant. Thus, for such a system

$$\sum_n{}' |(\rho_{\mathbf{q}}^\dagger)_{n0}|^2 \omega_{n0} = \frac{nq^2}{2m}, \qquad q \to 0; \tag{1.3.133}$$

the prime on the sum again indicates that multipair states are excluded. Now ω_{n0} must be less than the maximum frequency of any single pair state or zero sound mode; thus one finds immediately from (1.3.133) and (1.3.127) that

$$S(q) \geq \frac{q}{2mc_{\max}}, \qquad q \to 0, \tag{1.3.134}$$

where $c_{\max}$ is the largest* longitudinal zero sound velocity ($> v_f$) or, if there are no collective modes, v_f.

*Recall that the zero sound dispersion relation for a given m value may have more than one root.

One immediate consequence of the inequalities (1.3.132) and (1.3.134) is that

$$c_{\max} \geq c_1, \tag{1.3.135}$$

or, in other words, the velocity of the fastest longitudinal zero sound mode, or if there is no such mode, the Fermi velocity, must exceed the first sound velocity. Combining the inequalities (1.3.134) and (1.3.132) one finds

$$\frac{nq}{2m\,c_{\max}} \leq S(q) \leq \frac{nq}{2mc_1}, \quad q \to 0. \tag{1.3.136}$$

For liquid ^{3}He the observed zero sound velocity differs from the first sound velocity by no more than a few percent, and therefore the inequalities (1.3.136) pin down $S(q)$ rather precisely, if one is prepared to make the not unreasonable assumption that there is no longitudinal zero sound mode with a velocity greater than that of the observed zero sound mode.

1.4. SCATTERING OF QUASIPARTICLES AND FINITE TEMPERATURE EFFECTS

1.4.1. Landau Parameters and Scattering Amplitudes

We turn now to studying the relation between the quasiparticle scattering amplitude and the Landau parameters. There are two reasons for doing this: first, to learn how to generate reasonable approximations for the scattering amplitude for use in the transport theory described in Section 1.2, and, second, to relate the finite temperature properties of a Fermi liquid to the quasiparticle properties.

Let us consider a Fermi liquid initially in a state $|i\rangle$ to which we apply a perturbation

$$\int n_{\mathbf{p}'}(\mathbf{r}) U_{\mathbf{p}'}(\mathbf{r}, t)\, d^3r, \tag{1.4.1}$$

where $U_{\mathbf{p}'}(\mathbf{r}, t)$ is a real external potential coupled to the local quasiparticle number operator $n_{\mathbf{p}'}(r)$. In terms of quasiparticle creation and annihilation operators, $a^\dagger$ and a, we have [cf. (1.2.3)]

$$n_{\mathbf{p}}(\mathbf{r}) = \int \frac{d^3q}{(2\pi)^3}\, e^{-i\mathbf{q}\cdot\mathbf{r}} a^\dagger_{\mathbf{p}+\mathbf{q}/2} a_{\mathbf{p}-\mathbf{q}/2}. \tag{1.4.2}$$

A single Fourier component, $e^{i\mathbf{q}\cdot\mathbf{r}-i(\omega+i\eta)t} U_{\mathbf{p}'}(\mathbf{q}, \omega)$, of $U_{\mathbf{p}'}(\mathbf{r}, t)$ in the perturbation, when acting on the initial state, scatters a quasiparticle from state $\mathbf{p} - \mathbf{q}/2$, if occupied, to state $\mathbf{p} + \mathbf{q}/2$, if unoccupied; in other words, it creates a quasiparticle–quasihole excitation. Here the factor $e^{\eta t}$, where η is a positive infinitesimal, slowly turns the perturbation on. If we denote (in the interaction representation for U) the state of the system after a while by

$$|\Psi(t)\rangle = |i\rangle + |\delta\Psi(t)\rangle, \tag{1.4.3}$$

then the amplitude for finding the system in a final state $|\mathbf{p}+\mathbf{q}/2, \mathbf{p}-\mathbf{q}/2;f\rangle$ that differs from $|i\rangle$ by the presence a quasiparticle of momentum $\mathbf{p} + \mathbf{q}/2$ and a quasihole of momentum $\mathbf{p} - \mathbf{q}/2$ (neither present in the initial state), is given to first order in U by elementary time dependent perturbation theory as

$$\begin{aligned}&\langle \mathbf{p}+\mathbf{q}/2,\mathbf{p}-\mathbf{q}/2;f|\delta\Psi(t)\rangle \\ &\quad = \frac{e^{-i\omega t}\langle \mathbf{p}+\mathbf{q}/2,\mathbf{p}-\mathbf{q}/2;f|a^\dagger_{\mathbf{p}'+\mathbf{q}/2}a_{\mathbf{p}'-\mathbf{q}/2}|i\rangle}{\omega + i\eta - \varepsilon_{\mathbf{p}+\mathbf{q}/2} + \varepsilon_{\mathbf{p}-\mathbf{q}/2}}\, U_{\mathbf{p}'}(\mathbf{q},\omega).\end{aligned} \tag{1.4.4}$$

In the absence of quasiparticle interactions, the matrix element is simply

$$\langle \mathbf{p}+\mathbf{q}/2,\mathbf{p}-\mathbf{q}/2;f|a^\dagger_{\mathbf{p}'+\mathbf{q}/2}a_{\mathbf{p}'-\mathbf{q}/2}|i\rangle = \delta_{\mathbf{p},\mathbf{p}'}n^0_{\mathbf{p}-\mathbf{q}/2}(1 - n^0_{\mathbf{p}+\mathbf{q}/2}), \tag{1.4.5}$$

where n^0 denotes the quasiparticle distribution function in the initial state. In general, however, when quasiparticle interactions are nonnegligible, the initial quasiparticle–quasihole pair ($\mathbf{p}' + \mathbf{q}/2$, $\mathbf{p}' - \mathbf{q}/2$) can scatter into the pair state ($\mathbf{p} + \mathbf{q}/2$, $\mathbf{p} - \mathbf{q}/2$). Such scattering might occur directly, or through several intermediate momenta ($\mathbf{p}'' + \mathbf{q}/2$, $\mathbf{p}'' - \mathbf{q}/2$), and the matrix element will no longer vanish if $\mathbf{p} \neq \mathbf{p}'$.

Let us denote the general matrix element in (1.4.4) by

$$\langle \mathbf{p}+\mathbf{q}/2,\mathbf{p}-\mathbf{q}/2;f|a^\dagger_{\mathbf{p}'+\mathbf{q}/2}a_{\mathbf{p}'-\mathbf{q}/2}|i\rangle \equiv s_{\mathbf{p}\mathbf{p}'}(\mathbf{q},\omega + i\eta)n^0_{\mathbf{p}-\mathbf{q}/2}(1 - n^0_{\mathbf{p}+\mathbf{q}/2}), \tag{1.4.6}$$

where we have extracted explicitly the quasiparticle distribution factors that guarantee that the initial state $\mathbf{p} - \mathbf{q}/2$ is occupied and the final state $\mathbf{p} + \mathbf{q}/2$ is unoccupied. The operators $a^\dagger_{\mathbf{p}'+\mathbf{q}/2}$ and $a_{\mathbf{p}'-\mathbf{q}/2}$ acting to the right on $|i\rangle$ create and annihilate quasiparticles of momentum $\mathbf{p}' + \mathbf{q}/2$ and $\mathbf{p}' - \mathbf{q}/2$. However, because of repeated intermediate particle-hole scatterings, the action of $a^\dagger_{\mathbf{p}'+\mathbf{q}/2}$ and $a_{\mathbf{p}'-\mathbf{q}/2}$ to the left on the final state is more complicated. It is necessary to distinguish $a^\dagger$ and a from the creation and annihilation operators $a^\dagger_{\mathbf{p}'+\mathbf{q}/2,f}$ and $a_{\mathbf{p}'-\mathbf{q}/2,f}$ that create and annihilate quasiparticles of momentum $\mathbf{p}' + \mathbf{q}/2$ and $\mathbf{p}' - \mathbf{q}/2$ when acting on the final state.* For example, the quasiparticle distribution function in the final state is given by

$$\delta n_{\mathbf{p}}(\mathbf{r},t) = \int \frac{d^3q}{(2\pi)^3}\, e^{i\mathbf{q}\cdot\mathbf{r}}\langle\Psi(t)|a^\dagger_{\mathbf{p}-\mathbf{q}/2,f}\, a_{\mathbf{p}+\mathbf{q}/2,f}|\Psi(t)\rangle, \tag{1.4.7}$$

while the state $|\mathbf{p}+\mathbf{q}/2, \mathbf{p}-\mathbf{q}/2;f\rangle$ is given by

$$|\mathbf{p}+\mathbf{q}/2, \mathbf{p}-\mathbf{q}/2;f\rangle = a^\dagger_{\mathbf{p}+\mathbf{q}/2,f}\, a_{\mathbf{p}-\mathbf{q}/2,f}|i\rangle. \tag{1.4.8}$$

*In the language of scattering theory, the $a^\dagger$ and a are the "in" operators, while the $a_f^\dagger$ and a_f are the "out" operators. See, for example, Reference [54].

We can use the kinetic equation to determine the effective matrix element

$$t_{\mathbf{p}}(\mathbf{q}, \omega + i\eta) \equiv \sum_{\mathbf{p}'} s_{\mathbf{p}\mathbf{p}'}(\mathbf{q}, \omega + i\eta) U_{\mathbf{p}'}(\mathbf{q}, \omega) \tag{1.4.9}$$

in a Fermi liquid for a general external perturbation. The kinetic equation determines $\delta n_{\mathbf{p}}(\mathbf{r}, t)$, the variation in the quasiparticle distribution function due to the perturbation. Using the relations (1.4.3), (1.4.7), and (1.4.8), we see that the spatial Fourier transform of the distribution function is given by

$$\delta n_{\mathbf{p}}(\mathbf{q}, t) = \left\langle \mathbf{p}+\frac{\mathbf{q}}{2}, \mathbf{p}-\frac{\mathbf{q}}{2}; f \,\middle|\, \delta\Psi(t) \right\rangle + \left\langle \delta\Psi(t) \,\middle|\, \mathbf{p}-\frac{\mathbf{q}}{2}, \mathbf{p}+\frac{\mathbf{q}}{2}; f \right\rangle. \tag{1.4.10}$$

The first term in (1.4.10) will have a component $\propto e^{-i\omega t}$ arising from the term $U_{\mathbf{p}}(\mathbf{q},\omega) e^{-i(\omega+i\eta)t} a^{\dagger}_{\mathbf{p}+\mathbf{q}/2} a_{\mathbf{p}-\mathbf{q}/2}$ in the perturbation acting on the initial state, while the second term will have such a component arising from the Hermitean conjugate of this perturbation acting on the initial state. If we use (1.4.4), (1.4.6), and (1.4.9), together with the fact that for a Hermitean external perturbation $t_{\mathbf{p}}(\mathbf{q},\omega + i\eta) = (t_{\mathbf{p}}(-\mathbf{q}, -\omega + i\eta))^*$, then we see that the component of (1.4.10) $\propto e^{-i\omega t}$ is given by

$$\begin{aligned} \delta n_{\mathbf{p}}(\mathbf{q},t) = & \frac{e^{-i\omega t}}{\omega + i\eta - \varepsilon_{\mathbf{p}+\mathbf{q}/2} + \varepsilon_{\mathbf{p}-\mathbf{q}/2}} \\ & \times \left[n^0_{\mathbf{p}-\mathbf{q}/2}(1 - n^0_{\mathbf{p}+\mathbf{q}/2}) - n^0_{\mathbf{p}+\mathbf{q}/2}(1 - n^0_{\mathbf{p}-\mathbf{q}/2})\right] t_{\mathbf{p}}(\mathbf{q},\omega + i\eta). \end{aligned} \tag{1.4.11}$$

This is, however, precisely the result one finds from the linearized kinetic equation (1.2.22) in the absence of collisions, provided we make the identification

$$t_{\mathbf{p}}(\mathbf{q}, \omega + i\eta) = \delta\varepsilon_{\mathbf{p}}(\mathbf{q}, \omega), \tag{1.4.12}$$

where $\delta\varepsilon_{\mathbf{p}}(\mathbf{q}, \omega)$ is the space and time Fourier transform of $\delta\varepsilon_{\mathbf{p}}(\mathbf{r}, t)$ (the frequency ω is assumed to approach the real axis from the upper half plane). In other words, the amplitude for scattering a quasiparticle from state $\mathbf{p} - \mathbf{q}/2$ to state $\mathbf{p} + \mathbf{q}/2$ is simply the variation $\delta\varepsilon_{\mathbf{p}}(\mathbf{q}, \omega)$ of the quasiparticle energy computed from the kinetic equation.

In the case that the perturbation on the system is the external scalar potential in (1.4.1) with Fourier transform $U_{\mathbf{p}}(\mathbf{q}, \omega)$ we find that $\delta\varepsilon_{\mathbf{p}}(\mathbf{q}, \omega)$ is given, for small $\mathbf{q}$, as the solution of

$$\begin{aligned} \delta\varepsilon_{\mathbf{p}} &= \delta U_{\mathbf{p}} + \sum_{\mathbf{p}'} f_{\mathbf{p}\mathbf{p}'} \delta n_{\mathbf{p}'} \\ &= \delta U_{\mathbf{p}} - \sum_{\mathbf{p}'} f_{\mathbf{p}\mathbf{p}'} \frac{1}{\omega + i\eta - \mathbf{q}\cdot\mathbf{v}_{\mathbf{p}'}} (\mathbf{q}\cdot\nabla_{p'} n^0_{p'})\delta\varepsilon_{\mathbf{p}'}, \end{aligned} \tag{1.4.13}$$

since from (1.2.21)

$$\delta n_{\mathbf{p}'} = -\frac{\mathbf{q}\cdot\nabla_{p'} n^0_{p'}}{\omega + i\eta - \mathbf{q}\cdot\mathbf{v}_{\mathbf{p}'}}\,\delta\varepsilon_{\mathbf{p}'}. \tag{1.4.14}$$

The first term in (1.4.13) is the Born scattering amplitude, while the second term takes into account the scattering of the quasiparticle because of the modification of the quasiparticle distribution function caused by U. Note that ω is the increase in energy of the quasiparticle, in scattering from $\mathbf{p} - \mathbf{q}/2$ to $\mathbf{p} + \mathbf{q}/2$; and recall that only after long times must ω equal $\varepsilon_{\mathbf{p}+\mathbf{q}/2} - \varepsilon_{\mathbf{p}-\mathbf{q}/2}$.

To determine the amplitude for the scattering of two quasiparticles of initial momenta $\mathbf{p} - \mathbf{q}/2$ and $\mathbf{p}' + \mathbf{q}/2$ to final momenta $\mathbf{p} + \mathbf{q}/2$ and $\mathbf{p}' - \mathbf{q}/2$ we must find the modification of the effective potential $\delta\varepsilon_{\mathbf{p}}(\mathbf{q}, \omega)$ acting on the first quasiparticle as a result of the second quasiparticle making a transition from $\mathbf{p}' + \mathbf{q}/2$ to $\mathbf{p}' - \mathbf{q}/2$ (with energy loss ω). We denote the change in $\varepsilon_{\mathbf{p}}$ due to this change in the quasiparticle distribution $\delta n_{\mathbf{p}'}$ by

$$(\delta\varepsilon_{\mathbf{p}}/\delta n_{\mathbf{p}'})_{\mathbf{q},\omega} \equiv t_{\mathbf{pp}'}(\mathbf{q}, \omega + i\eta). \tag{1.4.15}$$

At first sight one might think that this amplitude, in the limit of small $\mathbf{q}$ and ω, is just $f_{\mathbf{pp}'}$. This would be the case were the quasiparticle distribution $n_{\mathbf{p}''}$, for $\mathbf{p}'' \neq \mathbf{p}'$, held fixed, as $n_{\mathbf{p}'}$ is varied. However, in general, the other quasiparticles are allowed to adjust to the presence of the modified quasiparticle distribution $\delta n_{\mathbf{p}'}$, and so in the limit of small $\mathbf{q}$ and ω we can write

$$\begin{aligned}\delta\varepsilon_{\mathbf{p}} &= \left(\frac{\partial\varepsilon_{\mathbf{p}}}{\partial n_{\mathbf{p}'}}\right)_{n_{p''}} \delta n_{\mathbf{p}'} + \sum_{\mathbf{p}''\neq\mathbf{p}'} \left(\frac{\partial\varepsilon_{\mathbf{p}}}{\partial n_{\mathbf{p}''}}\right)_{n_{\mathbf{p}}^{\cdots}} \delta n_{\mathbf{p}''} \\ &= f_{\mathbf{pp}'}\delta n_{\mathbf{p}'} + \sum_{\mathbf{p}''\neq\mathbf{p}'} f_{\mathbf{pp}''}\delta n_{\mathbf{p}''}.\end{aligned} \tag{1.4.16}$$

Substituting (1.4.14) in (1.4.16) we find an integral equation for the scattering amplitude:

$$t_{\mathbf{pp}'}(\mathbf{q}, \omega + i\eta) = f_{\mathbf{pp}'} - \sum_{\mathbf{p}''\neq\mathbf{p}'} f_{\mathbf{pp}''} \frac{\mathbf{q}\cdot\nabla_{p''} n^0_{p''}}{\omega + i\eta - \mathbf{q}\cdot\mathbf{v}_{\mathbf{p}''}}\, t_{\mathbf{p}''\mathbf{p}'}(\mathbf{q}, \omega + i\eta). \tag{1.4.17}$$

This equation is Landau's version of the Bethe–Salpeter equation for the repeated scattering of a quasiparticle–quasihole pair [55]. Diagramatically, we can represent (1.4.17) by Figure 1.6, where the shaded circles represent the full scattering amplitude $t_{\mathbf{p}''\mathbf{p}'}(\mathbf{q}, \omega + i\eta)$ and the open circle stands for the "bare" quasiparticle scattering amplitude $f_{\mathbf{pp}''}$. The second term of (1.4.17) describes how the scattering of a quasiparticle from $\mathbf{p}' + \mathbf{q}/2$ to $\mathbf{p}' - \mathbf{q}/2$, or equivalently the creation of a quasiparticle $(\mathbf{p}' - \mathbf{q}/2)$–quasihole $(\mathbf{p}' + \mathbf{q}/2)$ pair, produces, as shown in Figure 1.6, an amplitude $\delta\varepsilon_{\mathbf{p}''}$ for creating a quasiparticle–quasihole pair $(\mathbf{p}'' + \mathbf{q}/2, \mathbf{p}'' - \mathbf{q}/2)$ that then propagates, with propagator $-(\mathbf{q}\cdot\nabla_{p''} n^0_{p''})/(\omega + i\eta - \mathbf{q}\cdot\mathbf{v}_{\mathbf{p}''})$, and recombines creating a quasiparticle–quasihole pair $(\mathbf{p} + \mathbf{q}/2, \mathbf{p} - \mathbf{q}/2)$. Physically, this corresponds to the screening of the interaction between quasiparticles, or, equivalently,

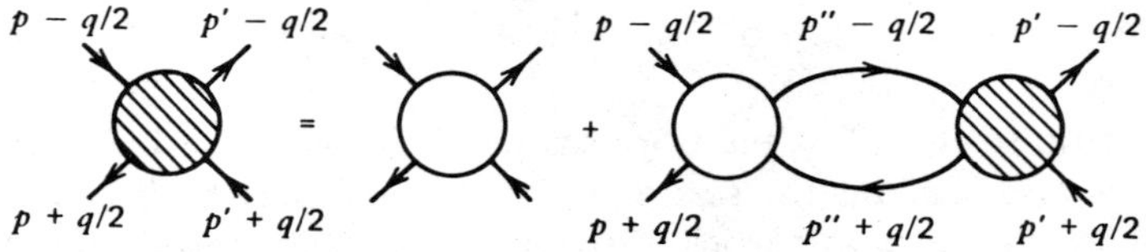

FIGURE 1.6. Equation 1.4.17 for repeated quasiparticle–quasihole scattering. The shaded circle represents the full scattering amplitude (or in Section 1.4.4(b) the K or T matrix) and the unshaded circle represents the bare quasiparticle–quasihole interaction.

the interaction between the dressing clouds associated with the quasiparticles **p** and **p**′; for a more detailed discussion of some of these effects, see [56].

An extensive discussion of (1.4.17) and its derivation from microscopic theory is given by Nozières [8]. Note that from a comparison of (1.4.17) and (1.4.13), using (1.4.9) and (1.4.12), we have

$$s_{\mathbf{pp}'} = \delta_{\mathbf{pp}'} - t_{\mathbf{pp}'} \frac{\mathbf{q}\cdot\nabla_{p'} n^0_{p'}}{\omega + i\eta - \mathbf{q}\cdot\mathbf{v}_{\mathbf{p}'}}, \tag{1.4.18}$$

which relates the response of the system to an external potential to the quasiparticle–quasihole scattering amplitude.

In the high frequency limit (called the "ω-limit") in which $\omega \gg qv_f$, the second term in (1.4.17) vanishes, and

$$t_{\mathbf{pp}'}(0, \omega) = f_{\mathbf{pp}'}, \qquad (\omega \ll \varepsilon_f). \tag{1.4.19}$$

In this limit the effective scattering amplitude reduces to the bare interaction between quasiparticles, because the disturbance $\delta n_{\mathbf{p}'}$, for frequencies $\omega \gg qv_f$, cannot modify the quasiparticle distribution for $\mathbf{p}'' \neq \mathbf{p}, \mathbf{p}'$.

In the opposite low frequency limit in which $qv_f \gg \omega$ (called the "k-limit") (1.4.17) reduces to

$$t_{\mathbf{pp}'}(\mathbf{q}, 0) = f_{\mathbf{pp}'} + \sum_{\mathbf{p}''} f_{\mathbf{pp}''} \frac{\partial n^0_{p''}}{\partial \varepsilon_{p''}} t_{\mathbf{p}''\mathbf{p}'}(\mathbf{q}, 0). \tag{1.4.20}$$

While $f_{\mathbf{pp}'} = (\delta\varepsilon_{\mathbf{p}}/\delta n_{\mathbf{p}'})_{n_{p''}}$, we can conclude from (1.4.17) that

$$\lim_{q\to 0} t_{\mathbf{pp}'}(\mathbf{q}, 0) = (\delta\varepsilon_{\mathbf{p}}/\delta n_{\mathbf{p}'})_{\mu}, \tag{1.4.21}$$

where μ is the chemical potential. To see this we note that at fixed μ, $n_{\mathbf{p}''}$ depends only on $\varepsilon_{\mathbf{p}''}$ and thus from (1.4.16)

$$\left(\frac{\delta\varepsilon_{\mathbf{p}}}{\delta n_{\mathbf{p}'}}\right)_{\mu} = f_{\mathbf{pp}'} + \sum_{\mathbf{p}''\neq\mathbf{p}'} f_{\mathbf{pp}''}\left(\frac{\partial n^0_{p''}}{\partial \varepsilon_{p''}}\right)\left(\frac{\delta\varepsilon_{\mathbf{p}''}}{\delta n_{\mathbf{p}'}}\right)_{\mu}. \tag{1.4.22}$$

Comparing with (1.4.20) we see that in the k-limit the scattering amplitude describes the variation in the quasiparticle energy at fixed chemical potential.

Expanding $t_{\mathbf{pp}'}(\mathbf{q}, 0)$ in Legendre polynomials $P_l(\cos\theta)$, where θ is the angle between $\mathbf{p}$ and $\mathbf{p}'$ (both close to the Fermi surface):

$$t_{\mathbf{pp}'}(\mathbf{q}, 0) = \sum_l t_l(\mathbf{q}, 0)P_l(\cos\theta), \tag{1.4.23}$$

we find from (1.4.19) that

$$t_l^s = \frac{f_l^s}{1 + F_l^s/(2l+1)}, \quad q \ll p_f \tag{1.4.24a}$$

and

$$t_l^a = \frac{f_l^a}{1 + F_l^a/(2l+1)}, \quad q \ll p_f. \tag{1.4.24b}$$

Here the spin indices s and a refer, as previously, to (quasiparticle–quasihole) scattering in spin symmetric and spin antisymmetric states. The factors $1 + F_l/(2l + 1)$ in the denominators act as "screening" factors for the various harmonics of the quasiparticle interaction in essentially the same way as $1 + F_0^s$ does in the response of the system to a shift in the chemical potential [as in (1.1.48)], or $1 + F_0^a$ in the response to a static external magnetic field [as in (1.1.55)]. In ^{3}He at zero pressure, the largest screening effects occur in the $l = 0$ channels. Since $F_0^s \simeq 10$, the low frequency scattering amplitude t_0^s is reduced by a factor $1 + F_0^s \simeq 11$ from its high frequency value. Similarly since $F_0^a \simeq -2/3$, the low frequency scattering amplitude t_0^a is *enhanced* by a factor 3 from its high frequency value. This enhancement has, as we shall see, an important effect on the finite temperature corrections to the transport coefficients and the specific heat.

Equation (1.4.17) can be solved, for arbitrary values of $s = \omega/qv_f$, in exactly the same way as we solved for zero sound in Section 1.3.1(b). Note that (1.3.10), when written in terms of the variable $\nu_{\mathbf{p}}(\omega - \mathbf{q}\cdot\mathbf{v}_{\mathbf{p}})/\mathbf{q}\cdot\mathbf{v}_{\mathbf{p}}$, has exactly the same structure as (1.4.17) for $t_{\mathbf{pp}'}$ except that U in (1.3.10) is replaced by $f_{\mathbf{pp}'}$ in (1.4.17). For example, if only f_0^s and f_0^a are nonzero, then from (1.4.17) we see that $t_{\mathbf{pp}'}^i(\mathbf{q}, \omega)$ for small q and ω (where $i = s$ or a), is independent of the angle between $\mathbf{p}$ and $\mathbf{p}'$ and is given by

$$t_{\mathbf{pp}'}^i(\mathbf{q}, \omega) = \frac{f_0^i}{1 + F_0^i\Omega_{00}(s)}. \tag{1.4.25}$$

When F_0^s is > 0 the system has a zero sound mode that appears as a singularity in $t_{\mathbf{pp}'}^s$ at $\omega = s_0 q$. This pole in t corresponds to the process in which the quasiparticle that scatters from $\mathbf{p}' + \mathbf{q}/2$ to $\mathbf{p}' - \mathbf{q}/2$ creates a zero sound oscillation which then scatters a quasiparticle from $\mathbf{p} - \mathbf{q}/2$ to $\mathbf{p} + \mathbf{q}/2$. The denominator in (1.4.25) is an example of the *dynamical screening* of the effective interaction by the other quasiparticles in the system. One point to notice is that quite generally $t_{\mathbf{pp}'}(\mathbf{q}, \omega)$ depends only on the variable $s = \omega/v_f q$ and the angles between the vectors $\mathbf{p}$, $\mathbf{p}'$ and $\mathbf{q}$, even if one includes arbitrarily many Landau parameters.

The collision term in the kinetic equation (1.2.53) used in calculating transport properties depends on the scattering amplitude $\langle \mathbf{p}_3\mathbf{p}_4|t|\mathbf{p}_1\mathbf{p}_2\rangle$ for quasiparticles whose energies lie within κT of the Fermi surface. Quasiparticle energy is conserved in the scattering and so ω, the energy transfer, equals $\varepsilon_3 - \varepsilon_1$. In terms of the variables $\mathbf{p} \pm \mathbf{q}/2$, $\mathbf{p}' \pm \mathbf{q}/2$ we have

$$\begin{aligned}&\langle \mathbf{p} + \mathbf{q}/2, \mathbf{p}' - \mathbf{q}/2|t|\mathbf{p} - \mathbf{q}/2, \mathbf{p}' + \mathbf{q}/2\rangle \\ &\quad = t_{\mathbf{pp}'}(\mathbf{q}, \omega = \varepsilon_{\mathbf{p}+\mathbf{q}/2} - \varepsilon_{\mathbf{p}-\mathbf{q}/2}).\end{aligned} \tag{1.4.26}$$

The q values of importance range up to $2p_f$, while $\omega \lesssim \kappa T$. Thus, to a good approximation in computing low temperature transport properties, one can approximate the amplitude by $t_{\mathbf{pp}'}(\mathbf{q}, 0)$; as we shall see in detail, inclusion of the ω dependence produces terms of order T^{-1} in τ (compared with the basic T^{-2} behavior of τ at low T).

In describing transport in Section 1.2 we introduced the variables θ, the angle between the momenta of the incident quasiparticles, and ϕ, the angle between the plane containing the momentum vectors of the incident quasiparticles and the plane containing the momentum vectors of the scattered quasiparticles. Small ϕ corresponds to nearly forward scattering, that is, small q, and in this limit the scattering amplitude can be expressed in terms of Landau parameters, as in (1.4.24):

$$t^i_{\mathbf{pp}'}(\mathbf{q} \to 0,0) \equiv t^i(\theta, \phi = 0) = \sum_l \frac{f^i_l P_l(\cos\theta)}{1 + F^i_l/(2l+1)}, \tag{1.4.27}$$

where $i = s$ or a.

From the Pauli exclusion principle, the scattering amplitude must be antisymmetric under the interchange of the spins and momenta of the final (or initial) two quasiparticles, that is,

$$\langle \mathbf{p}_4\sigma_4, \mathbf{p}_3\sigma_3|t|\mathbf{p}_1\sigma_1, \mathbf{p}_2\sigma_2\rangle = -\langle \mathbf{p}_3\sigma_3, \mathbf{p}_4\sigma_4|t|\mathbf{p}_1\sigma_1, \mathbf{p}_2\sigma_2\rangle. \tag{1.4.28}$$

Since in the absence of spin-orbit forces the total spin of the incident quasiparticles is conserved in the scattering, this amplitude, as far as the spin variables are concerned, depends only on two amplitudes, t_s for total spin zero (singlet) and t_t for total spin one (triplet). The scattering of two up-spin particles must be triplet and so

$$\langle \mathbf{p}_3\uparrow, \mathbf{p}_4\uparrow|t|\mathbf{p}_1\uparrow, \mathbf{p}_2\uparrow\rangle = \langle \mathbf{p}_3\mathbf{p}_4|t_t|\mathbf{p}_1\mathbf{p}_2\rangle. \tag{1.4.29}$$

Similarly, the amplitude for scattering of one up-spin with one down-spin quasiparticle is the average of the singlet and triplet amplitudes:

$$\langle \mathbf{p}_3 \uparrow, \mathbf{p}_4 \downarrow |t| \mathbf{p}_1 \uparrow, \mathbf{p}_2 \downarrow \rangle = \frac{1}{2}[\langle \mathbf{p}_3\mathbf{p}_4|t_s|\mathbf{p}_1\mathbf{p}_2\rangle + \langle \mathbf{p}_3\mathbf{p}_4|t_t|\mathbf{p}_1\mathbf{p}_2\rangle]. \quad (1.4.30)$$

Note that the decomposition into singlet and triplet amplitudes (in the particle–particle scattering channel) is not the same as the decomposition (in the particle–hole channel) into spin symmetric and spin antisymmetric components; the látter is in terms of the variables σ_1 and σ_3, and σ_2 and σ_4. Since the left side of (1.4.29) equals $t^s + t^a$, and the left side of (1.4.30) equals $t^s - t^a$, we see that

$$t_s = t^s - 3t^a, \qquad t_t = t^s + t^a \quad (1.4.31)$$

or

$$t^s = \frac{1}{4}(3t_t + t_s), \qquad t^a = \frac{1}{4}(t_t - t_s). \quad (1.4.32)$$

In fact the superscripts s and a refer to the total spin exchanged in the scattering, with s corresponding to zero spin exchange and a corresponding to exchange of spin one. In other words, if one views the scattering in the cross channel, where quasiparticle 1 and quasihole 3 scatter into quasiparticle 4 and quasihole 2, then a denotes relative triplet spin of quasiparticle 1 and quasihole 3, while s denotes relative singlet spin. It is easy to see that as a consequence of the definition (1.1.26) of spin-symmetric and spin-antisymmetric components, the amplitudes for scattering of pairs in singlet and triplet spin states are in fact *twice* t^s and t^a respectively.*

From (1.4.28) we see that the triplet amplitude must be antisymmetric under exchange of $\mathbf{p}_3$ and $\mathbf{p}_4$, while the singlet amplitude must be symmetric under this exchange. Since under this interchange $\phi \to \phi + \pi$, we see that

$$t_s(\theta, \phi) = t_s(\theta, \phi + \pi),$$
$$t_t(\theta, \phi) = -t_t(\theta, \phi + \pi). \quad (1.4.33)$$

The scattering amplitude in the neighborhood of $\phi = 0$ is given by (1.4.27), and, thus, by use of (1.4.33) and (1.4.32) we can express the scattering amplitude in the neighborhood of $\phi = \pi$ in terms of the Landau parameters as

$$t^s(\theta, \pi) = -\frac{1}{2}[t^s(\theta, 0) + 3t^a(\theta, 0)]$$

$$t^a(\theta, \pi) = -\frac{1}{2}[t^s(\theta, 0) - t^a(\theta, 0)]. \quad (1.4.34)$$

*We note that there is a second particle-hole channel, corresponding to the scattering of a quasiparticle 1 and a quasihole 4 to a quasiparticle 3 and a quasihole 2. The decomposition of the scattering amplitude into singlet and triplet components for the two particle-hole channels are of course different; this will be important when we discuss finite temperature contributions to the specific heat [(Section 1.4.4 (b)].

Scattering with $\phi = \pi$ corresponds to $\mathbf{p}_4 \simeq \mathbf{p}_1$ and thus a momentum transfer $\mathbf{q} \simeq \mathbf{p}_2 - \mathbf{p}_1$.

From (1.4.29) and (1.4.28) we see that for $\theta = 0$ the triplet amplitude must vanish:

$$t_t(0, \phi) = 0, \tag{1.4.35}$$

since then $\mathbf{p}_1 = \mathbf{p}_2 = \mathbf{p}_3 = \mathbf{p}_4$. Since $t_t = t^s + t^a$ we derive from (1.4.35) and (1.4.27) the "forward scattering sum rule" on the Landau parameters [57]

$$\sum_l (A_l^s + A_l^a) = 0, \tag{1.4.36}$$

where

$$A_l^i = \frac{F_l^i}{1 + F_l^i/(2l + 1)}. \tag{1.4.37}$$

Equation 1.4.36 places an important constraint on the magnitudes of the Landau parameters. The microscopic derivation of (1.4.36) has been discussed in detail by Mermin [39].

Except at $\phi = 0$ and $\phi = \pi$, one cannot express the scattering amplitudes in terms of Landau parameters. The symmetry requirements (1.4.33) do, however, suggest a very simple first approximation for the scattering amplitudes in terms of the F_l^i, namely, to take [58]

$$\begin{aligned} t_s(\theta, \phi) &\simeq t_s(\theta, 0) \\ t_t(\theta, \phi) &\simeq t_t(\theta, 0) \cos \phi. \end{aligned} \tag{1.4.38}$$

This corresponds to assuming that the singlet scattering takes place only in the channel in which the component of the relative angular momentum of the incident particles along their total momentum is zero (*s*-wave), while the triplet scattering is assumed to take place only in the channel in which this component is $\pm\hbar$ (*p*-wave). The approximation (1.4.38) is equivalent to writing

$$\begin{aligned} N(0)t_{\uparrow\uparrow}(\theta, \phi) &\simeq \sum_l (A_l^s + A_l^a) P_l(\cos\theta) \cos\phi, \\ N(0)t_{\uparrow\downarrow}(\theta, \phi) &\simeq \sum_l \frac{1}{2} [(A_l^s - 3A_l^a) + (A_l^s + A_l^a)\cos\phi]\, P_l(\cos\theta), \end{aligned} \tag{1.4.39}$$

where the arrows refer to the spins of the incident quasiparticles. As we shall see in the next section, this simple approximation has proven to be quite successful in calculating low temperature transport coefficients in terms of the Landau parameters. (Note that for parallel spin scattering, processes with scattering angle $\pi - \phi$ are indistinguishable from processes with angle ϕ, and therefore only ϕ between 0 and $\pi/2$ should be counted. Compare the discussion in Section 1.2.4(a).)

to the leading terms. However, as we shall see in this section, the form of the density of scattering states, together with certain nonanalytic structure in the scattering amplitude, as a function of $\omega/v_f q$, due to dynamical screening, leads to contributions to the scattering rate, and hence to the transport coefficients, of relative order T compared with the leading behavior. These finite temperature corrections give rise to deviations from the leading terms that appear at lower temperatures than one would expect offhand. This behavior was first predicted on the basis of the spin fluctuation model for almost ferromagnetic Fermi liquids [61–65] but, as was stressed in later work [18, 19, 58, 66, 67], is in fact a general property of all normal Fermi liquids and does not depend on their being almost ferromagnetic.

Let us first, to understand the phase space in the scattering at finite temperatures, consider the lifetime τ_p of a quasiparticle of momentum $\mathbf{p}$ due to two-quasiparticle scattering processes, $\mathbf{p}, \mathbf{p}' \rightarrow \mathbf{p} - \mathbf{q}, \mathbf{p}' + \mathbf{q}$, in which the momentum transfer q is smaller than some cutoff momentum $q_c \ll p_f$. (For a simple discussion of the phase space see Reference [56].) We assume for now that the scattering amplitude is spin independent; then

$$\begin{aligned}\frac{1}{\tau_p} &= 2\pi \int_0^{q_c} \frac{d^3q}{(2\pi)^3} 2 \int \frac{d^3p'}{(2\pi)^3} |\langle \mathbf{p} - \mathbf{q}, \mathbf{p}' + \mathbf{q}|t|\mathbf{p}, \mathbf{p}'\rangle|^2 \\ &\times \delta(\varepsilon_{\mathbf{p}} + \varepsilon_{\mathbf{p}'} - \varepsilon_{\mathbf{p}-\mathbf{q}} - \varepsilon_{\mathbf{p}'+\mathbf{q}}) \\ &\times [n^0_{\mathbf{p}'}(1 - n^0_{\mathbf{p}'+\mathbf{q}})(1 - n^0_{\mathbf{p}-\mathbf{q}}) + (1 - n^0_{\mathbf{p}'})n^0_{\mathbf{p}'+\mathbf{q}}n^0_{\mathbf{p}'-\mathbf{q}}].\end{aligned} \tag{1.4.40}$$

The first term in square brackets represents the rate at which quasiparticles of momentum $\mathbf{p}$ are scattered to new states, while the second term represents the blocking due to the presence of a quasiparticles of momentum $\mathbf{p}$, of processes that scatter quasiparticles into momentum $\mathbf{p}$ (See Reference [12], Section 4.4.)

For simplicity we shall assume that the scattering amplitude depends only on the ratio of the energy transfer $\omega = \varepsilon_{\mathbf{p}} - \varepsilon_{\mathbf{p}-\mathbf{q}}$ to the momentum transfer q; we denote the amplitude by $t(s)$, where $s = |\omega|/qv_f$. Then doing the $\mathbf{p}'$ and $\mathbf{q}$ angular integrals in (1.4.38) and substituting for the variable q in terms of s we have

$$\begin{aligned}\frac{1}{\tau_p} = \frac{N(0)}{4\pi v_f^3[1 - n(\varepsilon)]} \int_{-\infty}^{\infty} d\omega|\omega| \int_{-\infty}^{\infty} d\varepsilon'[1 - n(\varepsilon - \omega)]n(\varepsilon')[1 - n(\varepsilon' + \omega)] \\ \times \int_{|\omega|/v_f q_c}^{1} \frac{ds}{s^2}|t(s)|^2.\end{aligned} \tag{1.4.41}$$

If we write $|t(s)|^2$ as $|t(0)|^2 + [|t(s)|^2 - |t(0)|^2]$, we can do the $|t(0)|^2$ integral over s explicitly; since $|t(s)|^2 - |t(0)|^2$ will in general vanish as s^2, we may replace the lower limit in this term by 0. Using (A.1) and (A.9) for the integral I_3 we find then

$$\frac{1}{\tau_p} = \frac{\pi}{4v_f^2p_f^2}\left\{\frac{1}{2}[(\varepsilon_p - \mu)^2 + (\pi\kappa T)^2]v_f q_c|N(0)t(0)|^2 - \Xi L(\varepsilon_p - \mu, T)\right\}, \tag{1.4.42}$$

where

$$L(\varepsilon_p - \mu, T) = \int_{-\infty}^{\infty} d\omega \int_{-\infty}^{\infty} d\varepsilon'|\omega| \frac{(1 - n(\varepsilon - \omega)n(\varepsilon')(1 - n(\varepsilon' + \omega))}{1 - n(\varepsilon)} \tag{1.4.43}$$

and

$$\Xi = N(0)^2\left[|t(0)|^2 + \int_0^1 \frac{ds}{s^2}(|t(0)|^2 - |t(s)|^2)\right]. \tag{1.4.44}$$

The ε' integral is simply $\kappa T I_2(\omega/\kappa T) = \omega/(1 - e^{\omega/\kappa T})$, and thus

$$L = (\kappa T)^3[4\zeta(3) + M(x)], \tag{1.4.45}$$

where $x = (\varepsilon_p - \mu)/\kappa T$, $\zeta(n)$ is the Riemann zeta function of order n, and

$$M(x) = \int_0^{\infty} y^2 dy\left[\frac{1}{e^{y+x} + 1} + \frac{1}{e^{y-x} + 1}\right]. \tag{1.4.46}$$

As $T \to 0$, $L \to \frac{1}{3}|\varepsilon_p - \mu|^3$.

The first term in (1.4.42), quadratic in $\varepsilon_p - \mu$ and T, is part of the leading contribution to the lifetime for quasiparticle and quasihole excitations near the Fermi surface. When averaged over $\varepsilon_p - \mu$ within κT of the Fermi surface, this term produces part of the leading T^2 behavior of τ^{-1} [cf. (1.2.91)]. Finite angle scattering contributes as well to the full T^2 term. The second term in (1.4.42), proportional to L, is nonanalytic in $(\varepsilon_p - \mu)$ and is independent of the cutoff q_c. This nonanalytic behavior comes entirely from small momentum transfers where $q \ll q_c$. At finite temperature this term, when averaged over $\varepsilon_p - \mu$ within κT of the Fermi surface, will produce a T^3 term in τ^{-1}. Were the L term analytic in $\varepsilon_p - \mu$, its thermal average over ε_p would produce a correction term to $\tau^{-1} \sim T^4$. Such a correction would lead to finite temperature corrections to the transport coefficients of relative order T^2, and until the calculations of Rice on almost ferromagnetic Fermi liquids [61–63] it was generally believed that the finite temperature corrections were of order T^2. Note that the nonanalytic $|\varepsilon_p - \mu|^3$ terms in τ_p^{-1} are present even if the scattering amplitude is independent of s; these terms were in fact contained implicitly in Galitskii's work on the dilute Fermi gas [68].

In liquid ^{3}He the nonanalytic terms are greatly enhanced by the strong frequency dependence of the effective quasiparticle interaction. To see how this comes about we consider a model in which only a single Landau parameter F_0^a is nonzero; it is in this channel that the largest dynamical screening effects occur in ^{3}He. We shall assume also that $1 + F_0^a$ is a small positive number. Recall that according to the discussion in Section 1.1.3(e), the system is unstable if F_0^a is $\leqslant -1$; this instability corresponds to a ferromagnetic spin ordering. Thus, when $1 + F_0^a$ is small and positive, the system is nearly ferromagnetic in the sense that a small external magnetic field will produce a large spin alignment.

The dynamically screened interaction in this case is [cf. (1.4.25)]

$$t_0^a(q, \omega) = \frac{f_0^a}{1 + F_0^a \Omega_{00}(s)}. \tag{1.4.47}$$

Detailed consideration of the spin sum changes the result (1.4.42) by a factor* 3.

From (1.3.16) for Ω_{00} we see that for small s and small $1 + F_0^a$,

$$N(0) t_0^a(q, \omega) \simeq \frac{A_0^a}{1 + i(\pi/2) s A_0^a}, \tag{1.4.48}$$

where $A_0^a = F_0^a/(1 + F_0^a)$ is large and negative. The effect of the denominator in (1.4.48) is to cut off the s integral in (1.4.42) when $s \gtrsim 2/(\pi|A_0^a)|)$; the screening, which enhances the low frequency (small s) scattering amplitude, is negligible at frequencies above $\sim 2v_f q/\pi|A_0^a|$. Using the form (1.4.48) in the s integral in (1.4.42) we find that to leading order in an expansion in $|A_0^a|^{-1}$ the s integral in the nonanalytic term in τ_p^{-1} is given by

$$N(0)^2 \int_0^1 \frac{ds}{s^2}\left(|t(0)|^2 - |t(s)|^2\right) = \frac{\pi^2}{4} |A_0^a|^3. \tag{1.4.49}$$

In fact this result is exact for all $A_0^a < 0$, as can be seen if one does the exact integral, with t given by (1.4.47), by the method of contours in the s plane [67]. Thus, for $A_0^a < 0$,

$$\Xi = \frac{\pi^2}{4} |A_0^a|^3 + (A_0^a)^2. \tag{1.4.50}$$

Notice how for large $|A_0^a|$ the frequency dependence in $t(s)$ amplifies the nonanalytic term by a factor $\pi^2 |A_0^a|/4$. The nonanalytic terms become more and more important, relative to the leading quadratic terms, as the ferromagnetic instability is reached. In the limit of large $|A_0^a|$ the characteristic temperature at which the finite temperature corrections become important is $T_f/|A_0^a| \simeq T_f(1 + F_0^a)$. At $T = 0$ the nonanalytic term in τ_p^{-1} is given by

*The factor 3 arises as follows: when the scattering angle ϕ is near zero, then [cf. (1.4.39) for $l = 0$] the spin sum gives in general $2(|t^s|^2 + |t^a|^2)$. In the case of scattering of quasiparticles of antiparallel spin, scattering with $\phi \simeq \pi$ is also a small angle scattering process, but with spin flip. This scattering contributes a term [cf(1.4.39)] $4|t^a|^2$. The total amplitude for small angle scattering is thus $2(|t^s|^2 + 3|t^a|^2)$; the factor 3 is the relative contribution of t^a compared with t^s.

$$(\tau_p^{-1})_{\text{non-an}} = \frac{\pi}{4v_f^2p_f^2}\left((A_0^a)^2 + \frac{\pi^2}{4}|A_0^a|^3\right)|\varepsilon_p - \mu|^3. \tag{1.4.51}$$

For liquid ^{3}He, A_0^a is only about -2 and it is a poor approximation to neglect other Landau parameters. Details of calculations of the quasiparticle lifetime using more general expressions for the effective interaction may be found in the work of Emery [66].

(b) Transport Coefficients. The nonanalytic structure in the quasiparticle lifetime is reflected, through the relaxation time, in the transport coefficients at finite temperature. As we shall now see, the first corrections to the low temperature transport coefficients are terms of relative order T. We consider first the thermal conductivity. The simplest method of computing the leading finite temperature correction is first to write the transport equation (1.2.81) in the symbolic form

$$|X\rangle = I|\Phi\rangle, \tag{1.4.52}$$

where $|X\rangle$ is the driving term on the left side, $|\Phi\rangle$ is the deviation of the distribution function (written as $\bar{\Phi}_{\mathbf{p}}$ in momentum space), and I is the linearized collision operator acting on $\bar{\Phi}$. From (1.2.67), (1.2.75), and (1.2.76) we see that

$$\langle\Phi|X\rangle = \mathbf{j}_T\cdot\frac{\nabla T}{T} = -K\frac{(\nabla T)^2}{T}, \tag{1.4.53}$$

where the scalar product simply denotes integration over $\mathbf{p}$ and summation over spin; $\mathbf{j}_T$ is the thermal current and K the thermal conductivity.

In the low temperature calculations of Section 1.2 we assumed, when evaluating the collision integral, that the energy transfer ω in quasiparticle collisions could be neglected. This enabled us to decouple the angular and energy integrals in the collision integral. Such a procedure gives the correct low T behavior. However, as we have seen in the calculation above of the nonanalytic terms in the quasiparticle lifetime, the finite temperature corrections arise from processes in which the energy transfer cannot be neglected.

To see the structure of these corrections let us write the collision operator as [58]

$$I = I^{(0)} + I^{(1)}, \tag{1.4.54}$$

where $I^{(0)}$ denotes the low temperature form of the collision operator in which the energy transfer is neglected, and $I^{(1)}$ represents the corrections arising from the energy dependence of the quasiparticle scattering. We also write

$$|\Phi\rangle = |\Phi^0\rangle + |\Phi^1\rangle \tag{1.4.55}$$

where $|\Phi^0\rangle$ denotes the exact low temperature solution for $|\Phi\rangle$ computed in

Section 1.2. If we let $|X^0\rangle$ denote the low temperature form of the left side of the transport equation, then

$$|X^0\rangle = I^{(0)}|\Phi^0\rangle. \tag{1.4.56}$$

To compute the first finite temperature corrections to the transport coefficients, which are of relative order T, we may neglect the finite temperature corrections to $|X\rangle$, which are of order T^2, and assume in (1.4.52) that $|X\rangle = |X^0\rangle$. If we subtract (1.4.56) from (1.4.52) and keep only the first temperature corrections we find

$$0 = I^{(1)}|\Phi^0\rangle + I^{(0)}|\Phi^1\rangle. \tag{1.4.57}$$

The correction to K is determined from $\langle X^0|\Phi^1\rangle$, which, from (1.4.57), is given by

$$\langle X^0|\Phi^1\rangle = \langle\Phi^0|I^{(0)}|\Phi^1\rangle = -\langle\Phi^0|I^{(1)}|\Phi^0\rangle. \tag{1.4.58}$$

The first temperature variation of the thermal conductivity is, from (1.4.53), thus

$$KT - (KT)_{T=0} = \frac{\langle\Phi^0|I^{(1)}|\Phi^0\rangle}{(\nabla T/T)^2}. \tag{1.4.59}$$

By simple algebra, using (1.4.56), we find the result for the thermal conductivity, valid to first order:

$$\frac{1}{KT} - \left(\frac{1}{KT}\right)_{T=0} = -\frac{\langle\Phi^0|I^{(1)}|\Phi^0\rangle}{\langle\Phi^0|X^0\rangle^2}\left(\frac{\nabla T}{T}\right)^2,$$

or

$$\frac{1}{KT} = -\frac{\langle\Phi^0|I|\Phi^0\rangle}{\langle\Phi^0|X^0\rangle^2}\left(\frac{\nabla T}{T}\right)^2. \tag{1.4.60}$$

This form, unlike (1.4.59), is independent of the choice of normalization of $|\Phi^0\rangle$. To compute the first temperature correction to K we need only know the low temperature distribution function. This is because of the variational property of the kinetic equation [16]: the exact solution for $|\Phi\rangle$ minimizes the right side of (1.4.60), and thus the first order correction to $|\Phi\rangle$ produces only a second order correction to $1/KT$.

Equations similar to (1.4.60) may be derived for the finite temperature spin diffusion coefficient and viscosity.

The details of the evaluation of (1.4.60) are complicated and unilluminating. The essential physics is illustrated if we consider the simplified case in which we keep only the $\bar{\Phi}_1$ term in—for example (1.2.86) for I, so that $I \rightarrow (\partial n_p^0/\partial\varepsilon_p)/\tau_p$, where τ_p is the quasiparticle lifetime. Then

$$X_{\mathbf{p}}^0 = -(\varepsilon_p - \mu)\mathbf{v}_{\mathbf{p}} \cdot \frac{\nabla T}{T} \frac{\partial n_p^0}{\partial \varepsilon_p}$$

$$\bar{\Phi}_{\mathbf{p}}^0 = -(\varepsilon_p - \mu)\mathbf{v}_{\mathbf{p}} \cdot \frac{\nabla T}{T} \tau_p^0, \tag{1.4.61}$$

where τ_p^0 is the low temperature form of the lifetime, without the nonanalytic terms; it is of the form [cf. Section 1.2.4(b)]

$$(\tau_p^0)^{-1} = \left(\frac{x^2 + \pi^2}{2}\right)\tau^{-1}, \tag{1.4.62}$$

where $x = (\varepsilon_p - \mu)/\kappa T$ and τ is given by (1.2.91). Thus, from (1.4.60) we have

$$\frac{1}{KT} = \frac{1}{(KT)_{T=0}} - \frac{1}{(KT)_{T=0}^2} N(0) \frac{v_f^2}{3} (\kappa T)^2 \int_{-\infty}^{\infty} dx \frac{\partial n(x)}{\partial x} (\tau_p^0)^2 \left(\frac{1}{\tau_p} - \frac{1}{\tau_p^0}\right). \tag{1.4.63}$$

The latter integral represents the finite temperature correction; $1/\tau_p - 1/\tau_p^0$ is the nonanalytic term in the quasiparticle collision rate, and when averaged over in the integral it contributes a T^3. Since $\tau_p^0 \propto T^{-2}$ the entire finite temperature term is $\propto T$, and in this approximation it is independent of τ_p^0. If we use the form (1.4.47) for the scattering amplitude then the finite temperature correction to $1/KT$ is negative and proportional to $(A_0^a)^2 (1 + (\pi^2/4)|A_0^a|)$.

Another simple calculation that can be carried out is to evaluate (1.4.60) using the simplest variational trial function for Φ^0. One finds, assuming that Landau parameters other than F_0^a may be neglected [67]

$$\frac{1}{KT} - \left(\frac{1}{KT}\right)_{T=0} = -2430 \frac{\zeta(5)}{\pi} \frac{m^{*3}}{p_f^3} \hbar^2 \kappa (A_0^a)^2 \left[1 + \frac{\pi^2}{4} |A_0^a|\right] T, \tag{1.4.64}$$

for $A_0^a \leqslant 0$, where $\zeta(n)$ is the Riemann zeta function of order n. The analogous result for the spin diffusion coefficient is

$$\frac{1}{DT^2} - \left(\frac{1}{DT^2}\right)_{T=0} = -36\pi\zeta(3) \frac{m^{*4}}{p_f^6} \frac{\kappa^3}{\hbar} \frac{(A_0^a)^2}{(1 + F_0^a)} \left[1 + \frac{\pi^2}{4} |A_0^a|\right] T, \tag{1.4.65}$$

for $A_0^a \leqslant 0$, and the linear correction to $1/\eta T^2$ vanishes.

Detailed expressions for the leading finite temperature corrections to the thermal conductivity, as well as the spin diffusion coefficient and first viscosity are given by Dy and Pethick [58], and by Emery and Cheng [19]. In all cases the leading corrections are of order T relative to the extreme low-temperature behavior, and the coefficients of these terms may be expressed as a sum of products of two sorts of factors. The first are functions only of the Landau parameters and are given by expressions similar to that for M in

(1.4.46), and the second are functions only of the parameters λ_K, λ_η, and λ_D that determine the form of the solution to the transport equation in the extreme low-temperature limit. Emery and Cheng evaluated the second factors using their approximate solution of the transport equation and give convenient analytical expressions for them; Dy and Pethick showed that these approximate expressions give results that for reasonable values of λ differ from the exact results by only a few percent.* Because the expressions are rather complicated we shall not repeat them here.

On the basis of the above calculations one finds that the contributions to $1/\eta T^2$ linear in T are considerably smaller than those to $1/KT$ and $1/DT^2$. (Recall that in the variational calculation the finite temperature contribution to $1/\eta T^2$ vanishes.) Physically, the reason for this is that small momentum transfer collisions are very ineffective in changing the momentum flux tensor. On the other hand, small momentum transfer processes do change the energy flux directly, while small momentum transfer processes accompanied by a spin flip change the spin current.

In Figure 1.7 we show some experimental data for liquid ^{3}He at low pressure; the thermal conductivity is from the measurements of Abel, Johnson, Wheatley, and Zimmermann [15] and the spin diffusion coefficient is from the work of Anderson, Reese, Sarwinski, and Wheatley [69]. $T^* + \Delta$, where T^* is the magnetic temperature, is an effective temperature that is close to the Kelvin temperature in the range of temperatures of interest [60]; we shall in fact neglect any difference between $T^* + \Delta$ and T. The straight lines shown are fits to the data. To appreciate the magnitude of the finite-temperature contribution, recall that the Fermi temperature $p_f^2/2m^*\kappa$ of liquid ^{3}He at low pressure is $\sim 1700\ mK$. More recently Tyler [70] has performed measurements of the spin diffusion coefficient of liquid ^{3}He at low pressure and found results in accord with those of Reference [69]; also Lawson et al. [71] have reported a measurement of the linear contribution to $1/\eta T^2$ for liquid ^{3}He at pressures near the melting curve.

Dy and Pethick [58] have applied their exact calculation to liquid ^{3}He determining the scattering amplitude $t_{\mathbf{pp}'}(q, \omega + i\eta)$ for small q and ω (which determines the finite temperature correction) from (1.4.17), with only the Landau parameters F_0^s, F_0^a, F_1^s and F_1^a nonzero. Since F_1^a is not well determined, they fitted the experimental finite T correction to K by adjusting F_1^a and found, at low pressure, $F_1^a = -0.46 \pm 0.14$ (where the errors include only the contribution from statistical uncertainties in fitting a straight line to the experimental finite T correction to $1/KT$). By fitting F_1^a to the finite T spin diffusion data instead they found $F_1^a = -0.39 \pm 0.14$. By comparison, to satisfy the forward scattering sum rule (1.4.36) with just $l = 0$ and $l = 1$

*However, see footnote 32 of Reference [58].

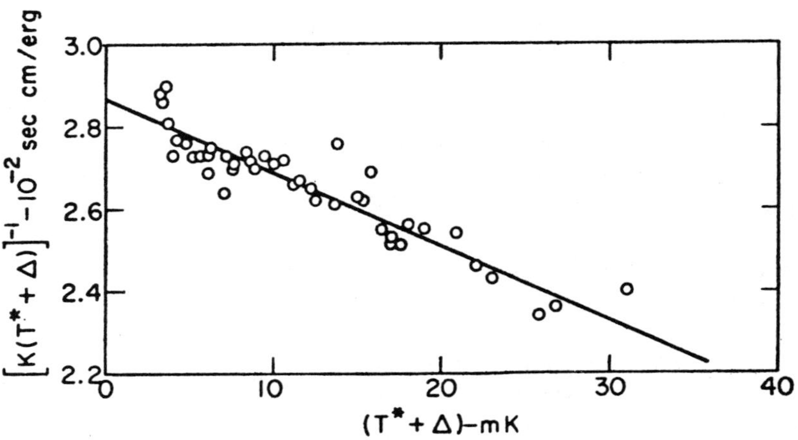

(a)

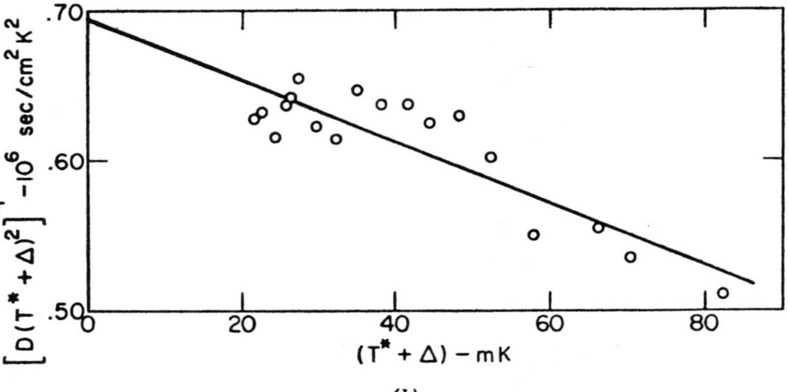

(b)

FIGURE 1.7. (a) The thermal conductivity data of Abel et al. [15] for liquid ^{3}He at low pressure, as analyzed by Abel and Wheatley [60]. The line is a fit to the data and is given by $[K(T^* + \Delta)]^{-1} = [(2.87 \pm 0.1) - (18 \pm 3)\,(T^* + \Delta)\,(K^{-1})] \times 10^{-2}$ sec cm/erg. ($\Delta = 0.3$ mK.) (b) The spin-diffusion coefficient data of Anderson et al. [69] for liquid ^{3}He at low pressure, as analyzed by Abel and Wheatley [3]. The line is a fit to the data and is given by $[D(T^* + \Delta)^2]^{-1} = [(0.695 \pm 0.012) - (2.0 \pm 0.3)$ $(T^* + \Delta)\,(K^{-1}) \times 10^6$ sec/cm^2 K^2. ($\Delta = 0.3$ mK.)

Landau parameters, one requires $F_1^a = -0.70$; at low pressures the spin echo experments of Corruccini et al. [44] indicate $F_1^a = -0.15 \pm 0.3$, but in view of the uncertainties in this determination one can view these various estimates of F_1^a as being in fair agreement.

So far we have not considered whether there are processes other than small momentum transfer ones that can give rise to finite temperature contributions ($\sim T$) to the transport coefficients. It has been suggested that processes in which the momentum transfer is $\sim 2p_f$ could give rise to such terms, since it was thought that such processes gave rise to some of the leading finite temperature contributions to the specific heat. However, in the light of Pickles' recent work [72], discussed at the end of Section 1.4.4(b), which indicates that there may be no such large-momentum transfer contributions to the specific heat, the question of whether such processes give contributions to the transport coefficients should be looked into more thoroughly. Should such contributions exist, it seems likely that in the case of liquid ^{3}He their contribution will be small compared with the contribution from small momentum transfer processes, which is greatly enhanced by the large negative value of F_0^a that leads to important spin fluctuation effects.

Finally a word about the earlier spin fluctuation or paramagnon calculations [62, 63]; in these calculations the collision integral in the transport equation is evaluated from the imaginary part of the diagrams shown in Figure 1.11. Physically this corresponds to taking into account scattering of two quasiparticles by exchange of quasiparticle-quasihole pairs, which is precisely the process we have considered. However, in the spin fluctuation calculations the interaction between fermions is assumed to be a simple contact interaction, and the renormalization of the fermion effective mass is neglected. The calculations contain the same basic physics as the Landau theory calculations and lead to the same qualitative behavior for the transport coefficients; for the case of an almost ferromagnetic Fermi liquid the paramagnon results essentially agree with the Landau theory calculations, but for ^{3}He, F_0^a is not sufficiently close to -1 for the "almost ferromagnetic" approximation to be valid. Contributions of higher order in the temperature than the ones considered here cannot be determined using Landau theory, since scattering processes other than exchange of long-wavelength fluctuations become important; models of the paramagnon type may be useful in estimating these contributions [62, 63, 65, 73].

1.4.4. Finite Temperature Contributions to the Specific Heat and Magnetic Susceptibility

(a) Introduction. Before describing recent calculations of the specific heat at finite temperatures we first give a brief account of the development of the subject. In the previous section we described the leading finite temperature contributions to the transport coefficients, and the nonanalytic behavior of the quasiparticle scattering amplitude that is responsible for them; these calculations are somewhat simpler than those of the specific heat,

and hence we considered them first. Historically, however, the nonanalytic behavior was first brought to light as a result of measurements of the specific heat, and not the transport coefficients.

As we have seen in Section 1.1.3(a) one expects the low-temperature specific heat of a Fermi liquid to be linear in T; however, when measurements of c_V were first carried out at temperatures below 100 mK it became clear that c_V/T was rather strongly temperature dependent, even though the temperature was well below the Fermi temperature $T_f = p_f^2/2m\kappa$ ($\sim$1700 mK for liquid ^{3}He at low pressure) that one would expect to set the characteristic temperature scale for the system. Also, the data for c_V could not be fitted by a series containing only *odd* powers of the temperature, as one would expect on the basis of Landau theory if the quasiparticle energy were analytic near the Fermi surface. To show the size of the finite-temperature effects, we give in Figure 1.8 some recent data on the specific heat of liquid ^{3}He at low pressure [59].

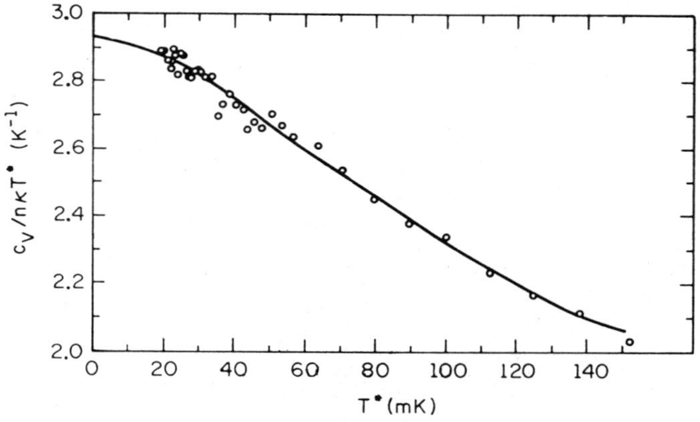

FIGURE 1.8. Specific heat of liquid ^{3}He at a pressure of 0.24 atm as a function of magnetic temperature T^* taken from the work of Mota et al. [59]. The smooth curve is a fit to the data, assuming c_V has only T^* and $T^{*3} \ln T^*$ terms, and has the form $c_V/n\kappa T^* = [2.93 - 57.4\ (T^*/K)^2 \ln(T^*/0.293\ K)]\ K^{-1}$.

Anderson [74] pointed out that the data at that time could be fitted quite well by a $T \ln T$ dependence, and suggested that coupling between quasiparticles and collective modes might be responsible for this nonanalytic behavior. This idea was taken up by Balian and Fredkin who explored the effects of coupling between quasiparticles and zero sound [75]. They considered the contribution to the quasiparticle energies due to the emission and reabsorption of a virtual zero sound phonon; this process may be represented diagrammatically by the graph shown in Figure 1.9. Taking this process into

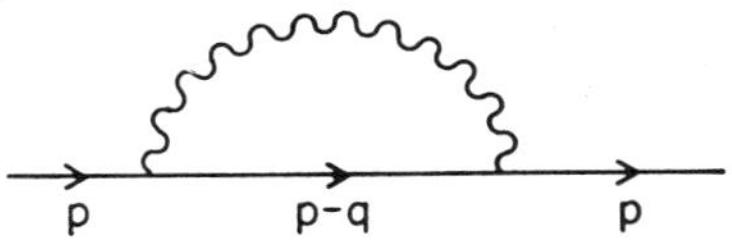

FIGURE 1.9. Diagram for the contribution to the fermion self-energy due to emission and reabsorption of a virtual zero sound phonon. The wavy line represents the phonon propagator.

account in the self-energy is essentially equivalent to allowing for the contribution to the quasiparticle interaction from exchange of a zero sound phonon (see Figure 1.10). Balian and Fredkin's simplest perturbation theory calculation did give a $T \ln T$ behavior for the specific heat, but a self-consistent solution of the problem gave a specific heat varying as $T(\ln T)^{1/2}$. In these calculations assumptions were made that are equivalent to the statement that the coupling between quasiparticles is long range, similar to the coupling between electrons and phonons in a piezoelectric crystal. As Engelsberg and Platzman [76], among others, showed, this assumption is not correct; as demonstrated in Section 1.3.3, the coupling is in fact of a deformation potential type ($\sim q^{1/2}$), like that between electrons and acoustic phonons in solids, and the leading contribution to the specific heat at low temperatures behaves as $T^3 \ln T$, just as Buckingham and Schafroth [77], and Eliashberg [78] had earlier shown to be the case for the electron-phonon interaction in solids. As we shall see in more detail later, the $T^3 \ln T$ term produced by this process is much too small to account for the effect observed experimentally.

While the consideration of the coupling to zero sound did not produce a sufficiently large effect, it did show how nonanalytic structure could be produced. The next step was the investigation by Doniach, Engelsberg, and Rice [61, 79], and by Berk and Schrieffer [80] of the coupling of quasiparticles to incoherent spin fluctuations; this too gives a $T^3 \ln T$ contribution to the specific heat. The magnetic susceptibility of liquid ^{3}He is considerably larger than that of a free Fermi gas; this leads to a strong coupling between quasiparticles and low-lying states containing spin fluctuations—that is, particle-

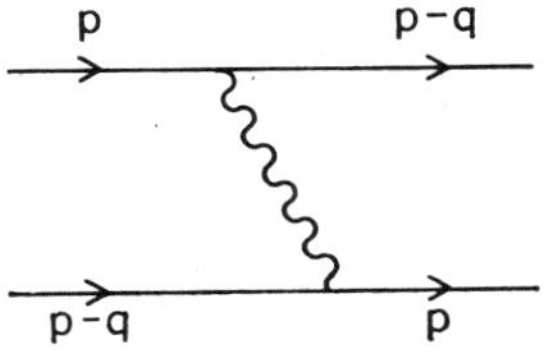

FIGURE 1.10. Diagram for the zero sound exchange contribution to the quasiparticle interaction.

hole pairs in triplet spin states. These spin fluctuations are not true collective modes of the system, like spin waves (magnons) in ferromagnets, or spin zero sound [see Section 1.3.1(f)], but are more closely related to the critical fluctuations that occur in the paramagnetic phase at temperatures above the Curie temperature of a ferromagnet. For this reason, the name "paramagnon" was coined for these fluctuations.

The spin fluctuation contribution to the specific heat was first calculated in a model in which fermions interact via a repulsive contact interaction. The diagrams for the self-energy taken into account are shown in Figure 1.11; these diagrams are essentially the same as the diagrams for the zero sound contribution, Figure 1.9, except that the zero sound propagator is replaced by the spin fluctuation propapator. Doniach and Engelsberg calculated the contribution from spin fluctuations with magnetic quantum number $m = \pm 1$ (Figure 1.11*a*), and the importance of the $m = 0$ contribution (Figure 1.11*b*) was pointed out by Bucher, Brinkman, Maita, and Williams [81], by Penn [82], by Nozières [83], and by Brenig [83]. We note in passing that the diagram in Figure 1.11*b* contains contributions from both triplet (spin) fluctuations and singlet (density) fluctuations; the contribution to the self-energy from the zero sound pole in the singlet part of the propagator corresponds physically to the modification of the quasiparticle energies by emission and reabsorption of zero sound phonons. In the early work it was assumed that the quasiparticle energy defined as a functional derivative (1.1.1), which following Balian

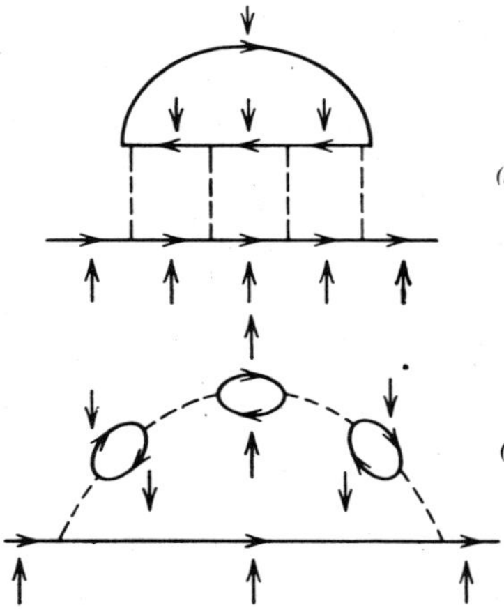

FIGURE 1.11. Diagrams for spin fluctuation contributions to the self-energy. (*a*) is the contribution from fluctuations with $m = \pm 1$, and (*b*) is that from fluctuations with $m = 0$. (*b*) also includes a term from fluctuations with spin zero, part of which gives the zero sound contribution; however for liquid ^{3}He this is small compared with the spin fluctuation term and is usually neglected.

and De Dominicis [10] we refer to as the *statistical* quasiparticle energy, could be identified with the *dynamical* quasiparticle energy given by the pole of the single particle propagator; the entropy was therefore calculated by evaluating the usual quasiparticle expression (1.1.3) using dynamical quasiparticle energies. The theoretical estimate of the spin-fluctuation contribution to the specific heat was of the right order of magnitude to account for the experimental data, and the $T^3 \ln T$ behavior was not inconsistent with the experimental data then available. (In Figure 1.8 the solid line is a fit to the data using T and $T^3 \ln T$ terms.)

The paramagnon calculations described above suffered from a number of difficulties; first, it was not clear how one could take into account the fact that the true interaction is more complicated than the simple contact interaction, and second, it was not obvious whether one should use the bare mass or the effective mass in the fermion propagators. These problems were resolved when it was realized that in the paramagnon calculations the $T^3 \ln T$ contributions to the specific heat all came from very long wavelength, low frequency spin fluctuations, whose properties can be computed exactly by Landau Fermi-liquid theory. Additional advantages of using the Landau theory here are that it treats spin fluctuations and density fluctuations on exactly the same footing, and that it may be applied to systems that are not almost ferromagnetic. Such calculations of the dynamical quasiparticle energies were carried out by Brenig and Mikeska [84], by Amit, Kane, and Wagner [85], and by Emery [66]; the $T^3 \ln T$ contribution to the specific heat was then estimated, as in the earlier calculations, by inserting the dynamical quasiparticle energies into the quasiparticle expression for the entropy.

Further work by Brenig, Mikeska, and Riedel [86], by Riedel [87], and by Brinkman and Engelsberg [88] on models of almost ferromagnetic systems showed that the method of calculating the entropy used in the above calculations, while giving the term linear in T correctly, does not give the correct $T^3 \ln T$ term; for the models considered, the method gave a result *three* times as large as the true answer.

Riedel [87] showed that the correct result for the entropy could be expressed as the usual quasiparticle expression evaluated using dynamical quasiparticle energies, plus an additional "Bose" contribution coming from interacting particle-hole pairs. The Bose contribution has two terms. The first comes from true collective modes (such as zero sound, in liquid ^{3}He) that correspond to isolated poles in the particle-hole propagator; it has the same form as the entropy of a system of noninteracting bosons whose energies are given by the poles of the particle-hole propagator. However, at low temperatures it is of order T^3, and can therefore be neglected when one is concerned only with $T^3 \ln T$ terms. The remaining part of the Bose contribution to the entropy comes from the particle-hole continuum and it includes a contribution from incoherent spin fluctuations, which are not true collective modes of the system. Thus, in Riedel's calculation the entropy to order $T^3 \ln T$ is expressed as a sum of the *dynamical* quasiparticle contribution and the Bose contribution coming from the particle-hole continuum. However, since Balian and De Dominicis [10] and Luttinger [89] have established that the entropy is given by the quasiparticle expression evaluated using *statistical* quasiparticle energies,* the $T^3 \ln T$ term in the specific heat can be expressed in terms of the nonanalytic terms in the statistical quasiparticle energies alone. From this point of view, Riedel's Bose contribution from the particle-hole continuum may be regarded as resulting from the difference between dynamical and statistical quasiparticle energies.

A direct calculation of the $T^3 \ln T$ term in the specific heat in terms of the *statistical* quasiparticle energies was given by Pethick and Carneiro [90] who showed that the long-wavelength fluctuations lead to nonanalytic behavior of the quasiparticle interaction $f_{\mathbf{p},\mathbf{p}+\mathbf{q}}$ for small $\mathbf{q}$, which in turn gives rise to $T^3 \ln T$ terms in the specific heat. The coefficient of the $T^3 \ln T$ term was calculated in terms of Landau parameters, and it was shown how the difference between statistical and dynamical quasiparticle energies is closely related to the difference between energies of interaction and forward scattering amplitudes. We now describe these calculations and indicate how they relate to earlier work.

(b) The Specific Heat. Consider two neighboring quasiparticle states, of momenta $\mathbf{p}$ and $\mathbf{p} + \mathbf{q}$, where $q \ll p_f$, whose energies lie within $\sim \kappa T$ of the Fermi surface; then, in the limit of vanishing temperature ($\kappa T \ll v_f q \ll v_f p_f$), $\hat{\mathbf{p}} \cdot \hat{\mathbf{q}}$ must be very small, since

$$\hat{\mathbf{p}} \cdot \hat{\mathbf{q}} = (\varepsilon_{\mathbf{p}+\mathbf{q}} - \varepsilon_{\mathbf{p}})/v_f q \sim \kappa T/v_f q.$$

(Note that $\hat{\mathbf{p}} \cdot \hat{\mathbf{q}}$ is the familiar variable $s = \omega/v_f q$, evaluated on the energy shell.) As we shall see below, for such quasiparticles $f_{\mathbf{p},\mathbf{p}+\mathbf{q}}$ has the form

$$f_{\mathbf{p},\mathbf{p}+\mathbf{q}} = f(0) + b(\hat{\mathbf{p}} \cdot \hat{\mathbf{q}})^2 + \mathcal{O}((\hat{\mathbf{p}} \cdot \hat{\mathbf{q}})^4), \quad q \ll p_f. \tag{1.4.66}$$

Here $f(0)$ is the value of $f_{\mathbf{p},\mathbf{p}+\mathbf{q}}$ for $\hat{\mathbf{p}} \cdot \hat{\mathbf{q}} = 0$. Thus, even for $\mathbf{q} \to 0$, $f_{\mathbf{p},\mathbf{p}+\mathbf{q}}$ depends critically on the angle between $\mathbf{p}$ and $\mathbf{q}$.

*We note that in the calculations in References [10] and [89], which are based on perturbation theory, it is assumed that the thermal averaging may be performed before summing to infinite order in the perturbing potential. While this procedure properly treats processes involving virtual collective modes, such as the exchange of a virtual zero sound phonon by particles, it omits contributions to the thermodynamic properties coming from the real excitation of collective modes. The leading contribution to the specific heat due to the presence of real collective modes is of order T^3, and it is therefore unimportant here since we are concerned only with the $T^3 \ln T$ contribution.

We shall first show that if the quasiparticle interaction has the form (1.4.66), then the quasiparticle energies* have contributions that behave as $(p - p_f)^3 \ln|p - p_f|$, giving rise to $T^3 \ln T$ terms in the specific heat. Later we shall show that $f_{\mathbf{p},\,\mathbf{p}+\mathbf{q}}$ has the form (1.4.66).

To compute the contribution $\Delta\varepsilon_{\mathbf{p}}$ to the energy of a given quasiparticle, of momentum $\mathbf{p}$, due to its interaction with its neighboring quasiparticles, of momentum $\mathbf{p} + \mathbf{q}$, we must in general solve the functional integral equation

$$\frac{\delta\varepsilon_{\mathbf{p}}}{\delta n_{\mathbf{p}+\mathbf{q}}} = f_{\mathbf{p},\,\mathbf{p}+\mathbf{q}}, \tag{1.4.67}$$

since f is itself a functional of the distribution function. However, here we are interested only in the contribution to $\varepsilon_{\mathbf{p}}$ resulting from interactions between the quasiparticle and the small number of neighboring quasiparticles, whose momenta differ from $\mathbf{p}$ by less than some small amount q_c, where $|p - p_f| \ll q_c \ll p_f$. Thus, the dependence of f on $n_{\mathbf{p}'}$ does not have to be taken into account. For an unpolarized Fermi system the quasiparticle energy is independent of spin, and therefore one can integrate (1.4.67) immediately to give $\Delta\varepsilon_{\mathbf{p}}$:

$$\Delta\varepsilon_{\mathbf{p}} = 2 \sum_{\mathbf{q}(q\leqslant q_c)} f^s_{\mathbf{p},\,\mathbf{p}+\mathbf{q}} n^0_{\mathbf{p}+\mathbf{q}}, \tag{1.4.68}$$

where $f^s_{\mathbf{p},\,\mathbf{p}+\mathbf{q}}$ is evaluated using the equilibrium distribution function. To the order to which we are working the temperature dependence of f does not have to be taken into account. The interesting contribution to $\Delta\varepsilon_{\mathbf{p}}$ is that due to the $(\hat{\mathbf{p}}\cdot\hat{\mathbf{q}})^2$ term in f^s (1.4.66), and, thus, neglecting other contributions, one has

$$\Delta\varepsilon_{\mathbf{p}} = 2b^s \sum_{\mathbf{q}(q\leqslant q_c)} (\hat{\mathbf{p}}\cdot\hat{\mathbf{q}})^2 n^0_{\mathbf{p}+\mathbf{q}}. \tag{1.4.69}$$

The sum over $\mathbf{q}$ in (1.4.69) is easily performed in cylindrical polar coordinates with the polar axis in the direction of $\mathbf{p}$. Then $(\hat{\mathbf{p}}\cdot\hat{\mathbf{q}})^2 = q_\parallel^2/(q_\parallel^2 + q_\perp^2)$, where $q_\parallel$ is the component of $\mathbf{q}$ parallel to $\mathbf{p}$ and $q_\perp$ is the perpendicular component. At zero temperature the quasiparticle distribution function is simply a step function; for small q, the restriction $|\mathbf{p} + \mathbf{q}| < p_f$ is equivalent to $q_\parallel < p_f - p$. Thus (1.4.69) becomes

$$\Delta\varepsilon_{\mathbf{p}}(T = 0) = \frac{2b^s}{(2\pi)^3}\int_{-q_c}^{p_f-p} dq_\parallel \int_0^{q_c^2-q^2} \pi d(q_\perp^2)\frac{q_\parallel^2}{q_\parallel^2 + q_\perp^2} \tag{1.4.70}$$

$$= -\frac{b^s}{2\pi^2}\int_{-q_c}^{p_f-p} dq_\parallel q_\parallel^2 \ln\left|\frac{q_\parallel}{q_c}\right| \tag{1.4.71}$$

$$= \frac{b^s}{6\pi^2}(p - p_f)^3 \ln\left|\frac{p - p_f}{q_c}\right|, \tag{1.4.72}$$

*In this section the phrase "quasiparticle energy" will always refer to the *statistical* quasiparticle energy, unless explicitly stated otherwise.

where in (1.4.72) we have neglected a constant term, as well as a term of order $(p - p_f)^3$.

One can easily understand the existence of a $(p - p_f)^3 \ln|p - p_f|$ term in the *dynamical* quasiparticle energy. We saw in Section 1.4.3(a) [cf. (1.4.43) as $T \to 0$] that the quasiparticle collision rate has contributions varying as $|p - p_f|^3$; microscopically these come from $|\omega|^3$ contributions to the imaginary part of the self-energy [12] $\Sigma(p, \omega)$ (and related contributions involving $\varepsilon_p - \mu$); since the real and imaginary parts of Σ satisfy a simple dispersion relation $\mathrm{Re}\Sigma(\omega) = \Sigma_0 - \int(d\omega'/\pi)\,\mathrm{Im}\Sigma(\omega')/(\omega' - \omega)$, it follows that $\mathrm{Re}\Sigma$ has $\omega^3 \ln|\omega|$ terms, which lead to $(p - p_f)^3 \ln|p - p_f|$ terms in the dynamical quasiparticle energy. However, since the statistical quasiparticle energy is not simply related to the self-energy, it is not at all obvious from this argument that the *statistical* quasiparticle energy should also contain such terms.

To include the leading finite temperature contributions in $\Delta\varepsilon_{\mathbf{p}}$, from (1.4.69), one may neglect the dependence of $n^0_{\mathbf{p}+\mathbf{q}}$ on both the temperature-dependent and the nonanalytic contributions to the quasiparticle energy; that is, $n^0_{\mathbf{p}+\mathbf{q}}$ may be replaced by $[\exp((\varepsilon_{\mathbf{p}+\mathbf{q}} - \mu)/\kappa T) + 1]^{-1}$, where for small q and $p - p_f$, $\varepsilon_{\mathbf{p}+\mathbf{q}} - \mu = (|\mathbf{p} + \mathbf{q}| - p_f)v_f = (p + q_\parallel - p_f)v_f$. Substituting this expression for $n^0_{\mathbf{p}+\mathbf{q}}$ into (1.4.69) and performing the $q_\perp$ integration we find

$$\Delta\varepsilon_{\mathbf{p}}(T) = -\frac{b^s}{2\pi^2}\int_{-q_c}^{q_c} dq_\parallel q_\parallel^2 \ln\left|\frac{q_\parallel}{q_c}\right| \frac{1}{e^{(p+q_\parallel-p_f)v_f/\kappa T} + 1}. \tag{1.4.73}$$

The upper limit on the integral is effectively the maximum of $(p - p_f)v_f$ and κT. Thus

$$\Delta\varepsilon_{\mathbf{p}}(T) = \frac{b^s}{6\pi^2 v_f^2}(p - p_f)[(p - p_f)^2 v_f^2 + \pi^2\kappa^2 T^2]\ln\left|\frac{\max[(p - p_f)v_f; \kappa T]}{\kappa T_c}\right| \tag{1.4.74}$$

to within a constant, plus higher order terms; here $T_c \sim v_f q_c/\kappa$ is a cutoff temperature.

To evaluate the contribution to the entropy we insert this expression for $\Delta\varepsilon_{\mathbf{p}}$ into the quasiparticle expression (1.1.3); to first order in $\Delta\varepsilon_{\mathbf{p}}$ the change in the entropy is given by

$$\Delta S = -\sum_{\mathbf{p}\sigma} \Delta\varepsilon_{\mathbf{p}}(T)\frac{\partial n_p^0}{\partial T} \tag{1.4.75}$$

$$= -\frac{2}{15}\pi^2\kappa B^s\left(\frac{\kappa T}{v_f}\right)^3 \ln\left(\frac{T}{T_c}\right) \tag{1.4.76}$$

$$= -\frac{1}{20}\pi^4 n\kappa B^s\left(\frac{T}{T_f}\right)^3 \ln\left(\frac{T}{T_c}\right) \tag{1.4.77}$$

where

$$B^s = N(0)b^s, \tag{1.4.78}$$

$n = p_f^3/3\pi^2$ is the density of particles, and $\kappa T_f = p_f^2/2m^*$. The corresponding contribution to the specific heat per unit volume is

$$\Delta c_V = -\frac{3}{20}\pi^4 n\kappa B^s\left(\frac{T}{T_f}\right)^3 \ln\left(\frac{T}{T_c}\right). \tag{1.4.79}$$

Since the $\hat{\mathbf{p}}\cdot\hat{\mathbf{q}}$ term in the quasiparticle interaction has as its source the exchange of long-wavelength spin and density fluctuations between quasiparticles of almost equal momenta (1.4.79) is an exact expression for the $T^3 \ln T$ contribution to the specific heat coming from long-wavelength fluctuations; its magnitude is proportional to the coefficient of the $(\hat{\mathbf{p}}\cdot\hat{\mathbf{q}})^2$ term in f. Next we show that f has the assumed form (1.4.66) and evaluate the coefficient b^s in terms of Landau parameters.

The contribution to the quasiparticle scattering amplitude due to the exchange of long-wavelength fluctuations is easily evaluated by the methods of Section 1.4.1. Our problem is to find the contribution of such fluctuations to $f_{\mathbf{p},\,\mathbf{p}+\mathbf{q}}$ for small $\mathbf{q}$. This would be solved if we knew how to relate the Landau quasiparticle interaction $f_{\mathbf{p},\,\mathbf{p}+\mathbf{q}}$ to the scattering amplitude. Often the energy of interaction is assumed to be the same as the real part of the forward scattering amplitude. However, this assumption is not true in general, since it fails even for two particles interacting via a central potential—as is well known, in this case the energy of interaction is proportional to the phase shift δ, while the real part of the forward scattering amplitude is proportional to $\sin\delta\cos\delta$. Only for small δ do these agree. We shall first investigate the energy of interaction of a pair of particles, and then apply the results to the interaction of a quasiparticle and a quasihole.

Consider two particles with relative momentum $\mathbf{k}$ interacting via a central potential. From elementary quantum mechanics it is known that the interaction energy of the particles—that is, the shift of the energy of the state of the particles—is given by [91]

$$\Delta E_{\mathbf{k}} = -\frac{2\pi}{mk}\sum_l (2l+1)\delta_l, \tag{1.4.80}$$

where δ_l is the phase shift for the lth partial wave, and m is the reduced mass of the pair of particles. On the other hand, the forward scattering amplitude is given by

$$T_{\mathbf{kk}} = -\frac{2\pi}{mk}\sum_l \frac{(2l+1)}{2i}(e^{2i\delta_l} - 1). \tag{1.4.81}$$

The difference between ΔE and T arises from terms in perturbation theory that have vanishing energy denominators. To see how this difference arises it is convenient to express these quantities in terms of the reactance matrix, or K matrix, whose angular momentum components are related to the phase shift by

$$K_l = -\frac{2\pi}{mk}(2l+1)\tan\delta_l. \tag{1.4.82}$$

$\Delta E_{\mathbf{k}}$ and $T_{\mathbf{kk}}$ may therefore be written as

$$\Delta E_{\mathbf{k}} = \frac{2\pi}{mk}\sum_l (2l+1)\arctan\left(\frac{mk}{2\pi}\frac{K_l}{2l+1}\right) \tag{1.4.83}$$

and

$$T_{\mathbf{kk}} = \sum_l \frac{K_l}{1 + i\dfrac{mk}{2\pi}\dfrac{K_l}{2l+1}}. \tag{1.4.84}$$

The importance of vanishing energy denominators, which correspond to on-energy-shell intermediate states, is well brought out by writing the equations in a more general notation, which is also very convenient for applying the results to the quasiparticle case. The K matrix satisfies the operator equation

$$K(E) = V + VP(E - H_0)^{-1}K(E), \tag{1.4.85}$$

where H_0 is the kinetic energy operator, E is the energy, V is the interaction potential, and P denotes a principal value integral. The scattering amplitude is given by

$$T(E) = V + V(E - H_0 + i\eta)^{-1}T(E), \qquad \eta \to +0, \tag{1.4.86}$$

which when written in terms of the K matrix is

$$T(E) = K(E) - i\pi K(E)\delta(E - H_0)T(E) \tag{1.4.87}$$

$$= K(E)[1 + i\pi\delta(E - H_0)K(E)]^{-1}. \tag{1.4.88}$$

The real part of $T(E)$ is given by

$$\mathrm{Re}T(E) = [1 - i\pi K(E)\delta(E - H_0)]^{-1}K(E)[1 + i\pi\delta(E - H_0)K(E)]^{-1} \tag{1.4.89}$$

$$= K(E)[1 + (\pi\delta(E - H_0)K(E))^2]^{-1} \tag{1.4.90}$$

$$= K(E)[1 + \sum_{n=1}^{\infty}(-1)^n(\pi\delta(E - H_0)K(E))^{2n}] \tag{1.4.91}$$

$$= K(E) - \pi^2 K(E)\delta(E - H_0)K(E)\delta(E - H_0)K(E) + \dots . \tag{1.4.92}$$

On the other hand, the interaction energy ΔE_a in a state a is given by the diagonal matrix elements of the operator

$$
\begin{aligned}
\Delta E &= \arctan K(E) \\
&= K(E)\left[1 + \sum_{n=1}^{\infty} \frac{(-1)^n}{2n+1} [\pi\delta(E - H_0)K(E)]^{2n}\right] \qquad (1.4.93) \\
&= K(E) - \tfrac{1}{3}\pi^2 K(E)\delta(E - H_0)K(E)\delta(E - H_0)K(E) + \cdots \qquad (1.4.94)
\end{aligned}
$$

evaluated on the energy shell, that is, for $E = E_a$, the unperturbed energy of the state. Note the factor 1/3 in (1.4.94), compared with (1.4.92).

We shall now take these results over and apply them to the calculation of the energy of a quasiparticle–quasihole pair with small total momentum. The physical process taken into account in the two-body problem is the repeated interaction of the pair of particles; the equation for the T or K matrices may be represented by the graphical equation shown in Figure 1.12. In the case of a quasiparticle–quasihole pair the analogous equation in shown in Figure 1.6; this takes into account repeated scatterings of a pair by the *bare* quasiparticle–quasihole interaction, which as we saw in Section 1.4.1 is just the Landau f-function. The algebraic equations for the T and K matrices, obtained by applying (1.4.85) and (1.4.86) to a quasiparticle $(\mathbf{p}' + \mathbf{q}/2)$–quasihole $(\mathbf{p}' - \mathbf{q}/2)$ pair are

$$
k^{\lambda}_{\mathbf{pp}'}(\mathbf{q}, \omega) = f^{\lambda}_{\mathbf{pp}'} + 2P\sum_{\mathbf{p}''} f^{\lambda}_{\mathbf{pp}''} \frac{n^0_{\mathbf{p}''-\mathbf{q}/2} - n^0_{\mathbf{p}''+\mathbf{q}/2}}{\omega - \varepsilon_{\mathbf{p}''+\mathbf{q}/2} + \varepsilon_{\mathbf{p}''-\mathbf{q}/2}} k^{\lambda}_{\mathbf{p}''\mathbf{p}'}(\mathbf{q}, \omega), \qquad (1.4.95)
$$

and

$$
t^{\lambda}_{\mathbf{pp}'}(\mathbf{q}, \omega + i\eta) = f^{\lambda}_{\mathbf{pp}'} + 2\sum_{\mathbf{p}''} f^{\lambda}_{\mathbf{pp}''} \frac{n^0_{\mathbf{p}''-\mathbf{q}/2} - n^0_{\mathbf{p}''+\mathbf{q}/2}}{\omega + i\eta - \varepsilon_{\mathbf{p}''+\mathbf{q}/2} + \varepsilon_{\mathbf{p}''-\mathbf{q}/2}} t^{\lambda}_{\mathbf{p}''\mathbf{p}'}(\mathbf{q}, \omega + i\eta), \qquad (1.4.96)
$$

where $\lambda = s$ and a denotes the symmetric or antisymmetric combination of the two quasiparticle spins. The factor of 2 is from the spin sum, and P means that in the sum terms with vanishing energy denominators are excluded. Compare with (1.4.17). The combination of occupation numbers $n^0_{\mathbf{p}''-\mathbf{q}/2} - n^0_{\mathbf{p}''+\mathbf{q}/2}$ takes into account the exclusion principle in intermediate states. It is the difference of $n^0_{\mathbf{p}''-\mathbf{q}/2}(1 - n^0_{\mathbf{p}''+\mathbf{q}/2})$ and $n^0_{\mathbf{p}''+\mathbf{q}/2}(1 - n^0_{\mathbf{p}''-\mathbf{q}/2})$; the first factor corresponds physically to the initial pair being scattered to a state with a

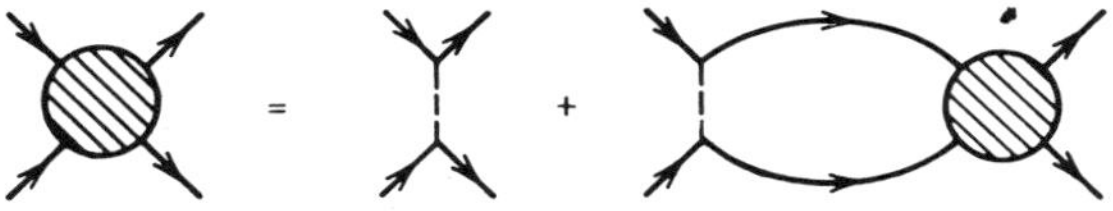

FIGURE 1.12. Equation for the K (or T) matrix for particle–particle scattering. The dashed line represents the bare two body potential while the shaded circle represents the full K (or T) matrix.

quasiparticle of momentum $\mathbf{p}'' + \mathbf{q}/2$ and a quasihole of momentum $\mathbf{p}'' - \mathbf{q}/2$, while the second term corresponds to the annihilation of the initial pair by a quasiparticle of momentum $\mathbf{p}'' - \mathbf{q}/2$ and a quasihole of momentum $\mathbf{p}'' + \mathbf{q}/2$ present in the system. The two factors correspond to the two possible time orders for the quasiparticle–quasihole bubble, the closed $\mathbf{p}''$ loop in Figure 1.6.

For the two particle problem there is only one term, since the pair of particles must always move forward in time; in the absence of a background medium the pair of particles cannot annihilate against a pair of holes. In Fermi liquid theory the terms coming from the two time directions for the bubble diagram are equally important and they must both be included. On the other hand, for the scattering of two particles in a medium, the problem considered in the pair approximation of the Brueckner–Bethe theory of nuclear matter [92, 93] (as opposed to the quasiparticle–quasihole scattering considered here), processes in which a pair of particles is scattered to particle states outside the Fermi sea, are generally much more important than processes in which the initial pair of particles is annihilated by two holes. For this reason, in the Bethe–Goldstone equation for the G matrix only repeated scattering of a pair of particles to states outside the Fermi surface is taken into account, and therefore the intermediate state carries a factor $(1 - n_{\mathbf{p}_1}) \times (1 - n_{\mathbf{p}_2})$, where $\mathbf{p}_1$ and $\mathbf{p}_2$ are the momenta of the particles in the intermediate state; the term $-n_{\mathbf{p}_1}n_{\mathbf{p}_2}$ coming from annihilation of the initial pair of particles by two holes is not included. The factor $n^0_{\mathbf{p}+\mathbf{q}/2} - n^0_{\mathbf{p}-\mathbf{q}/2}$ is also familiar from the quantum kinetic equation (1.2.22); in fact, had we used the quantum kinetic equation rather than its classical limit (1.2.21), then (1.4.96) would agree precisely with (1.4.17) for the t matrix derived in Section 1.4.1. Since f does not depend sensitively on the momentum variables, the solutions of (1.4.96) and (1.4.17) are identical for $\omega \ll p_f v_f$, $q \ll p_f$ irrespective of the relative magnitudes of ω, $v_f q$, and κT.

The energy of interaction of a quasiparticle with a quasihole is given in terms of the K matrix by the quasiparticle–quasihole version of (1.4.94):

$$\begin{aligned}\Delta\omega_{\mathbf{pq}} = k^{\lambda}_{\mathbf{pp}}(\mathbf{q}, \omega_{\mathbf{qp}}) &- \frac{1}{3}\pi^2(2)^2 \sum_{\mathbf{p}'\mathbf{p}''} \dot{k}^{\lambda}_{\mathbf{pp}'}(\mathbf{q}, \omega_{\mathbf{pq}}) \\ &\times (n^0_{\mathbf{p}'-\mathbf{q}/2} - n^0_{\mathbf{p}'+\mathbf{q}/2})\delta(\omega_{\mathbf{pq}} - \omega_{\mathbf{p}'\mathbf{q}})k^{\lambda}_{\mathbf{p}'\mathbf{p}''}(\mathbf{q}, \omega_{\mathbf{pq}}) \\ &\times (n^0_{\mathbf{p}''-\mathbf{q}/2} - n^0_{\mathbf{p}''+\mathbf{q}/2})\delta(\omega_{\mathbf{pq}} - \omega_{\mathbf{p}''\mathbf{q}})k^{\lambda}_{\mathbf{p}''\mathbf{p}}(\mathbf{q}, \omega_{\mathbf{pq}}) + \dots, \end{aligned} \tag{1.4.97}$$

where

$$\omega_{\mathbf{pq}} = \varepsilon_{\mathbf{p}+\mathbf{q}/2} - \varepsilon_{\mathbf{p}-\mathbf{q}/2}. \tag{1.4.98}$$

By contrast, the real part of the forward scattering amplitude, given by the quasiparticle version of (1.4.92), is

$$\begin{aligned}\mathrm{Re}(t^{\lambda}_{\mathbf{pp}}(\mathbf{q}, \omega_{\mathbf{pq}})) = k^{\lambda}_{\mathbf{pp}}(\mathbf{q}, \omega_{\mathbf{pq}}) - \pi^2(2)^2 \sum_{\mathbf{p}'\mathbf{p}''} k^{\lambda}_{\mathbf{pp}'}(\mathbf{q}, \omega_{\mathbf{pq}}) \\ \times (n^0_{\mathbf{p}'-\mathbf{q}/2} - n^0_{\mathbf{p}'+\mathbf{q}/2})\delta(\omega_{\mathbf{pq}} - \omega_{\mathbf{p}'\mathbf{q}})k^{\lambda}_{\mathbf{p}'\mathbf{p}''}(\mathbf{q}, \omega_{\mathbf{pq}}) \\ \times (n^0_{\mathbf{p}''-\mathbf{q}/2} - n^0_{\mathbf{p}''+\mathbf{q}/2})\delta(\omega_{\mathbf{pq}} - \omega_{\mathbf{p}''\mathbf{q}})k^{\lambda}_{\mathbf{p}''\mathbf{p}}(\mathbf{q}, \omega_{\mathbf{pq}}) + \dots\end{aligned} \quad (1.4.99)$$

The T and K matrices for scattering of a quasiparticle–quasihole pair with total momentum $\mathbf{q}$ and energy ω are $2t^s$ and $2k^s$ for the singlet spin state of the pair and $2t^a$ and $2k^a$ for triplet spin. We can see this as follows. Let us compactly denote a quasiparticle–quasihole state in which the quasiparticle has spin σ and the quasihole has spin σ' by $|\sigma\sigma'\rangle$. Then, for example,

$$\begin{aligned} k^s_{\mathbf{pp}'} &= \frac{\langle\uparrow\uparrow|K|\uparrow\uparrow\rangle + \langle\uparrow\uparrow|K|\downarrow\downarrow\rangle}{2} \\ k^a_{\mathbf{pp}'} &= \frac{\langle\uparrow\uparrow|K|\uparrow\uparrow\rangle - \langle\uparrow\uparrow|K|\downarrow\downarrow\rangle}{2}. \end{aligned} \quad (1.4.100)$$

A singlet spin quasiparticle–quasihole state has the form $(|\uparrow\uparrow\rangle + |\downarrow\downarrow\rangle)/\sqrt{2}$, while the antisymmetric combination $(|\uparrow\uparrow\rangle - |\downarrow\downarrow\rangle)/\sqrt{2}$ is a triplet state; the sign in these combinations is opposite what it would be for quasiparticle–quasiparticle spin states as a consequence of the antisymmetry of fermion states. The matrix element between singlet (or triplet) quasiparticle–quasihole states is thus

$$\left(\frac{\langle\uparrow\uparrow| \pm \langle\downarrow\downarrow|}{\sqrt{2}}\right)K\left(\frac{|\uparrow\uparrow\rangle \pm |\downarrow\downarrow\rangle}{\sqrt{2}}\right) = 2k^{s(a)}_{\mathbf{pp}'}. \quad (1.4.101)$$

The interaction energies of a quasiparticle and a quasihole in singlet and triplet spin states are, respectively, $2\Delta\omega^s_{\mathbf{pp}}$ and $2\Delta\omega^a_{\mathbf{qq}}$. A pair consisting of a quasiparticle $(\mathbf{p} + \mathbf{q}/2, \uparrow)$ and a quasihole $(\mathbf{p} - \mathbf{q}/2, \downarrow)$ is in a triplet state, and therefore its energy is

$$\Delta\omega_{\mathbf{pq}\uparrow\downarrow} = 2\Delta\omega^a_{\mathbf{pq}}; \quad (1.4.102)$$

a pair consisting of a quasiparticle $(\mathbf{p} + \mathbf{q}/2, \uparrow)$ and a quasihole $(\mathbf{p} - \mathbf{q}/2, \uparrow)$ is half singlet and half triplet, and therefore its energy is

$$\Delta\omega_{\mathbf{pq}\uparrow\uparrow} = \Delta\omega^s_{\mathbf{pq}} + \Delta\omega^a_{\mathbf{pq}}. \quad (1.4.103)$$

The contribution to the total energy from long wavelength fluctuations with total momentum less than q_c is obtained by summing the interaction energies over all possible pairs and is

$$\Delta E = \frac{1}{2} \sum_{\mathbf{p},\, \mathbf{q}(q<q_c)} \Delta\omega_{\mathbf{pq}\sigma\sigma'} n^0_{\mathbf{p}-\mathbf{q}/2,\sigma}(1 - n^0_{\mathbf{p}+\mathbf{q}/2,\sigma'}). \quad (1.4.104)$$

The factor of $\frac{1}{2}$ in (1.4.104) is necessary since $\Delta\omega_{\mathbf{pq}\sigma\sigma'}$ is in fact the energy of interaction of a quasiparticle $(\mathbf{p} - \mathbf{q}/2, \sigma)$ and a quasihole $(\mathbf{p} + \mathbf{q}/2, \sigma')$

plus the energy of interaction of a pair with the quasiparticle $(\mathbf{p}+\mathbf{q}/2, \sigma')$ and quasihole $(\mathbf{p}-\mathbf{q}/2, \sigma)$. This is most easily seen by expressing $\Delta\omega$ in terms of Goldstone graphs. The basic structure of graphs for $\Delta\omega$ is shown in Figure 1.13*a*, where the double wavy line represents a chain of quasiparticle–quasihole bubbles. Because the quasiparticle–quasihole propagator we have used corresponds to the sum of two Goldstone graphs that differ only in the time-ordering of the initial and final vertices, graphs for $\Delta\omega$ may be divided into two classes according to which of the two vertices, 1 or 2, in Figure 1.13*a* occurs earlier in time. To construct graphs for contributions to the total energy one ties together the free ends of graphs for $\Delta\omega$, as shown in Figure 1.13*b* and *c*. The fermion lines in Figure 1.13*b* give rise to a factor $n^0_{\mathbf{p}-\mathbf{q}/2}(1 - n^0_{\mathbf{p}+\mathbf{q}/2})$, and therefore this term corresponds to the energy of a pair consisting of a quasiparticle $\mathbf{p}-\mathbf{q}/2$ and a quasihole $\mathbf{p}+\mathbf{q}/2$. On the other hand, the graph shown in Figure 1.13*c* carries a factor $n^0_{\mathbf{p}+\mathbf{q}/2}(1 - n^0_{\mathbf{p}-\mathbf{q}/2})$, and therefore corresponds to the energy of a pair consisting of a quasiparticle $\mathbf{p}+\mathbf{q}/2$ and a quasihole $\mathbf{p}-\mathbf{q}/2$. The factor of $\frac{1}{2}$ in (1.4.104) is needed to avoid overcounting contributions; for further details consult Carneiro's thesis [94]. Equation 1.4.104 also agrees with the result obtained by treating the Landau quasiparticle Hamiltonian in the random phase approximation.

Equation 1.4.104 is very similar to the expression for the interaction energy in the pair approximation to the Brueckner–Bethe theory of nuclear matter:

$$E_{\text{int}} = \frac{1}{2}\sum_{\mathbf{p},\mathbf{p}'} G_{\mathbf{p}\mathbf{p}'} n_{\mathbf{p}} n_{\mathbf{p}'}. \tag{1.4.105}$$

Here $G_{\mathbf{p}\mathbf{p}'}$ is the two-particle G matrix that takes into account repeated scattering of two particles in the medium.

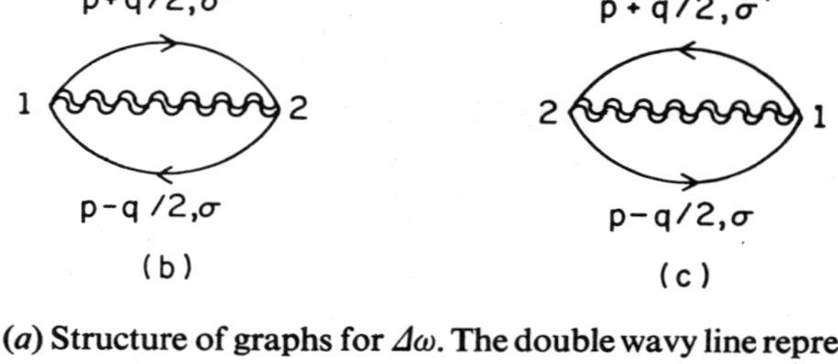

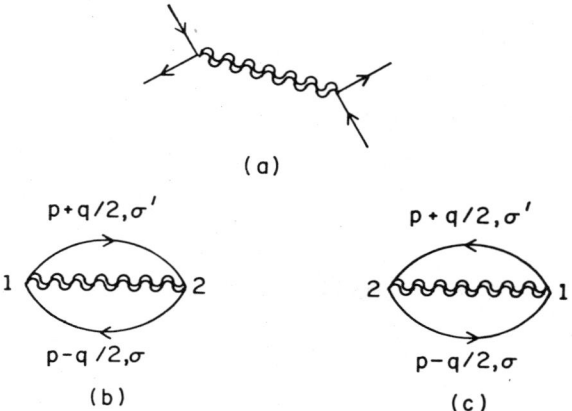

FIGURE 1.13. (*a*) Structure of graphs for $\Delta\omega$. The double wavy line represents a chain of quasiparticle–quasihole bubbles. (*b*) and (*c*) Diagrams for ΔE, equation 1.4.104.

To calculate fluctuation contributions to $f_{pp'}$ we differentiate the energy expression with respect to the distribution function as in (1.1.15). If $\Delta\omega$ were independent of the distribution function, then

$$\Delta f_{\mathbf{p}-\mathbf{q}/2,\sigma,\ \mathbf{p}+\mathbf{q}/2,\sigma'} = -\Delta\omega_{\mathbf{pq}\sigma\sigma'}. \tag{1.4.106}$$

This contribution may be represented by the graph shown in Figure 1.14, which corresponds to the direct exchange of a long-wavelength quasiparticle–quasihole pair. There are, however, additional contributions to $f_{pp'}$ coming

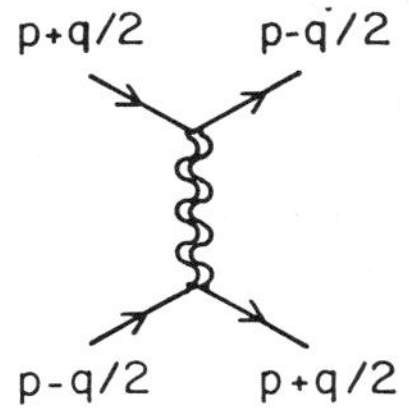

FIGURE 1.14. Diagram for the contribution to $f_{\mathbf{p}'-\mathbf{q}/2,\mathbf{p}+\mathbf{q}/2}$ not involving rearrangement terms. The double wavy line represents a chain of quasiparticle–quasihole bubbles.

from the dependence of $\Delta\omega$ on the distribution function. In calculations of effective interactions in nuclei one finds similar contributions; the effective interaction is equal to the G matrix plus additional contributions, often called *rearrangement terms*, resulting from the dependence of the G matrix on the distribution function. Following the nuclear terminology we refer to contributions to the quasiparticle energy and Landau f-function that involve derivatives of $\Delta\omega$ as rearrangement terms. The rearrangement contribution to f may be represented by diagrams having the structure shown in Figure 1.15. Whereas the direct term gives rise to interactions between quasiparticles with only slightly different momenta, the rearrangement terms yield interactions between quasiparticles whose momenta are very different, even though

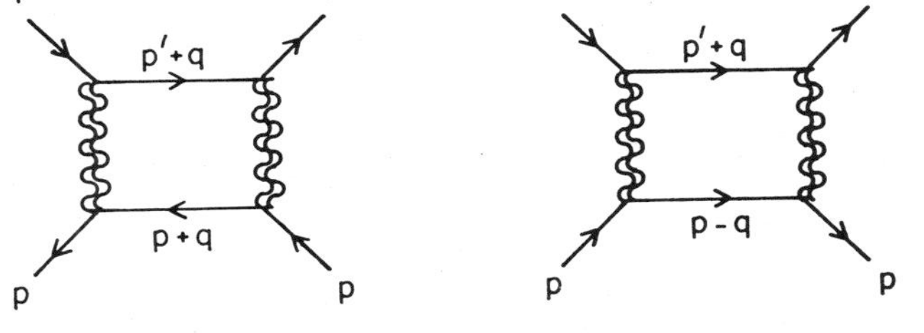

FIGURE 1.15. Diagrams for rearrangement contributions to $f_{pp'}$.

the quasiparticle–quasihole pairs exchanged have small momenta. By explicit calculation it may be shown that the rearrangement terms give no contribution to $f^s_{\mathbf{pp}'}$, and therefore do not affect the specific heat. From (1.4.102) to (1.4.104) and (1.4.106) one therefore sees that*

$$\Delta f^s_{\mathbf{p}-\mathbf{q}/2,\,\mathbf{p}+\mathbf{q}/2} = -\tfrac{1}{2}(\Delta\omega_{\mathbf{pq}\uparrow\uparrow} + \Delta\omega_{\mathbf{pq}\uparrow\downarrow}) = -\tfrac{1}{4}(\Delta\omega^s_{\mathbf{pq}} + 3\Delta\omega^a_{\mathbf{pq}}). \quad (1.4.107)$$

[Compare with the footnote below (1.4.42).] Although rearrangement terms do not contribute to f^s, they do contribute to f^a, and are important in determining finite temperature contributions to the magnetic susceptibility.

To determine how $f_{\mathbf{p},\,\mathbf{p}+\mathbf{q}}$ depends on $\mathbf{q}$ we need to know how $\Delta\omega$ (1.4.89) depends on $\mathbf{q}$. First we note that for small $\mathbf{q}$ the K-matrix element, $k_{\mathbf{pp}'}(\mathbf{q}, \omega)$, like $t_{\mathbf{pp}'}(\mathbf{q}, \omega)$, depends only on the variable $s = \omega/v_f q$ and the angles between $\mathbf{p}$, $\mathbf{p}'$, and $\mathbf{q}$. The sum over intermediate states in (1.4.97) may be performed using the fact that for small $\mathbf{q}$

$$\int p'^2 dp'(n^0_{\mathbf{p}'-\mathbf{q}/2} - n^0_{\mathbf{p}'+\mathbf{q}/2})\delta(\omega - \omega_{\mathbf{p}'\mathbf{q}}) = \pi^2 N(0) s\delta(s - \hat{\mathbf{p}}'\cdot\hat{\mathbf{q}}); \quad (1.4.108)$$

each sum over intermediate states in expression (1.4.97) for $\Delta\omega$ therefore gives a factor s. Thus we see that for small $\mathbf{q}$, $f_{\mathbf{p}-\mathbf{q}/2,\,\mathbf{p}+\mathbf{q}/2}$ and hence $f_{\mathbf{p},\,\mathbf{p}+\mathbf{q}}$ is indeed a function of $\hat{\mathbf{p}}\cdot\hat{\mathbf{q}}$; since the K matrix is an even function of s, f has an expansion of the form assumed (1.4.66).

The K matrix may be calculated taking into account any number of Landau parameters; however, here we shall just consider the case when all Landau parameters except F^s_0 and F^a_0 can be neglected. The K matrix is then given by

$$\mathscr{K}^\lambda_0(s) = N(0)k^\lambda_{\mathbf{pp}'}(\mathbf{q}, \omega) \equiv \frac{F^\lambda_0}{1 + F^\lambda_0 \mathrm{Re}\, \Omega_{00}(s)} \quad (1.4.109)$$

where $\Omega_{00}(s)$ is given by (1.3.16). From (1.4.97) with (1.4.108) we find $N(0)\Delta\omega_{\mathbf{pq}} = \mathscr{K} - (\pi^2/12)s^2\mathscr{K}^3$, and thus from (1.4.66) and (1.4.107), using the fact that on the energy shell $s = \hat{\mathbf{p}}\cdot\hat{\mathbf{q}}$, we derive

$$B^s = N(0)b^s = -\frac{1}{2}\sum_{\lambda=s,\,a} w_\lambda \left(\frac{1}{2}\frac{d^2\mathscr{K}^\lambda_0(s)}{ds^2}\bigg|_{s=0} - \frac{\pi^2}{12}[\mathscr{K}^\lambda_0(0)]^3\right), \quad (1.4.110)$$

$$= -\frac{1}{2}\left[(A^s_0)^2\left(1 - \frac{\pi^2}{12}A^s_0\right) + 3(A^a_0)^2\left(1 - \frac{\pi^2}{12}A^a_0\right)\right], \quad (1.4.111)$$

where the spin multiplicity factors are $w_s = 1$ and $w_a = 3$. If one neglects all but the $(A^a_0)^3$ term in B^s, and uses (1.4.76) and (1.4.111), one finds for the specific heat

*One side of this equation involves a spin-symmetric quantity, and the other side both spin-symmetric and spin antisymmetric quantities because the labels refer to the two different particle-hole channels. Deriving (1.4.105) we essentially use the crossing symmetry of the vertex.

$$c_V = \frac{\pi^2}{2} n\kappa \frac{T}{T_f}\left(1 - \frac{3\pi^4}{80}(A_0^a)^3\left(\frac{T}{T_f}\right)^2 \ln\left(\frac{T}{T_c}\right)\right). \quad (1.4.112)$$

The characteristic temperature that sets the scale of the finite temperature contributions is therefore $\sim T_f/|A_0^a|^{3/2}$, which for an almost ferromagnetic Fermi liquid is considerably less than T_f. The results of these calculations are consistent with those of Riedel [87], if the vertex functions in his calculation are replaced by fully renormalized ones.

Results for the case when Landau parameters with $l \leqslant 2$ are taken into account are given in Reference [90].

Specific heat data for liquid ^{3}He at low pressure are shown in Figure 1.8 [59]. The solid line is a fit containing only T and $T^3 \ln T$ terms. As Mota et al. [59] point out, the parameters in the fit cannot be determined precisely because of the scatter in the data at low temperatures, and this makes it impossible to arrive at quantitative conclusions by comparing theory and experiment. However, we note that the theoretical values of the magnitude of the T^3 ln T term [90] are consistent with the experimental value for liquid ^{3}He at low pressure for reasonable values of the Landau parameters F_1^a and F_2^a, but in view of the uncertainty in the experimental value, and ignorance about F_1^a and F_2^a it is not possible to arrive at any quantitative conclusions. We note, however, that if one neglects the Landau parameter F_2^a, the theoretical and experimental values of the T^3 ln T term in the specific heat are consistent if $F_1^a = 0.27$ ($A_1^a = 0.25$). Other estimates of A_1^a are -0.91 (from the forward scattering sum rule), -0.55 (from finite temperature contributions to the thermal conductivity), and -0.15 (from spin echo experiments), but because of the uncertainties the differences between the various estimates are not particularly significant.

Empirically, the fit to the data using only T and $T^3 \ln T$ terms is remarkably good. It is in fact good to temperatures much higher than one would expect on the basis of Brinkman and Engelsberg's [88] calculations for the paramagnon model. At the moment it is not clear whether the paramagnon model calculations underestimate the temperature at which the low temperature expansion breaks down, or whether the good fit to the experimental data is fortuitous (but see Section 3.1.1.)

For the case of liquid ^{3}He it is found that the contribution to B^s from exchange of zero spin excitations, which includes the zero sound contribution, is very small, typically of the order of one percent of the total. The reasons for this are twofold: first, the spin multiplicity factor, and, second, large contributions to B^s result from the coupling of quasiparticles to low-lying excitations. The zero spin excitations such as zero sound have high energies, due to the large value of F_0^s, and therefore contribute little to B^s. On the other hand, quasiparticles couple strongly to the low-lying pairs with spin 1, due to the large negative value of F_0^a.

How do the above results differ from those obtained by inserting dynamical quasiparticle energies into the entropy expression? By comparing the results of the calculations of Emery [66] with those above, one can see that the *dynamical* quasiparticle energy is related to the real part of the *forward scattering amplitude* in exactly the same way as the *statistical* quasiparticle energy is related to the *energy of interaction* $\Delta\omega$. The dynamical quasiparticle energy therefore has the same form as the statistical quasiparticle energy (1.4.74), but with the coefficient b^s replaced by b_D^s, which is related to the real part of the forward scattering amplitude in the same way as b^s is related to $\Delta\omega_{\mathbf{pq}}$. If only the Landau parameters F_0^s and F_0^a are taken into account, one finds

$$B_D^s = N(0)b_D^s = -\frac{1}{2}\sum_{\lambda=s,a} w_\lambda\left(\frac{1}{2}\left.\frac{d^2\mathcal{K}_0^\lambda(s)}{ds^2}\right|_{s=0} - \frac{\pi^2}{4}\,[\mathcal{K}_0^\lambda(0)]^3\right). \quad (1.4.113)$$

Note that the $\mathcal{K}^3$ term here is three times larger than in the expression for B^s (1.4.110); this reflects the same factor of 3 difference in the K^3 contributions to ΔE(1.4.94) and ReT (1.4.92). In the limit of an almost ferromagnetic Fermi liquid ($A_0^a \to -\infty$) the K^3 term is the dominant one, and this explains the factor of 3 differences found in the earlier paramagnon calculations. If one calculates B^s and B_D^s for values of the Landau parameters appropriate for liquid ^{3}He at low pressure one finds that B_D^s is two to four times as large as B^s.

We have shown how to calculate the $T^3 \ln T$ term in the specific heat coming from small momentum transfer processes, and one may ask whether any other processes give rise to $T^3 \ln T$ terms. In their field-theoretic calculations of the fermion self-energy Amit, Kane, and Wagner [85] found an inconsistency if they assumed that all the logarithmic terms came from small momentum-transfer processes, and suggested that the inconsistency could be resolved if there were additional logarithmic terms coming from processes in which $q \sim 2p_f$. However Pickles [72] has recently shown that the inconsistency is a consequence of the neglect of certain low-momentum transfer processes, and that it can be resolved without invoking high-momentum-transfer processes. Some calculations by Pickles suggest that $q \sim 2p_f$ processes do not give rise to $T^3 \ln T$ terms in the specific heat if one takes into account repeated scattering of pairs of quasiparticles with small total momentum; as yet the calculations have been carried through in detail only for a simplified form of the quasiparticle interaction, and we must await further calculations before any definite conclusions can be reached. However, it seems likely that even if such contributions exist, their magnitude in the case of liquid ^{3}He will be much less than the large long-wavelength contribution we derived above.

(c) The Magnetic Susceptibility. If the quasiparticle energy has con-

tributions $\sim (p - p_f)^3 \ln|p - p_f|$ one might expect there to be $T^2 \ln T$ contributions to the magnetic susceptibility. Although such terms have been found in calculations based on Fermi liquid theory [96], the "paramagnon" calculations of Béal-Monod, Ma, and Fredkin [97] gave no such terms. The later calculation started from the expression for the thermodynamic potential in a magnetic field and the magnetic susceptibility was evaluated by direct differentiation. The reason for the discrepancy between these two calculations has recently been considered by Carneiro [94] and Carneiro and Pethick [118] who showed that the Landau theory calculations when consistently carried out give no $T^2 \ln T$ term from long-wavelength fluctuations; the essential point is that in calculating the spin-antisymmetric Landau parameter, one cannot neglect the rearrangement terms. The rearrangement terms were taken into account (implicitly) in Reference [97], but not in Reference [96]. Misawa [98] has suggested that there may be $T^2 \ln T$ terms in the magnetic susceptibility because of processes involving large momentum transfers. However, it is not yet clear whether such terms survive when one takes into account repeated scattering of a pair of particles with small total momentum (cf. the discussion above of the $q \sim 2p_f$ contribution to the $T^2 \ln T$ term in the specific heat). It should also be noted that the experimental data for liquid ^{3}He at low pressure can be accounted for quite well by a T^2 dependence [99].

1.5. CONCLUDING REMARKS

We have based our discussions here on phenomenological considerations. Essentially all the results of the phenomenological calculations have also been derived using microscopic theory. We have already mentioned the results for *static* equilibrium properties obtained by Balian and De Dominicis [10] and others. Both equilibrium and nonequilibrium properties have been discussed using field-theoretic methods, and extensive descriptions of the earlier work on this subject have been given by Nozières [8], and by Abrikosov, Gorkov, and Dzyaloshinskii [100]. More recently, Leggett has discussed in detail properties of correlation functions, and has given a microscopic derivation of the stability conditions [46], and inequalities [41]. Amit, Kane, and Wagner [85] investigated in detail the nonanalytic contributions to the quasiparticle energy, and Carneiro [94, 115] and Pethick have shown how to calculate the leading finite-temperature contributions to static properties.

Finally, we give references to a number of extensions of Landau theory to ore complicated systems than the one-component uncharged normal Fermi system considered here. (i) Solutions of liquid ^{3}He in liquid ^{4}He are described in the following chapter. (ii) Elecctrons in metals have been discussed by a number of authors [46, 101, 102, 103]. Here both the electron-phonon and the electron–lattice interaction must be taken into account. Properties of collective excitations in metals have been considered in detail in the monograph of Platzman and Wolff [104]. (iii) Fermi-liquid effects in superfluid Fermi systems have been considered by Leggett [46, 105], and by Betbeder-Matibet and Nozières [106]. (See also Leggett's [113] theoretical review of the properties of superfluid ^{3}He). (iv) Landau theory has been applied to real nuclei by Migdal and collaborators [107]. (See also Reference [95].)

APPENDIX A: SOME USEFUL FERMI INTEGRALS

First we describe here a convenient and systematic method for calculating integrals of the form

$$I_\nu(y) = \int_{-\infty}^{\infty} dx_1 \ldots dx_\nu n(x_1) n(x_2) \ldots n(x_\nu) \delta(x_1 + x_2 + \ldots + x_\nu + y), \quad \text{(A.1)}$$

which arise in evaluation of transport properties of a Fermi liquid [108]. In (A.1), $n(x) = (e^x + 1)^{-1}$. We first use the algebraic relation

$$\left[\prod_{i=1}^{\nu} n(x_i) - e^y \prod_{i=1}^{\nu} (1 - n(x_i))\right] \delta\left(\sum_{i=1}^{\nu} x_i + y\right) = 0 \quad \text{(A.2)}$$

to write (A.1) as

$$[1 - (-1)^\nu e^{-y}]\, I_\nu(y) \equiv J_\nu(y)$$
$$= \int_{-\infty}^{\infty} dx_1 \ldots dx_\nu \left[\prod_i n(x_i) - \prod_i (n(x_i) - 1)\right] \delta(\textstyle\sum_i x_i + y). \quad \text{(A.3)}$$

If we then represent the delta function as $\delta(x) = \int_{-\infty}^{\infty} e^{itx}\, dt/2\pi$, we find

$$J_\nu(y) = \int_{-\infty}^{\infty} \frac{dt}{2\pi} e^{ity} [(b(t))^\nu - (c(t))^\nu], \quad \text{(A.4)}$$

where

$$b(t) = \int_{-\infty}^{\infty} dx \frac{e^{i(t - i\eta)x}}{e^x + 1}$$

$$c(t) = -\int_{-\infty}^{\infty} dx \frac{e^{i(t + i\eta)x}}{e^{-x} + 1}, \quad \text{(A.5)}$$

and η is a positive infinitesimal number added to make the integrals converge. The integrals (A.5) are given by [109]

$$b(z) = \frac{\pi}{i \sinh \pi z}, \qquad 0 > \operatorname{Im} z > -1$$

$$c(z) = \frac{\pi}{i \sinh \pi z}, \qquad 1 > \operatorname{Im} z > 0. \quad \text{(A.6)}$$

The net result is then

$$J_\nu(y) = \oint \frac{dz}{2\pi} e^{izy} \left(\frac{\pi}{i \sinh \pi z} \right)^\nu, \tag{A.7}$$

where the contour goes from $-\infty$ to $+\infty$ just under the real axis and from $+\infty$ to $-\infty$ just above the real axis. Since the only singularity of the integral on the real axis is at $z = 0$, the contour can be collapsed to encircle the pole of order ν at the origin; the integral then equals the νth order polynomial in y:

$$J_\nu(y) = \frac{1}{i^{\nu-1}(\nu - 1)!} \frac{\partial^{(\nu-1)}}{\partial z^{(\nu-1)}} \left[\left(\frac{\pi z}{\sinh \pi z} \right)^\nu e^{izy} \right]_{z=0}. \tag{A.8}$$

The first few I_ν are

$$\begin{aligned} I_1(y) &= \frac{1}{1 + e^{-y}} \\ I_2(y) &= \frac{y}{1 - e^{-y}} \\ I_3(y) &= \frac{1}{2} \frac{y^2 + \pi^2}{1 + e^{-y}} \\ I_4(y) &= \frac{1}{6} \frac{y(y^2 + 4\pi^2)}{1 - e^{-y}}. \end{aligned} \tag{A.9}$$

In calculating the transport coefficients we encounter the integrals

$$H_\nu = \int_{-\infty}^{\infty} dx \frac{\phi_\nu(x)}{\cosh \frac{x}{2}}, \qquad \nu \text{ odd} \tag{A.10}$$

and

$$L_\nu = \int_{-\infty}^{\infty} dx \frac{x\, \phi_\nu(x)}{\cosh \frac{x}{2}}, \qquad \nu \text{ even}, \tag{A.11}$$

where $\phi_\nu(x)$, of parity $(-1)^{\nu+1}$, is the eigenfunction of the collision operator, defined by (1.2.100). These integrals are evaluated in terms of the Fourier transforms of the integrands. From (A.5) and (A.6) we see that

$$\int_{-\infty}^{\infty} dx \frac{e^{-ikx}}{\cosh \frac{x}{2}} = 2b(-k - i/2) = \frac{2\pi}{\cosh \pi k} \tag{A.12}$$

$$\int_{-\infty}^{\infty} dx \frac{xe^{-ikx}}{\cosh \frac{x}{2}} = i \frac{\partial}{\partial k} \frac{2\pi}{\cosh \pi k} = \frac{2\pi^2}{\mathrm{i}} \frac{\tanh \pi k}{\cosh \pi k}, \tag{A.13}$$

so that from (1.2.66):

$$\frac{\nu(\nu+1)H_\nu}{\sqrt{2\nu+1}} = \int_{-\infty}^{\infty} \frac{dk}{\cosh \pi k} P_\nu^1(\tanh \pi k)$$
$$= \int_{-1}^{1} \frac{dz}{\pi} \frac{P_\nu^1(z)}{(1-z^2)^{1/2}} = -\int_{-1}^{1} \frac{dz}{\pi} \frac{d}{dz} P_\nu(z) = -\frac{2}{\pi}.$$

Here we have used

$$P_\nu^1(z) = -(1-z^2)^{1/2}\, dP_\nu(z)/dz. \tag{A.14}$$

Thus

$$\int_{-\infty}^{\infty} dx \frac{\phi_\nu(x)}{\cosh \frac{x}{2}} = -\frac{2}{\pi} \frac{\sqrt{2\nu+1}}{\nu(\nu+1)}, \qquad \nu \text{ odd} \tag{A.15}$$

and 0 if ν is even. Similarly,

$$\frac{\nu(\nu+1)L_\nu}{\sqrt{\nu+1}} = i\pi \int_{-\infty}^{\infty} \frac{dk}{\cosh \pi k} \tanh \pi k \, P_\nu^1(\tanh \pi k)$$
$$= -i \int_{-1}^{1} dz\, z \frac{dP_\nu(z)}{dz} = -2i,$$

so that

$$\int_{-\infty}^{\infty} dx \frac{x\, \phi_\nu(x)}{\cosh \frac{x}{2}} = -\frac{2i\sqrt{2\nu+1}}{\nu(\nu+1)}, \qquad \nu \text{ even}, \tag{A.16}$$

and 0 for ν odd.

APPENDIX B: PROPERTIES OF $\Omega_{ll'}$

$\Omega_{ll'}$, (1.3.15), may be expressed directly in terms of Legendre functions of the second kind, $Q_l(s)$ [110]

$$\Omega_{l'l} = \Omega_{ll'} = \frac{\delta_{ll'}}{2l+1} - sP_{l'}(s)Q_l(s), \qquad (l' \leqslant l). \tag{B.1}$$

$sQ_l(s)$ may be written in the form [110]

$$sQ_l(s) = -P_l(s)\Omega_{00}(s) + P_l(s) - sW_{l-1}(s), \tag{B.2}$$

where

$$\begin{aligned} W_{l-1}(s) &= \sum_{k=1}^{2E[(l-1)/2]} \frac{2(l-2k)-1}{(2k+1)(l-k)} P_{l-2k-1}(s), \qquad (l \geqslant 1) \\ &= \sum_{k=1}^{l} \frac{1}{k} P_{k-1}(s) P_{l-k}(s), \qquad (l \geqslant 1) \end{aligned} \tag{B.3}$$

$$W_{-1} = 0. \tag{B.4}$$

Here $E[x]$ denotes the integer part of x. The first few $q_{ll'}$ are given by

$$\Omega_{00} = 1 - \frac{s}{2}\ln\frac{s+1}{s-1},$$

$$\Omega_{10} = s\,\Omega_{00}, \qquad \Omega_{11} = \tfrac{1}{3} + s^2\Omega_{00},$$

$$\Omega_{20} = \tfrac{1}{2} + P_2(s)\Omega_{00}, \qquad \Omega_{21} = s\left[\tfrac{1}{2} + P_2(s)\Omega_{00}\right],$$

$$\Omega_{22} = \tfrac{3}{4}s^2 - \tfrac{1}{20} + P_2(s)^2 \qquad \Omega_{00} = \tfrac{1}{5} + P_2(s)\Omega_{20}(s). \quad \text{(B.5)}$$

APPENDIX C: FERMI LIQUID PARAMETERS FOR LIQUID ^{3}He[a]

Pressure	0 bar	27 bar
m^*/m	2.80	5.17
F_0^s	9.28	68.17
F_1^s	5.39	12.79
F_0^a	−0.696	−0.760
F_1^a (from Γ)	−0.54	−1.00
F_1^a (from sum rule)	−0.46	−0.27
$p_f(10^{-20}$ g-cm/sec)	8.28	9.234
v_f (cm/sec)	5.90×10^3	3.57×10^3
c_1 (cm/sec)	1.829×10^4	3.893×10^4

[a]The values of m^*/m and F_1^s are from D. S. Greywall, *Phys. Rev. B* **33**, 7520 (1986), where values at other pressures can also be found. The values of F_0^a, v_f, and c_1 are based on Wheatley [112], with m^*/m adjusted. The first set of estimates for F_1^a are from Greywall's measurements of the $T^3 \log T$ term in the specific heat, while the second set are from the forward scattering sum rule; in both estimates, Landau parameters with $l > 1$ were neglected.

REFERENCES

1. L. D. Landau, *Zh. Eksp. Teor. Fiz.* **30,** 1058 (1956) [English transl.: *Sov. Phys.—JETP* **3,** 920 (1957)].
2. L. D. Landau, *Zh. Eksp. Teor. Fiz.* **32,** 59 (1957) [English transl.: *Sov. Phys.—JETP* **5,** 101 (1957)].
3. J. C. Wheatley, in *Progress in Low Temperature Physics*, Vol. VI, edited by C. J. Gorter, (North-Holland Publishing, Amsterdam, 1970), p. 77.
4. A. A. Abrikosov and I. M. Khalatnikov, *Rept. Progr. Phys.* **22,** 329 (1959).
5. J. C. Wheatley, in *Quantum Fluids,* edited by D. F. Brewer (North-Holland Publishing, Amsterdam, 1966), p. 183.
6. D. Pines and P. Nozières, *The Theory of Quantum Liquids* (Benjamin, New York, 1966) Vol. I.
7. J. Wilks, *The Properties of Liquid and Solid Helium* (Clarendon Press, Oxford, 1967).

8. P. Nozières, *Theory of Interacting Fermi Systems* (Benjamin, New York, 1964).
9. D. D. Osheroff, R. C. Richardson, and D. M. Lee, *Phys. Rev. Lett.* **28,** 885 (1972); H. Kojima, D. N. Paulson, and J. C. Wheatley, *Phys. Rev. Lett.* **32,** 141 (1974).
10. R. Balian and C. De Dominicis, *Ann. Phys. (N. Y.)* **62,** 229 (1971).
11. V. P. Silin, *Zh. Eksp. Teor. Fiz.* **33,** 495 (1957) [English transl.: *Sov. Phys.—JETP* **6,** 387 (1958)].
12. See, for example, L. P. Kadanoff and G. Baym, *Quantum Statistical Mechanics* (Benjamin, New York, 1962).
13. A. A. Abrikosov and I. M. Khalatnikov, *Zh. Eksp. Teor. Fiz.* **32,** 1083 (1957) [English transl.: *Sov. Phys.–JETP* **5,** 887 (1957)].
14. D. Hone, *Phys. Rev.* **121,** 669 (1961).
15. W. R. Abel, R. T. Johnson, J. C. Wheatley, and W. Zimmermann, Jr., *Phys. Rev. Lett.* **18,** 737 (1967).
16. See, for example, J. M. Ziman, *Electrons and Phonons* (Oxford University Press, New York, 1960).
17. G. Baym and C. Ebner, *Phys. Rev.* **170,** 346 (1968).
18. K. S. Dy and C. J. Pethick, *Phys, Rev. Lett.* **21,** 876 (1968).
19. V. J. Emery and D. Cheng, *Phys. Rev. Lett.* **21,** 533 (1968); V. J. Emery, *Phys. Rev.* **175,** 251 (1968).
20. G. A. Brooker and J. Sykes, *Phys, Rev. Lett.* **21,** 279 (1968).
21. H. Højgaard Jensen, H. Smith, and J. W. Wilkins, *Phys. Lett.* **27A,** 532 (1968); *Phys. Rev.* **185,** 323 (1969).
22. J. Sykes and G. A. Brooker, *Ann. Phys. (N. Y.)* **56,** 1 (1970).
23. See, for example, I. S. Gradshteyn and I. M. Ryzhik, *Tables of Integrals, Series, and Products* (Academic Press, New York, 1965), 8.700.1.
24. G. A. Brooker, *Proc. Phys. Soc. (London)* **90,** 397 (1967).
25. L. D. Landau and E. M. Lifshitz, *Fluid Mechanics* (Addison-Wesley, Reading, Mass., 1959) pp. 298–300.
26. C. J. Pethick, *Phys. Rev.* **185**, 384 (1969).
27. L. R. Corruccini, J. S. Clarke, N. D. Mermin, and J. W. Wilkins, *Phys. Rev.* **180,** 225 (1969).
28. B. S. Lukyanchuk, *Zh. Eksp. Teor. Fiz.* **56,** 1338 (1969) [English transl.: *Sov. Phys.—JETP* **29,** 719 (1969)].
29. G. A. Brooker and J. Sykes, *Ann. Phys. (N .Y.)* **61,** 387 (1970).
30. J. Sykes and G. A. Brooker, *Ann. Phys. (N. Y.)* **74,** 67 (1972).
31. B. E. Keen, P. W. Matthews, and J. Wilks, *Phys. Lett.* **5,** 5 (1963); *Proc. Roy. Soc. A* **284,** 125 (1965).
32. W. R. Abel, A. C. Anderson, and J. C. Wheatley, *Phys. Rev. Lett.* **17,** 74 (1966).
33. I. Bekarevich and I. M. Khalatnikov, *Zh. Eksp. Teor. Fiz.* **39,** 1699 (1960) [English transl.: *Sov. Phys.—JETP* **12,** 1187 (1961)].
34. J. Gavoret, *Phys. Rev.* **137,** A721 (1965).
35. D. S. Betts, B. E. Keen, and J. Wilks, *Proc. Roy. Soc. A* **289,** 34 (1965); I. J. Kirby and J. Wilks, *Phys. Lett.* **24A,** 60 (1967).
36. A. C. Anderson and W. L. Johnson, *J. Low Temp. Phys.* **7,** 1 (1972).

37. See for example, Reference [25], § 24.

38. I. A. Fomin, *Zh. Eksp. Teor. Fiz.* **54,** 1881 (1968) [English transl.: *Sov. Phys.—JETP* **27,** 1010 (1968)]; M. J. Lea, A. R. Birks, P. M. Lee, and E. R. Dobbs, *J. Phys. C* **6,** L226 (1973); I. A. Fomin, *Zh. Eksp. Teor. Fiz. Pis. Red.* **24,** 90 (1976) [English transl.: *JETP Lett.* **24,** 77 (1976)]; E. G. Flowers, R. W. Richardson, and S. J. Williamson, *Phys. Rev. Lett.* **37,** 309 (1976).

39. N. D. Mermin, *Phys. Rev.* **159,** 161 (1967).

40. A. J. Leggett and M. J. Rice, *Phys. Rev. Lett.* **20,** 586 (1968).

41. A. J. Leggett, *J. Phys. C* **12,** 447 (1970).

42. S. Doniach, *Phys. Rev.* **177,** 336 (1969).

43. See J. C. Wheatley, *Proceedings of the Eleventh International Conference on Low Temperature Physics*, St. Andrews, Scotland, 1968, edited by J. F. Allen, D. M. Finlayson and D. M. McCall, (St. Andrews University, 1969) Vol. I, p. 409.

44. L. R. Corruccini, D. D. Osheroff, D. M. Lee, and R. C. Richardson, *Phys. Rev. Lett.*, **27,** 650 (1971); *J. Low Temp. Phys.* **8,** 229 (1972).

45. G. Baym, in *Mathematical Methods in Solid State and Superfluid Theory*, edited by R. C. Clark and G. H. Derrick (Oliver and Boyd, Edinburgh, 1969).

46. A. J. Leggett, *Ann. Phys. (N.Y.)* **46,** 76 (1968).

47. E. Safier and A. Widom, *Phys. Rev. A* **7,** 252 (1973).

48. A. J. Leggett, *Phys. Rev.* **140,** *A* 1869 (1965).

49. H. T. Tan, *Phys. Rev. A* **4,** 256 (1971).

50. E. Safier and A. Widom, *J. Low Temp. Phys.* **6,** 397 (1972).

51. A. A. Abrikosov and I. M. Khalatnikov, *Zh. Eksp. Teor. Fiz.* **34,** 198 (1958) [English transl.: *Sov. Phys.—JETP* **7,** 135 (1958)].

52. L. D. Landau and E. M. Lifshitz, *Electrodynamics of Continuous Media* (Pergamon, Oxford, 1960), Chap. XIV.

53. E. Feenberg, *Theory of Quantum Fluids* (Academic Press, New York, 1969).

54. See, for example, R. Newton, *Scattering Theory of Waves and Particles* (McGraw-Hill, New York, 1966), p. 152 ff.

55. L. D. Landau, *Zh. Eksp. Teor. Fiz.* **35,** 97 (1958) [English transl., *Soviet Phys.—JETP* **8,** 70 (1959)].

56. C. J. Pethick, in *Lectures in Theoretical Physics,* edited by K. Mahanthappa and W. E. Brittin (Gordon and Breach, New York, 1969) p. 187.

57. D. Hone, *Phys. Rev.* **125,** 1494 (1962).

58. K. S. Dy and C. J. Pethick, *Phys. Rev.* **185,** 373 (1969).

59. A. C. Mota, R. P. Platzeck, R. Rapp, and J. C. Wheatley, *Phys. Rev.* **177,** 266 (1969).

60. W. R. Abel and J. C. Wheatley, *Phys. Rev. Lett.* **21,** 597 (1968).

61. S. Doniach, S. Engelsberg, and M. J. Rice, *Proceedings of the Tenth International Conference on Low Temperature Physics,* edited by M. P. Malkov, L. P. Pitaevskii, and Yu. D. Anufrieyev (Viniti, Moscow, 1967), Vol. I. p. 356.

62. M. J. Rice, *Phys. Rev.* **159,** 153 (1967).

63. M. J. Rice, *Phys. Rev.* **162,** 189 (1967).

64. S. Doniach and M. J. Rice, in *Methods and Problems of Theoretical Physics,* edited by J. E. Bowcock (North-Holland, Amsterdam, 1970), p. 101.

65. D. S. Betts and M. J. Rice, *Phys. Rev.* **166,** 159 (1968).
66. V. J. Emery, *Phys. Rev.* **170**, 205 (1968).
67. C. J. Pethick, *Phys. Lett.* **27A,** 219 (1968); *Phys. Rev.* **177,** 391 (1969).
68. V. M. Galitskii, *Zh. Eksp. Teor. Fiz.* **34,** 151 (1958) [English transl.: *Sov. Phys.—JETP* **7,** 104 (1958)].
69. A. C. Anderson, W. Reese, R. J. Sarwinski, and J. C. Wheatley, *Phys. Rev. Lett.* **7,** 220 (1961).
70. A. Tyler, *J. Phys. C* **4,** 1479 (1971).
71. D. T. Lawson, W. J. Gully, S. Goldstein, J. D. Reppy, D. M. Lee, and R. C. Richardson, *J. Low Temp. Phys.* **13,** 503 (1973).
72. J. Pickles, Ph.D. Thesis, Cambridge University (1973).
73. The spin diffusion coefficient has recently been calculated at higher temperatures by M. T. Béal-Monod, *Phys. Rev. Lett.* **31,** 513 (1973), using the spin-fluctuation model. A rather more sophisticated version of the model has also been used by S. Babu and G. E. Brown, *Ann. Phys. (N.Y.)* **78,** 1 (1973), to calculate the Landau parameters for liquid ^{3}He.
74. P. W. Anderson, *Physics* **2,** 1 (1965).
75. R. Balian and D. R. Fredkin, *Phys. Rev. Lett.* **15,** 480 (1965).
76. S. Engelsberg and P. M. Platzman, *Phys. Rev.* **148,** 103 (1966).
77. M. J. Buckingham and M. R. Schafroth, *Proc. Phys. Soc. (London) A* **67,** 828 (1954).
78. G. M. Eliashberg, *Zh. Eksp. Teor. Fiz.* **43,** 1105 (1962) [English transl.: *Sov. Phys.—JETP* **16,** 780 (1962)].
79. S. Doniach and S. Engelsberg, *Phys. Rev. Lett.* **17,** 750 (1966).
80. N. F. Berk and J. R. Schrieffer, *Phys. Rev. Lett.* **17,** 433 (1966).
81. E. Bucher, W. F. Brinkman, J. P. Maita, and H. J. Williams, *Phys. Rev. Lett.* **18,** 1125 (1967).
82. D. R. Penn, *Phys. Lett.* **25A,** 269 (1967).
83. Cited by J. R. Schrieffer, *J. Appl. Phys.* **39,** 642 (1968).
84. W. Brenig and H. J. Mikeska, *Phys. Lett.* **24A,** 332 (1967).
85. D. J. Amit, J. W. Kane, and H. Wagner, *Phys. Rev. Lett.* **19,** 425 (1967); *Phys. Rev.* **175,** 313, 326 (1968).
86. W. Brenig, H. J. Mikeska, and E. Riedel, *Z. Physik* **206,** 439 (1967).
87. E. Riedel, *Z. Physik* **210,** 403 (1968).
88. W. F. Brinkman and S. Engelsberg, *Phys. Rev.* **169,** 417 (1968).
89. J. M. Luttinger, *Phys. Rev.* **174,** 263 (1968).
90. C. J. Pethick and G. M. Carneiro, *Phys. Rev. A***7,** 304 (1973).
91. See, for example, K. Gottfried, *Quantum Mechanics* (W. A. Benjamin, New York, 1966), Vol. 1, Chap. VII.
92. See, for example, H. A. Bethe, *Ann. Rev. Nucl. Sci.* **21,** 93 (1971).
93. G. E. Brown, *Unified Theory of Nuclear Models and Forces* (North-Holland, Amsterdam, 1971), Chap. XI.
94. G. M. Carneiro, Ph.D. Thesis, University of Illinois (1973).
95. G. E. Brown, *Rev. Mod. Phys.* **43,** 1 (1971).

96. M. Ya. Amusia and V. E. Starodubsky, *Phys. Lett.* **35A,** 115 (1971).

97. M. T. Béal-Monod, S. K. Ma, and D. R. Fredkin, *Phys. Rev. Lett.* **20,** 929 (1968).

98. S. Misawa, *Phys. Lett.* **32A,** 541 (1970); and in *Proceedings of the Twelfth International Conference on Low Temperature Physics,* Kyoto, Japan, 1970, edited by E. Kanda (Keigaku, Tokyo, 1971), p. 151.

99. H. Ramm, P. Pedroni, J. R. Thompson, and H. Meyer, *J. Low Temp. Phys.* **2,** 539 (1970).

100. A. A. Abrikosov, L. P. Gorkov, and I. Ye. Dzyaloshinskii, *Quantum Field Theoretical Methods in Statistical Physics* (Pergamon, Oxford, 1965).

101. J. W. Wilkins, *Observable Many-Body Effects in Metals* (Nordita, Copenhagen, 1968).

102. R. E. Prange and L. P. Kadanoff, *Phys. Rev.* **134,** A566 (1964).

103. R. E. Prange and A. Sachs, *Phys. Rev.* **158,** 672 (1967).

104. P. M. Platzman and P. A. Wolff, *Solid State Phys., Suppl.* **13** (Academic Press, New York, 1973).

105. A. J. Leggett, *Phys. Rev.* **147,** 119 (1966).

106. O. Betbeder-Matibet and P. Nozières, *Ann. Phys. (N.Y.)* **51,** 392 (1969).

107. A. B. Migdal, *Theory of Finite Fermi Systems and Applications to Atomic Nuclei* (Interscience, New York, 1962).

108. P. Morel and P. Nozières, *Phys. Rev.* **126,** 1909 (1962).

109. Reference [23], 3.311.3.

110. Reference [23], 7.224 and 8.831.

111. J. C. Wheatley, *Phys. Rev.* **165,** 304 (1968).

112. J. C. Wheatley, *Rev. Mod. Phys.* **47,** 415 (1975).

113. A. J. Leggett, *Rev. Mod. Phys.* **47,** 331 (1975).

114. P. Bhattacharyya, C. J. Pethick, and H. Smith, *Phys. Rev. B* **15**, 3367, 3384 (1977); C. J. Pethick and H. Smith, Sussex Symposium of Superfluid ^{3}He, *Physica* **90B**, 107 (1977).

115. G. M. Carneiro and C. J. Pethick, *Phys. Rev. B***11,** 1106 (1975).

116. P. R. Roach and J. B. Ketterson, *Phys. Rev. Lett.* **36,** 736 (1976).

117. D. D. Osheroff, W. van Roosbroeck, H. Smith, and W. F. Brinkman, *Phys. Rev. Lett.* **38,** 134 (1977).

118. G. M. Carneiro and C. J. Pethick, *Phys. Rev. B***16,** 1933 (1977).

119. E. Egilsson and C. J. Pethick, *J. Low Temp. Phys.* **29, 99** (1977).

120. I. M. Khalatnikov, in *The Physics of Liquid and Solid Helium*, Part 1, edited by K. H. Bennemann and J. B. Ketterson (Wiley, New York, 1976), p. 1; also *An Introduction to the Theory of Superfluidity* (Benjamin, New York, 1965).

121. N. Viviandi, E. Buendia, S. Fantoni, and S. Rosati, *Phys. Rev. B* **38**, 4523 (1988).

122. D. M. Lee and R. C. Richardson, in *The Physics of Liquid and Solid Helium*, Part 2, edited by K. H. Bennemann and J. B. Ketterson (Wiley, New York, 1978), p. 287.

123. P. W. Anderson and W. F. Brinkman, in *The Physics of Liquid and Solid Helium*, Part 2, edited by K. H. Bennemann and J. B. Ketterson (Wiley, New York, 1978), p. 177.

2

LOW TEMPERATURE PROPERTIES OF DILUTE SOLUTIONS OF ^{3}He IN SUPERFLUID ^{4}He

INTRODUCTION

The study of dilute solutions of ^{3}He in superfluid ^{4}He received considerable impetus from the experimental discovery of Edwards *et al.* [1] in 1965 that ^{3}He is miscible in ^{4}He, at concentrations less than about 6%, down to absolute zero temperature. In this regime the ^{3}He in solution behaves as a Fermi liquid with the unique feature that its density and temperature can be varied experimentally, enabling it to be studied continuously from the degenerate quantum regime to the "high temperature" classical regime.

The phase diagram [2, 3] in the ^{3}He concentration $[x = N_3/(N_3 + N_4)]$, temperature plane is shown in Figure 2.1. The ^{4}He rich region at the left, marked II, is superfluid, while region I is normal. The λ line, which separates these two regions, intersects the T axis at the pure ^{4}He λ point. In the regime labelled I + II the system physically separates into two phases, a ^{4}He rich phase (in region II) below and a lighter ^{3}He rich phase (in I) above. The point labelled C, the meeting of the λ line and the phase separation curve, is the tricritical point; the study of the tricritical point is a subject in itself, which we will not discuss here, but refer the reader to the review by Ahlers [4]. The dashed line in Figure 2.1 is the ^{3}He degeneracy line, along which the temperature equals the ^{3}He degeneracy temperature T_f. Below this line the ^{3}He are degenerate.

The thermal and transport properties of dilute solutions at low temperatures—our main interest here—result from the interplay of the ^{3}He quasiparticles and the phonon and roton excitations of the superfluid. At very low temperatures the system behaves as a Fermi liquid of ^{3}He quasiparticles moving through a background ^{4}He ether; in this regime the full machinery of the Landau Fermi-liquid theory developed in Chapter 1

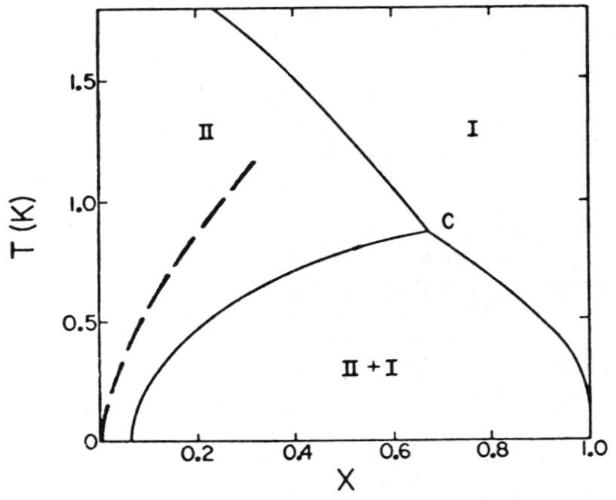

FIGURE 2.1. Phase diagram of solutions of ^{3}He in ^{4}He in the fractional ^{3}He concentration, temperature plane. Region I is normal, II superfluid, while II + I is the region of phase separation. Along the dashed line the ^{3}He Fermi temperature equals the temperature. *C* is the tricritical point.

is applicable. The freedom to vary the ^{3}He density experimentally enables one to deduce a great deal of information about effective interactions of ^{3}He quasiparticles. The ^{3}He-phonon system, which determines the properties of the system at temperatures below ~ 0.6 K, is rather analagous to nonrelativistic quantum electrodynamics in the sense that phonon velocities are large compared with characteristic ^{3}He velocities; the theory of the ^{3}He-phonon interaction can be developed by very general thermodynamic and Galilean invariance arguments, analogous to those used in the Landau Fermi-liquid theory.

Part of the original theoretical interest in dilute solutions centered on the possibility that the ^{3}He may themselves undergo a superfluid pairing transition [5, 6]. So far there has been no experimental evidence that the ^{3}He become superfluid. The effective interaction between ^{3}He quasiparticles is, as we shall see, only weakly attractive. Bardeen, Baym, and Pines estimated that the zero pressure transition temperature at which the ^{3}He become superfluid, paired as in a BCS superconductor in *s* states, is $\sim 2 \times 10^{-6}$ K. Landau et al. [7] estimate that the transition temperature under 10 to 20 atm pressure may be $\sim 10^{-4}$ K, while Hoffberg [8] has suggested that at high pressure and concentration *p* state pairing might occur at even higher temperatures. (But see [9].) We shall not consider properties of a possible ^{3}He superfluid phase here.

Other areas that we shall omit are surface and finite geometry properties of dilute solutions [10], the study of light and neutron scattering from dilute solutions [11–15] and the theory of dilute solutions of ^{4}He in ^{3}He [16–19]. Valuable further discussions of the properties of dilute solutions of ^{3}He in ^{4}He can be found in Khalatnikov's book [20], and in the review article by Ebner and Edwards [2]. The latter contains an extensive review of experiments on the thermal properties of dilute solutions, as well as a detailed bibliography. Additional reviews are found in References [21–26].

2.1. ELEMENTARY EXCITATIONS OF DILUTE SOLUTIONS

The theory of dilute solutions of ^3He in superfluid ^4He at low temperatures, below 1.5 K generally, is greatly simplified by the fact that one can, in the spirit of the Landau Fermi-liquid theory, describe the low-lying excited states of the system in terms of weakly interacting elementary excitations above the ground state. These excitations are of two types: phonon and roton excitations of the ^4He, whose parameters are nearly equal to those of pure ^4He; and ^3He quasiparticle excitations, characterized at low momentum by the Landau-Pomeranchuk spectrum [27]

$$\varepsilon_p = \varepsilon_0 + \frac{p^2}{2m^*}, \tag{2.1.1}$$

where $\mathbf{p}$ is the momentum of the excitation, and m^* is the quasiparticle effective mass. At low momentum the ^3He quasiparticles are kinematically stable against emission of ^4He excitations.

Consider a low-lying macroscopic state, in a box of volume V in which the density is uniform and the superfluid velocity $\mathbf{v}_s$ is everywhere zero. Such a state is characterized by the (smoothed) distribution functions $n_{\mathbf{p}\sigma}$, the mean number of ^3He quasiparticles of momentum $\mathbf{p}$ and spin σ, and $f_{\mathbf{q}}$ the mean number of ^4He excitations of momentum $\mathbf{q}$, and, in addition, the number density n_4 of ^4He atoms. The energy E (per unit volume) of the state is a functional only of the $n_{\mathbf{p}\sigma}$, $f_{\mathbf{q}}$ and n_4:

$$E = E(\{n_{\mathbf{p}\sigma}\}, \{f_{\mathbf{q}}\}, n_4). \tag{2.1.2}$$

Since the number of ^3He quasiparticles equals the number of ^3He atoms, we have

$$N_3 = n_3 V = \sum_{\mathbf{p}\sigma} n_{\mathbf{p}\sigma}, \tag{2.1.3}$$

where n_3 is the ^3He number density.

The ^3He quasiparticle energy $\varepsilon_{\mathbf{p}\sigma}$ is then given by the variation of the energy E with respect to $n_{\mathbf{p}\sigma}$, holding all other $n_{\mathbf{p}'\sigma'}$, all $f_{\mathbf{q}}$ and n_4 fixed:

$$\varepsilon_{\mathbf{p}\sigma} = V\left(\frac{\delta E}{\delta n_{\mathbf{p}\sigma}}\right)_{n_{\mathbf{p}'\sigma'},\, f_{\mathbf{q}},\, n_4} \tag{2.1.4}$$

Similarly the ^4He excitation energies are given by

$$\omega_{\mathbf{q}} = V\left(\frac{\delta E}{\delta f_{\mathbf{q}}}\right)_{n_{\mathbf{p}\sigma},\, f_{\mathbf{q}'},\, n_4}, \tag{2.1.5}$$

while the derivative of E with respect to n_4 defines μ_4, the ^{4}He chemical potential:

$$\mu_4 = \left(\frac{\delta E}{\delta n_4}\right)_{n_{\mathbf{p}\sigma},\, f_{\mathbf{q}}}. \tag{2.1.6}$$

Thus the total variation of the energy, at fixed volume and $\mathbf{v}_s$, is

$$\delta E = \mu_4 \delta n_4 + \frac{1}{V}\sum_{\mathbf{p}\sigma} \varepsilon_{\mathbf{p}\sigma}\, \delta n_{\mathbf{p}\sigma} + \frac{1}{V}\sum_{\mathbf{q}} \omega_{\mathbf{q}} \delta f_{\mathbf{q}}. \tag{2.1.7}$$

Note that μ_3, the ^{3}He chemical potential, does not occur in the variation of E, since in the quasiparticle picture n_3 does not occur as a variable independent of the $n_{\mathbf{p}\sigma}$. Note also that $\varepsilon_{\mathbf{p}\sigma}$ and $\omega_{\mathbf{q}}$ depend on the number of ^{3}He quasiparticles and ^{4}He excitations; this leads to interactions among the elementary excitations, and hence to the temperature dependence of $\varepsilon_{\mathbf{p}\sigma}$ and $\omega_{\mathbf{q}}$ in states of thermodynamic equilibrium.

It is often quite useful to regard the ^{4}He chemical potential μ_4 as the independent variable instead of n_4. Then, if we write

$$E'(\{n_{\mathbf{p}\sigma}\}, \{f_{\mathbf{q}}\}, \mu_4) = E(\{n_{\mathbf{p}\sigma}\}, \{f_{\mathbf{q}}\}, n_4) - \mu_4 n_4 \tag{2.1.8}$$

we have

$$\delta E' = -n_4 \delta\mu_4 + \frac{1}{V}\sum_{\mathbf{p}\sigma} \varepsilon_{\mathbf{p}\sigma} \delta n_{\mathbf{p}\sigma} + \frac{1}{V}\sum_{\mathbf{q}} \omega_{\mathbf{q}} \delta f_{\mathbf{q}}. \tag{2.1.9}$$

The quasiparticle energies are thus equivalently defined by

$$\varepsilon_{\mathbf{p}\sigma} = V\left(\frac{\delta E'}{\delta n_{\mathbf{p}\sigma}}\right)_{n_{\mathbf{p}'\sigma'},\, f_{\mathbf{q}},\, \mu_4} \tag{2.1.10}$$

$$\omega_{\mathbf{q}} = V\left(\frac{\delta E'}{\delta f_{\mathbf{q}}}\right)_{n_{\mathbf{p}\sigma},\, f_{\mathbf{q}'},\, \mu_4}.$$

For momenta p small compared with $m_4 s$, where m_4 is the ^{4}He atomic mass, and s($= 238$ m/sec at the vapor pressure) is the ^{4}He first sound velocity, the ^{3}He excitation spectrum has the form (2.1.1). At low T and q the ^{4}He excitations are phonons with energies ω_q given by

$$\omega_q = sq, \qquad q \ll q_D; \tag{2.1.11}$$

here $q_D = 1.09$ Å^{-1} is the ^{4}He Debye wavenumber, defined by $n_4^0 = q_D^3/6\pi^2$, where n_4^0 is the pure ^{4}He number density at $T = P = 0$. For $q \simeq q_0 = 1.91$ Å^{-1} the ^{4}He excitations are of the form

$$\omega_q = \Delta + \frac{(q - q_0)^2}{2\mu_r}, \qquad q \sim q_0, \tag{2.1.12}$$

which is the roton spectrum. In pure ^{4}He at $T = 0$ and zero pressure, $\Delta = 8.65$ K, $\mu_r = 0.16\ m_4$.

The entropy S of the macroscopic state depends only on the number of ways of distributing a given number of elementary excitations among the various possible momentum states. Thus, S is a sum of contributions from each individual mode of excitation, and we have [cf. (1.1.3)]

$$S = -\kappa \sum_{\mathbf{p}\sigma} [n_{\mathbf{p}\sigma} \ln n_{\mathbf{p}\sigma} + (1 - n_{\mathbf{p}\sigma}) \ln (1 - n_{\mathbf{p}\sigma})] - \kappa \sum_{\mathbf{q}} [f_{\mathbf{q}} \ln f_{\mathbf{q}} - (1 + f_{\mathbf{q}}) \ln (1 + f_{\mathbf{q}})], \tag{2.1.13}$$

where κ is Boltzmann's constant. The entropy has no explicit dependence on N_4 or $\mathbf{v}_s$; its total variation is given by

$$\delta S = -\kappa \sum_{\mathbf{p}\sigma} (\delta n_{\mathbf{p}\sigma}) \ln \left(\frac{n_{\mathbf{p}\sigma}}{1 - n_{\mathbf{p}\sigma}} \right) - \kappa \sum_{\mathbf{q}} (\delta f_{\mathbf{q}}) \ln \left(\frac{f_{\mathbf{q}}}{1 + f_{\mathbf{q}}} \right). \tag{2.1.14}$$

It should be stressed here that the "free particle" form of the entropy is completely consistent, as we shall see, with temperature dependent excitation energies, including a temperature dependent effective mass, and with weak interactions between all the elementary excitations.

In thermal equilibrium the smoothed distribution functions $n_{\mathbf{p}\sigma}$ and $f_{\mathbf{q}}$ are determined from the relation

$$\delta E = T\delta \frac{S}{V} + \mu_3 \delta n_3 + \mu_4 \delta n_4, \tag{2.1.15}$$

valid for any variation about thermodynamic equilibrium at temperature T. Using (2.1.7) and (2.1.14) we find, as in Chapter 1, that

$$n_{\mathbf{p}\sigma} = \frac{1}{e^{(\varepsilon_{\mathbf{p}\sigma} - \mu_3)/\kappa T} + 1} \tag{2.1.16}$$

and

$$f_{\mathbf{q}} = \frac{1}{e^{\omega_{\mathbf{q}}/\kappa T} - 1}. \tag{2.1.17}$$

The ^{3}He chemical potential μ_3, which enters the theory only as an equilibrium quantity, is determined by (2.1.16) and (2.1.3). The energy density E of a state of thermodynamic equilibrium depends on the temperature *only* through the ^{3}He and ^{4}He excitation distribution functions (2.1.16) and (2.1.17).

The total variation of the free energy density $F = E - TS/V$ is, according to (2.1.7), (2.1.15), (2.1.16) and (2.1.17),

$$\delta F = \mu_3 \delta\left[\frac{1}{V}\sum_{\mathbf{p}\sigma} n_{\mathbf{p}\sigma}\right] + \mu_4 \delta n_4 - \frac{S}{V}\delta T. \tag{2.1.18}$$

Also, in the quasiparticle picture the pressure P is given by

$$P = -\left(\frac{\partial EV}{\partial V}\right)_{n_{\mathbf{p}\sigma},\, f_{\mathbf{q}},\, n_4 V} \tag{2.1.19}$$

(this is equivalent to differentiating at constant entropy). Equation 2.1.18 agrees with the usual thermodynamic result, and leads in the limit of large volume to the Gibbs–Duhem relation

$$\delta P = n_3 \delta\mu_3 + n_4 \delta\mu_4 + \frac{S}{V}\delta T. \tag{2.1.20}$$

2.2. PROPERTIES OF ONE ^{3}He ATOM IN ^{4}He AT $T = 0$

The energy of a single ^{3}He quasiparticle in superfluid ^{4}He at $T = 0$ is given, for $p \ll m_4 s$, by $\varepsilon_p = \varepsilon_0 + p^2/2m$, where at zero pressure $\varepsilon_0 \simeq -2.785$ K, the ^{3}He effective mass m is $\simeq 2.34\ m_3$, and m_3 is the ^{3}He atomic mass. These values are determined experimentally from the heat-of-mixing [28] of ^{3}He in ^{4}He, and from the specific heat [29] of dilute solutions. In the limit of zero ^{3}He concentration ε_0 is just the ^{3}He chemical potential. The form of ε_p for larger p is discussed in References [30–32].

2.2.1. Volume Occupied by ^{3}He

The volume occupied by a single ^{4}He atom in pure ^{4}He is $n_4^{-1} \simeq 45.8$ Å^3. Let us write the volume occupied by a ^{3}He added to ^{4}He at constant *pressure*, as $v_3 = (1 + \alpha)n_4^{-1}$. Experimentally [33, 34], $\alpha \simeq 0.285$ at $T = 0$ and $P = 0$. If N_4 ^{4}He atoms occupy a volume V then at the same pressure N_4 ^{4}He atoms plus N_3 ^{3}He atoms will occupy a volume $V + N_3 v_3$. Thus, the change in the mean ^{4}He density on adding ^{3}He atoms at constant pressure is

$$\left(\frac{\partial n_4}{\partial n_3}\right)_P = -(1 + \alpha). \tag{2.2.1}$$

If we note from the Gibbs–Duhem relation (2.1.20) that in the limit of zero ^{3}He concentration and constant T, keeping P constant is equivalent to keeping μ_4 constant, we see that the left side of (2.2.1) is equivalently

$$\left(\frac{\partial n_4}{\partial n_3}\right)_{\mu_4} = -\left(\frac{\partial \mu_4}{\partial n_3}\right)_{n_4}\left(\frac{\partial n_4}{\partial \mu_4}\right)_{n_3}. \tag{2.2.2}$$

Since

$$\left(\frac{\partial \mu_4}{\partial n_3}\right)_{n_4} = \frac{\partial^2 E}{\partial n_3 \partial n_4} = \left(\frac{\partial \mu_3}{\partial n_4}\right)_{n_3}, \tag{2.2.3}$$

and as $n_3 \to 0$

$$\frac{\partial \mu_4}{\partial n_4} = \frac{m_4 s^2}{n_4} \tag{2.2.4}$$

(the relation between the compressibility and the sound velocity), we find the simple and useful relation between $\partial \varepsilon_0/\partial n_4$ and α:

$$\frac{\partial \varepsilon_0}{\partial n_4} = \frac{m_4 s^2}{n_4}(1 + \alpha). \tag{2.2.5}$$

Physically the reason a ^{3}He occupies a 30% larger volume than a ^{4}He is due to the smaller mass and hence the greater zero point motion of the ^{3}He atom. We can do a simple microscopic calculation that makes this idea quantitative [35]. The exact Hamiltonian of the mixture is

$$H = \sum_{^4\text{He}} \frac{p_i^2}{2m_4} + \frac{1}{2}\sum_{ij} u(\mathbf{r}_i - \mathbf{r}_j) + \sum_{^3\text{He}} \frac{p_j^2}{2m_3}, \tag{2.2.6}$$

where u is the bare interatomic potential, which is the same for all pairs of helium atoms, whether 4–4, 4–3 or 3–3; the sum in the middle term is over all pairs. Since $m_3 = 3m_4/4$ we can equivalently write H as

$$H = \sum_{i} \frac{p_i^2}{2m_4} + \frac{1}{2}\sum_{ij} u(\mathbf{r}_i - \mathbf{r}_j) + \sum_{^3\text{He}} \frac{p_i^2}{6m_4}, \tag{2.2.7}$$

where the first sum is now over both the ^{3}He and the ^{4}He atoms. The first two terms in (2.2.7) are formally the Hamiltonian for $N = N_3 + N_4$ ^{4}He atoms. The final term represents the additional kinetic energy of the ^{3}He atoms due to their lighter mass; it is essentially a small perturbation.

We may estimate the ground state energy of the system by choosing as a trial wave function the true ground state wave function of pure ^{4}He at the same particle density

$$n = n_3 + n_4 \tag{2.2.8}$$

as in the mixture. This trial wave function does not take into account the Pauli principle between ^{3}He of parallel spin, and thus our result will be valid for one ^{3}He atom, or for two ^{3}He atoms of opposite spin. With this trial function we find

$$E = E_0(n) + n_3 E_1(n), \tag{2.2.9}$$

where E_0 is the ground state energy density of pure ^{4}He at density n, and

$$E_1(n) = \left\langle \frac{p^2}{6m_4} \right\rangle, \tag{2.2.10}$$

that is, one-third the mean kinetic energy per particle in pure ^{4}He at density n at $T = 0$. For a single ^{3}He present we identify ε_0 as

$$\varepsilon_0 = \frac{\partial E}{\partial n_3} = \frac{\partial E_0}{\partial n_4} + E_1(n_4). \tag{2.2.11}$$

The value of $\partial E_0/\partial n_4 = \mu_4$ is -7.17 K, and McMillan [36] gives the theoretical value $\langle p^2/2m_4 \rangle \simeq 14.16$ K. We thus estimate in this simple model $\varepsilon_0 \simeq -2.45$ K. The result (1.27) is more generally an upper bound to ε_0. McMillan also estimates $n_4 \partial \langle p^2/2m_4 \rangle / \partial n_4 \simeq 25.1$ K, and so we find from (2.2.11) and (2.2.5) the estimate

$$\alpha \simeq \frac{1}{3} \frac{n_4}{m_4 s^2} \frac{\partial}{\partial n_4} \left\langle \frac{p^2}{2m_4} \right\rangle \simeq 0.31. \tag{2.2.12}$$

It is interesting to note that were $m_3 = m_4$, (2.2.9) with $E_1 \equiv 0$ would be exact because then the ground state wave function of the one ^{3}He, N_4 ^{4}He atom state would be the $N_4 + 1$ ^{4}He atom ground state wave function—that is, completely symmetric [37]. Similarly, the ground state of two ^{3}He of opposite spin and N_4 ^{4}He is the same as that of $N_4 + 2$ ^{4}He atoms.

Massey and Woo [38] have done a more complete calculation of α, as a function of density, which agrees remarkably well with experiment [33] (see Figure 5.5.2 of Woo's review [39]). They have also derived, using the work of Davison and Feenberg [40] the theoretical result $\varepsilon_0 = -2.70$ K.

2.2.2. ^{3}He Effective Mass

The effective mass of a ^{3}He in ^{4}He is larger than m_3 because as the ^{3}He atom moves through the fluid it creates a flow in the fluid which also carries momentum. The simplest estimate of the effective mass is from the classical result that a sphere of mass m_3 in a fluid of mass density $m_4 n_4$ behaves as if it has a mass [63]

$$m_{cl} = m_3 + \tfrac{1}{2} m_4 n_4 v_3, \tag{2.2.13}$$

where v_3 is the volume of the fluid displaced by the sphere; that is, the mass associated with the fluid flow is one-half the mass displaced by the sphere. From $n_4 v_3 = 1 + \alpha$ we find $m_{cl} = 1.85\, m_3$.

A microscopic calculation by Woo, Tan, and Massey [41], using the method of correlated basis functions, yields the result $m = 2.37\, m_3$ at $P = 0$. Earlier calculations of m are found in References [40] and [42]. Recently, Pandharipande and Itoh [43] have done a microscopic calculation that gives a good explanation of the pressure dependence of m. They write the energy of a ^{3}He quasiparticle of momentum p as

$$\varepsilon_p = \varepsilon_0(n_4) + \frac{p^2}{2m_3} + \Sigma(n_4) p^2, \tag{2.2.14}$$

and find that the coefficient of the momentum dependent interaction term,

$\Sigma(n_4)$, is to good accuracy a sum of attractive pairwise interactions of the ^{3}He with the individual ^{4}He and of the form

$$\Sigma(n_4) = -\frac{an_4}{\left(1 + \dfrac{m_3}{m_4}\right)^2}, \tag{2.2.15}$$

where a is a positive constant. Their result for the ^{3}He effective mass is thus

$$\frac{m}{m_3} = \frac{1}{1 - 2am_3n_4/(m_3 + m_4)^2}. \tag{2.2.16}$$

If one fits (2.2.16) to the experimental value $m = 2.34\, m_3$ at $P = 0$, one finds a form

$$\frac{m}{m_3} = \left[1 - 0.57\,\frac{n_4}{n_4^0}\right]^{-1}, \tag{2.2.17}$$

for the dependence of m on the ^{4}He density, where n_4^0 is the ^{4}He density at $T = P = 0$. Equation 2.2.17 is in excellent agreement with the experiments of Brubaker et al. [44] on second sound in dilute solutions, from which they deduce the dependence of m on pressure; their results at $P = 0$, 10 and 20 atm can be fit to (2.2.17) with a coefficient 0.56 instead of 0.57.

2.3. INTERACTIONS OF ^{3}He AT VERY LOW TEMPERATURE

At temperatures below $\sim$ 0.6K, the ^{4}He is essentially in its ground state; the excitations of phonons and rotons can be neglected. The ^{3}He in solution then behave as a pure Fermi liquid in the presence of a ^{4}He background "ether."

2.3.1. ^{3}He Landau Parameters

The ^{3}He Landau parameters, which describe the ^{3}He–^{3}He Fermi liquid interactions, are defined by

$$f_{\mathbf{p}\sigma,\mathbf{p}'\sigma'} = V^2\left(\frac{\delta^2 E'}{\delta n_{\mathbf{p}\sigma}\delta n_{\mathbf{p}'\sigma'}}\right)_{\mu_4} = V\left(\frac{\delta \varepsilon_{\mathbf{p}\sigma}}{\delta n_{\mathbf{p}'\sigma'}}\right)_{\mu_4}. \tag{2.3.1}$$

We can equivalently define another set of "direct" Landau parameters in terms of variations of E by

$$f^{\mathrm{dir}}_{\mathbf{p}\sigma,\mathbf{p}'\sigma'} \equiv V^2\left(\frac{\delta^2 E}{\delta n_{\mathbf{p}\sigma}\delta n_{\mathbf{p}'\sigma'}}\right)_{n_4} = V\left(\frac{\delta \varepsilon_{\mathbf{p}\sigma}}{\delta n_{\mathbf{p}'\sigma'}}\right)_{n_4}. \tag{2.3.2}$$

According to the rules for partial derivatives the two sets of Landau parameters are simply related by

$$f_{\mathbf{p}\sigma,\mathbf{p}'\sigma'} = f^{\mathrm{dir}}_{\mathbf{p}\sigma,\mathbf{p}'\sigma'} + \left(\frac{\partial \varepsilon_{\mathbf{p}\sigma}}{\partial n_4}\right)_{n_{\mathbf{p}'\sigma'}}\left(\frac{\delta n_4}{\delta n_{\mathbf{p}'\sigma'}}\right)_{\mu_4}. \tag{2.3.3}$$

Now

$$\left(\frac{\delta n_4}{\delta n_{\mathbf{p}'\sigma'}}\right)_{\mu_4} = -\frac{\delta^2 E'}{\delta\mu_4 \delta n_{\mathbf{p}'\sigma'}} = -\left(\frac{\partial \varepsilon_{\mathbf{p}'\sigma'}}{\partial \mu_4}\right)_{n_{\mathbf{p}\sigma}}$$
$$= -\left(\frac{\partial n_4}{\partial \mu_4}\right)_{n_{\mathbf{p}\sigma}}\left(\frac{\partial \varepsilon_{\mathbf{p}'\sigma'}}{\partial n_4}\right)_{n_{\mathbf{p}\sigma}}, \tag{2.3.4}$$

so that the relation becomes

$$f_{\mathbf{p}\sigma,\mathbf{p}'\sigma'} = f^{\text{dir}}_{\mathbf{p}\sigma,\mathbf{p}'\sigma'} - \left(\frac{\partial \varepsilon_{\mathbf{p}\sigma}}{\partial n_4}\right)_{n_{\mathbf{p}''\sigma''}}\left(\frac{\partial n_4}{\partial \mu_4}\right)_{n_{\mathbf{p}''\sigma''}}\left(\frac{\partial \varepsilon_{\mathbf{p}'\sigma'}}{\partial n_4}\right)_{n_{\mathbf{p}''\sigma''}}. \tag{2.3.5}$$

In the derivatives, the ^{3}He quasiparticle distribution is kept fixed. This equation has a simple physical interpretation: The interaction between two ^{3}He quasiparticles, described by $f_{\mathbf{p}\sigma,\mathbf{p}\sigma'}$, can take place either by the ^{3}He quasiparticles interacting at constant n_4, the "direct" term, or via processes in which the quasiparticle $\mathbf{p}'\sigma'$ creates a ^{4}He density fluctuation, described by (2.3.4), which then interacts with the quasiparticle $\mathbf{p}\sigma$. We shall refer to this latter process as the "phonon exchange" or "phonon induced" interaction, since it corresponds to the exchange of a phonon between the two ^{3}He quasiparticles; the full form of the phonon induced interaction is given in Section 2.4.2(b). See Figure 2.2. Then for small $\mathbf{p}$ and $\mathbf{p}'$ the contribution to $\mathbf{f}_{\mathbf{p}\sigma,\mathbf{p}'\sigma'}$, from the phonon-induced process is

$$-\frac{m_4 s^2}{n_4}(1+\alpha)^2, \tag{2.3.6}$$

where we have used (2.3.4) and (2.2.5).

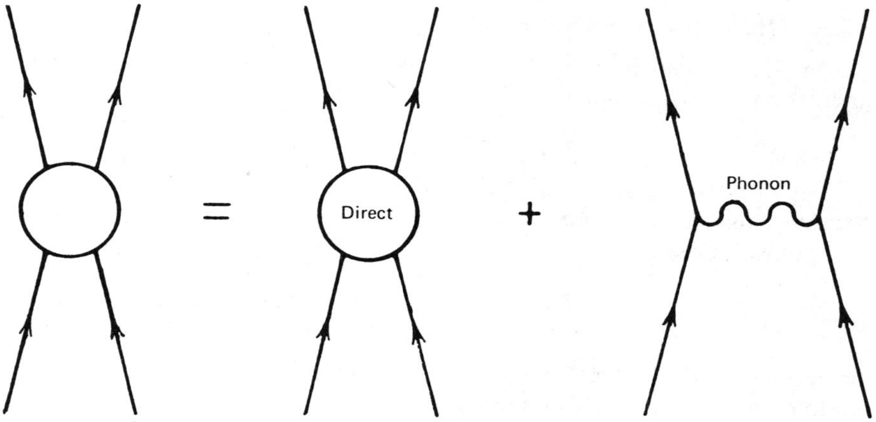

FIGURE 2.2. ^{3}He–^{3}He effective interaction as a sum of a direct scattering plus a phonon exchange scattering.

The consequences of Galilean invariance for $f_{\mathbf{p}\sigma,\mathbf{p}'\sigma'}$ are more complex than in a pure translationally invariant Fermi liquid, such a liquid ^{3}He. Even in the limit of weak interactions between quasiparticles, $f_{\mathbf{p}\sigma,\mathbf{p}'\sigma'}$ is not a function only of $\mathbf{p} - \mathbf{p}'$. This is because the rest frame of the background ^{4}He, the frame in which the superfluid velocity vanishes, establishes a preferred frame, and destroys Galilean invariance for the ^{3}He system alone. The spherical harmonic expansions of the Landau parameters for momenta on the Fermi surface, for very low T, are still given by (1.1.31) and (1.1.32).

In dilute solutions of ^{3}He it is sufficient to expand the total energy (2.1.2) to second order in $n_{\mathbf{p}\sigma}$. Thus, we write

$$E'(\{n_{\mathbf{p}\sigma}\}, \{f_{\mathbf{q}}\}, \mu_4) = E'(\{f_{\mathbf{q}}\}, \mu_4) + \frac{1}{V}\sum_{\mathbf{p}\sigma} \varepsilon^0_{\mathbf{p}\sigma} n_{\mathbf{p}\sigma} + \frac{1}{2}\frac{1}{V^2}\sum_{\mathbf{p}\sigma,\mathbf{p}'\sigma'} f_{\mathbf{p}\sigma,\mathbf{p}'\sigma'} n_{\mathbf{p}\sigma} n_{\mathbf{p}'\sigma'} + \dots, \tag{2.3.7}$$

where $\varepsilon^0_{\mathbf{p}}$ and $f_{\mathbf{p}\sigma,\mathbf{p}'\sigma'}$ depend only on $\{f_{\mathbf{q}}\}$ and μ_4, and $E'(\{f_{\mathbf{q}}\}, \mu_4)$ is the energy $E - \mu_4 n_4$ of pure ^{4}He. With (2.3.7) the ^{3}He quasiparticle energy is given by

$$\varepsilon_{\mathbf{p}\sigma} = \varepsilon^0_{\mathbf{p}} + \frac{1}{V}\sum_{\mathbf{p}'\sigma'} f_{\mathbf{p}\sigma,\mathbf{p}'\sigma'} n_{\mathbf{p}'\sigma'}. \tag{2.3.8}$$

At very low temperatures the ^{3}He form a completely degenerate Fermi liquid, filling a Fermi sea with radius the Fermi momentum

$$p_f = (3\pi^2 n_3)^{1/3}\hbar. \tag{2.3.9}$$

The ^{3}He effective mass, on the Fermi surface, is then defined by (1.1.9, 1.1.10); from (2.3.8) we have, for $p = p_f$,

$$\frac{\mathbf{p}}{m^*} = \frac{\mathbf{p}}{m} + \frac{1}{V}\sum_{\mathbf{p}'\sigma'} (\nabla_p f_{\mathbf{p}\sigma,\mathbf{p}'\sigma'}) n_{\mathbf{p}'\sigma'}. \tag{2.3.10}$$

Here we have taken $\varepsilon^0_p = \varepsilon_0 + p^2/2m$, where m is the effective mass of a single ^{3}He in pure ^{4}He. Multiplying both sides by $\nabla_p n_{\mathbf{p}\sigma}$, summing over all $\mathbf{p}\sigma$ and integrating this ∇_p by parts in the righthand terms we find

$$\frac{1}{m^*} = \frac{1}{m} + \frac{1}{3n_3} V^2 \sum_{\mathbf{p}\sigma,\mathbf{p}'\sigma'} (\nabla^2_p f_{\mathbf{p}\sigma,\mathbf{p}'\sigma'}) n_{\mathbf{p}\sigma} n_{\mathbf{p}'\sigma'}. \tag{2.3.11}$$

The density of states at the Fermi surface is, as before, $N(0) = m^* p_f/\pi^2\hbar^3$, while the ^{3}He degeneracy or Fermi temperature (the dashed line in Figure 2.1) is

$$T_f = \frac{p_f^2}{2m^*\kappa} \simeq 0.119\left(\frac{x}{0.01}\right)^{2/3} K. \tag{2.3.12}$$

The quantity

$$x = \frac{n_3}{n_3+n_4} = \frac{n_3}{n} \tag{2.3.13}$$

is the fractional concentration, by number, of ^{3}He in solution. The latter formula in (3.12), in Kelvin, is valid to lowest order in x, and the numerical coefficient is that for $P = 0$.

Were $f_{\mathbf{p}\sigma,\mathbf{p}'\sigma'}$ to depend only on $\mathbf{p} - \mathbf{p}'$, as would be the case if the ^{3}He themselves formed a Galilean invariant system, we could carry the ∇_p through to the $n_{\mathbf{p}'\sigma'}$, in (2.3.11), and use (1.1.8) to reduce (2.3.11) to the relation (1.1.56),

$$\frac{m^*}{m} = 1 + \frac{F_1^s}{3} \tag{2.3.14}$$

where F_1^s is defined by (1.1.33).

2.3.2. Low Temperature Properties of Dilute Solutions

We now describe briefly the low temperature properties of the dilute solutions and their dependence on the interaction between the ^{3}He quasiparticles. After this we discuss phenomenological fits to and theories of the interaction.

(a) Specific Heat. For ^{3}He concentrations $x \gtrsim 1\%$ the specific heat at $T \lesssim 0.5$ K is dominated entirely by the ^{3}He excitations. For $T \ll T_f$, the specific heat is linear in T and given by (1.1.40). Measurements [29] in this regime determine $m^*(x) \simeq 2.38\ m_3$ for $x = 1.3\%$ and $\simeq 2.47 m_3$ for $x = 5.0\%$. As the temperature is increased, one observes that the ^{3}He contribution to the specific heat is, as expected, essentially that of an ideal Fermi gas, approaching $\sim \frac{3}{2} n_3 \kappa$ for $T \gg T_f$. The effects of ^{3}He–^{3}He interactions on the high temperature specific heat have not been well studied, since it is difficult to sort out these corrections from the contribution to the specific heat of the ^{4}He excitations.

(b) Chemical Potentials and Molar Volumes. The ^{3}He chemical potential at $T = 0$ equals the energy of a quasiparticle at the top of the Fermi sea:

$$\mu_3(n_3, \mu_4) = \varepsilon_0(\mu_4) + \frac{p_f^2}{2m(\mu_4)} + \frac{2}{V}\sum_{\mathbf{p}'} (f^s_{\mathbf{p},\mathbf{p}'}(\mu_4))_{p=p_f} n_{\mathbf{p}'}. \tag{2.3.15}$$

The change in μ_3 with ^{3}He density, at constant μ_4, is computed exactly as in a pure Fermi liquid, and the result is (cf. 1.1.48)

$$\left(\frac{\partial \mu_3}{\partial n_3}\right)_{\mu_4} = \frac{1 + F_0^s}{N(0)}, \tag{2.3.16}$$

where F_0^s is defined by (1.1.33). More generally, at finite temperature T, the ^{3}He chemical potential is of the form

$$\mu_3(n_3, \mu_4, T) = \varepsilon_0(\mu_4, T) + \mu_f(n_3, \mu_4, T) + \varepsilon_{\text{int}}(n_3, \mu_4, T), \qquad (2.3.17)$$

where μ_f is the chemical potential of a free Fermi gas of mass m at temperature T and density n_3, and ε_{int} is a correction depending on the ^{3}He–^{3}He interaction. Note that μ_f has the structure

$$\mu_f = \kappa T_f^\circ g\left(\frac{T}{T_f^\circ}\right) \qquad (2.3.18)$$

where $T_f^\circ = p_f^2/2m\kappa$, $n_3 = p_f^3/3\pi^2\hbar^3$, and g is a known dimensionless function obeying $g(0) = 1$.

It is often more convenient experimentally to use the variables x, P, T in place of n_3, μ_4, T. As ^{3}He are added to ^{4}He at constant pressure the ^{4}He chemical potential changes; this change is most easily computed from the Gibbs–Duhem relation (2.1.20), which allows us to write

$$0 = n_3\left(\frac{\partial \mu_3}{\partial x}\right)_{T,P} + n_4\left(\frac{\partial \mu_4}{\partial x}\right)_{T,P}. \qquad (2.3.19)$$

Thus, dividing by n_4 and integrating with respect to x, we have

$$\mu_4(x, T, P) = \mu_4(0, T, P) - \int_0^x dx' \frac{x'}{1 - x'} \frac{\partial \mu_3(x', T, P)}{\partial x'}. \qquad (2.3.20)$$

This relation is the starting point for the calculation of the osmotic pressure Π defined by

$$\mu_4(x, T, P + \Pi) = \mu_4(0, T, P). \qquad (2.3.21)$$

Detailed calculations of the osmotic pressure are given by Ebner and Edwards [2].

In a dilute solution at $T = 0$, $\mu_3(x, P)$ is less than the chemical potential $\mu_3(1, P)$ of pure ^{3}He at the same pressure. As ^{3}He are added to the system they enter the solution until a limiting concentration x_s, where

$$\mu_3(x_s, P) = \mu_3(1, P). \qquad (2.3.22)$$

Beyond this point the system separates into two phases: one a dilute solution at concentration x_s and the other pure ^{3}He, which, because it is lighter, floats on top of the dissolved phase. The limiting concentration x_s is sensitive to the interaction between the ^{3}He quasiparticles. To see this we first note that at $T = 0$ the difference between $\mu_4(x, P)$ and $\mu_4(x = 0, P)$ is, from (2.3.20), at least of order $x^{5/3}$, and thus to terms of relative order x we can replace $\mu_4 = \mu_4(x, P)$ in (2.3.15) by $\mu_4(0, P) \equiv \mu_4(P)$. Now at $P = 0$, $\mu_3(x = 1, P = 0) = -2.47$ K while $\mu_3(x = 0, P = 0) = -2.79$ K. Neglecting the interaction term in (2.3.17), we would predict that phase separation occurs when $T_f(x) = \mu_3(1, 0) - \mu_3(0, 0)$, or from (2.3.12) at $x \simeq 4.3\%$. The experimental fact that x_s is $\simeq 6.5\%$ tells one immediately that a ^{3}He atom in solution at finite x is more tightly bound than it is as $x \to 0$; in other words, there must be a net attractive interaction between the ^{3}He. Experimental study [28] of the heat-of-mixing of dilute solutions shows that $\varepsilon_{\text{int}}(n_3, \mu_4(P = 0), T = 0)$ is $\sim -2x$ (in Kelvin).

The density $n = n_3 + n_4$, or molar volume $v = n^{-1}$, of the solution at given x, T, and P can be found from the thermodynamic relation

$$n^{-1} = \left(\frac{\partial \mu}{\partial \mathrm{P}}\right)_{x,T}, \tag{2.3.23}$$

where

$$\begin{aligned}\mu(x, T, P) &= (1 - x)\mu_4(x, T, P) + x\mu_3(x, T, P) \\ &= (1 - x)\mu_4(0, T, P) + (1 - x)\int_0^x dx' \frac{\mu_3(x', T, P)}{(1 - x')^2}\end{aligned} \tag{2.3.24}$$

is the Gibbs free energy per particle. To derive the latter form we use (2.3.20). Then evaluating $\mu_3(x, T, P)$ at zero temperature from (2.3.17), replacing $\mu_4(x, P)$ by $\mu_4(P)$, we find that to order x^2,

$$\begin{aligned}\mu(x, P) = (1 - x)\mu_4(P) + x\left[\varepsilon_0(\mu_4(P)) + \frac{3}{10}\frac{p_f^2}{m}\right] \\ + (1 - x)\int_0^x dx' \frac{\varepsilon_{\text{int}}(x'n_4(P), \mu_4(P))}{(1 - x')^2},\end{aligned} \tag{2.3.25}$$

where $n_4(P)$ is the density of pure ^{4}He at pressure P.

Let us write the molar volume in the form

$$n^{-1}(x, T, P) = n_4^{-1}(0, T, P)(1 + x\alpha(x, T, P)), \tag{2.3.26}$$

defining $\alpha(x, T, P)$. Using (2.3.25) and (2.3.23) we then derive the result for $\alpha(x, T = 0, P)$:

$$\begin{aligned}\alpha(x, P) = \alpha(x = 0, P) + \frac{p_f^2}{5m}\frac{1}{m_4 s^2}\left(1 - \frac{3}{2}\frac{n_4}{m}\frac{\partial m}{\partial n_4}\right) \\ + (1 - x)n_3\int_0^x dx' \frac{\partial \varepsilon_{\text{int}}(x'n_4(P), \mu_4(P))/\partial P}{(1 - x')^2},\end{aligned} \tag{2.3.27}$$

where

$$\alpha(x = 0, P) = \frac{\partial \varepsilon_0(\mu_4)}{\partial \mu_4} - 1 \tag{2.3.28}$$

is the fractional excess volume occupied by a single ^{3}He atom in pure ^{4}He at $T = 0$. We see from (2.3.27) that a calculation of the contribution of ^{3}He–^{3}He interactions to the molar volume requires a knowledge of the pressure dependence of the Landau parameters; these corrections to α at $T = 0$ are $\lesssim 1\%$ for $x \lesssim 5\%$. Measurements of the molar volume of dilute solutions are not delicate enough to give one information about the interactions of ^{3}He quasiparticles. More generally, at finite T

$$\alpha(x, T, P) = \alpha(0, T, P) + \frac{P_f}{n_3}\frac{\partial n_4}{\partial P}\left(1 - \frac{3}{2}\frac{n_4}{m}\frac{\partial m}{\partial n_4}\right) + \text{interaction terms}, \tag{2.3.29}$$

where $P_f(x, T, P)$ is the pressure of a free Fermi gas of mass m at temperature T and density n_3 equal to that of the ^{3}He in solution. A fuller discussion of the molar volume at finite T is given in Reference [34]. Note that since $\partial \ln m/\partial \ln n_4 \sim 1.2$, the coefficient of the "kinetic" term $\propto P_f$ in (2.3.29) is negative; this term tends to make dilute solutions have a negative coefficient of thermal expansion at low T.

(c) Spin Susceptibility. The ^{3}He spin susceptibility in dilute solutions of ^{3}He in ^{4}He at low temperatures is given by the Fermi liquid result (1.1.55):

$$\chi = \frac{\hbar^2}{4}\frac{\gamma^2 N(0)}{1 + F_0^a} \tag{2.3.30}$$

where γ is the gyromagnetic ratio, and F_0^a is defined by (1.1.33). F_0^a appears experimentally [2] to be positive and $\sim$ 0.05 to 0.10.

(d) Low Temperature ^{3}He Transport Properties. The specific heat, molar volume, and spin susceptibility are only weakly sensitive to the interactions among the ^{3}He quasiparticles. On the other hand, the very low temperature transport properties are determined entirely by the ^{3}He–^{3}He interactions, and these, together with heat-of-mixing data, provide the best handle by which to deduce the effective interactions from experiment.

At temperatures $T \ll T_f$ the spin diffusion coefficient of dilute solutions of ^{3}He is $\propto T^{-2}$ and is given explicitly by the exact Fermi liquid result (1.2.159). Thermal conduction in dilute solutions is by both the ^{4}He excitations and the ^{3}He quasiparticles; at low $T < 0.01$ K $\ll T_f$ conduction by ^{3}He, $\propto T^{-1}$, is the dominant mechanism. The ^{3}He thermal conductivity in this regime is given by the exact Fermi liquid result (1.2.113). Similarly, the dominant viscosity in this temperature regime is that of the ^{3}He, which is $\propto T^{-2}$ and is given by (1.2.129). The reason that the viscosity and thermal conductivity due to ^{4}He thermal excitations can be neglected at low T is that the number of ^{4}He excitations at very low T is $\propto T^3$, while phonon mean free paths are determined primarily by scattering from the walls. We discuss transport by ^{4}He excitations in Section 2.4.6.

We turn now to the description of the effective interactions between ^{3}He quasiparticles and how they can be deduced from experiment.

2.3.3. Phenomenological Effective Interaction

Imagine that we have just two ^{3}He quasiparticles in pure ^{4}He at $T = 0$, one of momentum $\mathbf{p}$ and spin $\uparrow$ and the other of momentum $\mathbf{p}'$ and spin $\downarrow$; and that they scatter together to a final state in which the two quasiparticles have momenta and spin $\mathbf{p} + \mathbf{q}, \uparrow$ and $\mathbf{p}' - \mathbf{q}, \downarrow$. Because of the weakness of microscopic ^{3}He interactions involving the spin, such as spin–orbit interactions or direct spin–spin interactions,* we can neglect the possibility of a spin flip in this process and regard the two quasiparticles as distinguishable. Let us denote the scattering amplitude for this process by $t_{\mathbf{p},\mathbf{p}'}(\mathbf{q})$, where we assume states to be normalized in unit volume. (Because of the lack of Galilean invariance this amplitude does not depend only on the difference $\mathbf{p} - \mathbf{p}'$.) Now if the two quasiparticles have parallel spin initially, then the scattering amplitude is the sum of a term $t_{\mathbf{p},\mathbf{p}'}(\mathbf{q})$ minus an exchange term $t_{\mathbf{p},\mathbf{p}'}(\mathbf{p}' - \mathbf{p} - \mathbf{q})$, in which $\mathbf{p} + \mathbf{q} \leftrightarrow \mathbf{p}' - \mathbf{q}$ in the final state; for the exchange process the momentum transfer is $\mathbf{p}' - \mathbf{p} - \mathbf{q}$. Thus, we may write the scattering amplitude for arbitrary initial spins (up or down) as

$$\langle \bar{\mathbf{p}}\sigma, \bar{\mathbf{p}}'\sigma' | t | \mathbf{p}\sigma, \mathbf{p}'\sigma' \rangle = t_{\mathbf{p},\mathbf{p}'}(\bar{\mathbf{p}} - \mathbf{p}') - \delta_{\sigma,\sigma'} t_{\mathbf{p},\mathbf{p}'}(\bar{\mathbf{p}}' - \mathbf{p}) \tag{2.3.31}$$

(from momentum conservation $\bar{\mathbf{p}} + \bar{\mathbf{p}}' = \mathbf{p} + \mathbf{p}'$ and from energy conservation $\varepsilon_{\bar{\mathbf{p}}} - \varepsilon_{\mathbf{p}} = \varepsilon_{\mathbf{p}'} - \varepsilon_{\bar{\mathbf{p}}'}$).

We may relate the scattering amplitude (2.3.31) to the Landau parameters $f_{\mathbf{p}\sigma,\mathbf{p}'\sigma'}$ by taking the "k-limit" of (2.3.31), that is, we first set the energy transfer $\omega = \varepsilon_{\bar{\mathbf{p}}} - \varepsilon_{\mathbf{p}}$ equal to zero and then let the momentum transfer $\bar{\mathbf{p}} - \mathbf{p}$ approach zero. Then

$$\lim_{\bar{\mathbf{p}}-\mathbf{p}\to 0} \lim_{\varepsilon_{\bar{\mathbf{p}}}-\varepsilon_{\mathbf{p}}\to 0} \langle \bar{\mathbf{p}}\sigma, \bar{\mathbf{p}}'\sigma' | t | \mathbf{p}\sigma, \mathbf{p}'\sigma' \rangle = f_{\mathbf{p}\sigma,\mathbf{p}'\sigma'}. \tag{2.3.32}$$

There are two important remarks to be made about this equation. First, if we look at the analogous relation (1.4.20) in a pure Fermi liquid we find an additional term $\sim f(\partial n/\partial \varepsilon)t$ on the right side, which represents the repeated excitation of particle-hole pairs. Such a term is absent in the low ^{3}He density limit, since the density of states $\Sigma_{\mathbf{p}''}\partial n^0_{p''}/\partial \varepsilon_{p''}$ vanishes as $x \to 0$. Second, the k-limit of t should be identified with the Landau parameter f defined at constand μ_4, because as the ^{3}He particles move through the ^{4}He they displace the ^{4}He so as to keep μ_4 constant. Compare with (1.4.21).

From the form of the right side of (2.3.31) we see that the Landau parameters obey

*As Leggett [45] has pointed out, such interactions do play an important role in determining the pairing state in pure superfluid ^{3}He.

$$f_{\mathbf{p}\sigma,\mathbf{p}'\sigma'} = t_{\mathbf{p},\mathbf{p}'}(0) - \delta_{\sigma,\sigma'} t_{\mathbf{p},\mathbf{p}'}(\mathbf{p}' - \mathbf{p}), \tag{2.3.33}$$

where in $t_{\mathbf{p},\mathbf{p}'}$ the energy transfer is set equal to zero, and then, in $t_{\mathbf{p},\mathbf{p}'}(0)$, the momentum transfer is taken to zero. Note that

$$f_{\mathbf{p}\uparrow,\mathbf{p}\uparrow} = 0. \tag{2.3.34}$$

Two basic approaches have been used in constructing a form for the scattering amplitude. The first is that of Emery [46], who assumes an effective potential $V(r)$ between ^{3}He atoms, and then determines the scattering phase shifts by solving the two-body Schrödinger equation. Emery pointed out that the parameters of the effective potential could be fitted directly to the spin diffusion data since there is no contribution to spin diffusion from the ^{4}He excitations. He specifically chose the potential to be of the form of a hard core of radius r_c inside an attractive square well of depth $V_0 = s_0(\pi\hbar/2b)^2/m$ between $r = r_c$ and $r = r_c + b$; outside $r_c + b$ the potential was taken to vanish. The parameters that gave the best fit to the measured spin diffusion at $x = 1.3\%$ and 5%, from the degenerate to the classical regime, are $r_c = 1.8$ Å, $b = 0.8$ Å, $s_0 = 0.81$, $m = 2.34\, m_3$; then $V_0 = 22$ K.

The second approach is that of Bardeen, Baym, and Pines [6] (BBP), who assume that $t_{\mathbf{p},\mathbf{p}'}(\mathbf{q})$ depends only on the momentum transfer $\mathbf{q}$, and write

$$t_{\mathbf{p},\mathbf{p}'}(\mathbf{q}) = v(q); \tag{2.3.35}$$

$v(\mathbf{q})$ is called the "effective interaction" between the ^{3}He. [One can imagine $v(q)$, of dimension energy $\times$ volume, to be the Fourier transform of an effective pseudopotential $v(r)$ acting between ^{3}He quasiparticles of opposite spin; because $v(q)$ is already the scattering amplitude, it would be double counting to attempt to solve the Schrödinger equation with $v(r)$.]

The effective interaction can be determined from experiment as follows. From (2.3.33) and (2.3.35) we have

$$f_{\mathbf{p}\sigma,\mathbf{p}'\sigma'} = v(0) - \delta_{\sigma,\sigma'} v(|\mathbf{p} - \mathbf{p}'|), \tag{2.3.36}$$

and

$$\begin{aligned} f^s_{\mathbf{p},\mathbf{p}'} &= v(0) - \tfrac{1}{2} v(|\mathbf{p} - \mathbf{p}'|) \\ f^a_{\mathbf{p},\mathbf{p}'} &= -\tfrac{1}{2} v(|\mathbf{p} - \mathbf{p}'|). \end{aligned} \tag{2.3.37}$$

Then from (2.3.17) and (2.3.15), we see that

$$\begin{aligned} \varepsilon_{\text{int}}(n_3, \mu_4, T = 0) &= 2 \int_0^{p_f} \frac{d^3p'}{(2\pi\hbar)^3} (f^s_{\mathbf{p},\mathbf{p}'})_{p=p_f} \\ &= 6n_3 \int_0^1 dy(1 - y)y^2[2v(0) - v(2p_f y)]. \end{aligned} \tag{2.3.38}$$

In the limit of small p_f, $\varepsilon_{\text{int}} \simeq \frac{1}{2} v(0) n_3$; comparing with the heat-of-mixing result that at $P = 0$, $\varepsilon_{\text{int}} \sim -2x(\text{K})$, we see that $v(0)$ is *negative* and $\sim -\frac{1}{10} m_4 s^2/n_4$ for zero pressure. Note that $m_4 s^2 = 27.3$ K sets a natural scale of energies in ^{4}He at $T = 0$. The effective interaction is thus relatively quite weak; we shall see later how this weakness results from that fact that ^{3}He and ^{4}He are isotopes.

The q dependence of v can be determined by fitting (1.2.159) and (1.2.113) to the measured ^{3}He spin diffusion coefficient D_σ and to the ^{3}He contribution to the thermal conductivity K in solution. To compute the transport coefficients in terms of $v(q)$ we note that in a quasiparticle scattering $\mathbf{p} + \mathbf{p}' \to \bar{\mathbf{p}} + \bar{\mathbf{p}}'$, the scattering amplitude (2.3.31) is given by $v(|\bar{\mathbf{p}} - \mathbf{p}|) - \delta_{\sigma,\sigma'} v(|\mathbf{p}' - \mathbf{p}|)$. In terms of the angles θ and ϕ (Figure 1.2),

$$|\bar{\mathbf{p}} - \mathbf{p}| = 2p_f \sin\frac{\theta}{2} \sin\frac{\phi}{2}$$

$$|\bar{\mathbf{p}}' - \mathbf{p}| = 2p_f \sin\frac{\theta}{2} \cos\frac{\phi}{2}, \tag{2.3.39}$$

where the scattering is assumed to take place on the Fermi surface. Then the scattering rates $W_{\sigma\sigma'}$ are given by

$$W_{\uparrow\uparrow}(\theta, \phi) = \frac{2\pi}{\hbar} \left| v\left(2p_f \sin\frac{\theta}{2} \sin\frac{\phi}{2}\right) - v\left(2p_f \sin\frac{\theta}{2} \cos\frac{\phi}{2}\right)\right|^2$$

$$W_{\uparrow\downarrow}(\theta, \phi) = \frac{2\pi}{\hbar} \left| v\left(2p_f \sin\frac{\theta}{2} \sin\frac{\phi}{2}\right)\right|^2. \tag{2.3.40}$$

These rates used in (1.2.88) and (1.2.91) for $\langle W \rangle$ and τ, together with (1.2.94) for λ_K, (1.2.113) for K, and (1.2.153) for λ_D and (1.2.159) for D_σ, completely specify the calculation of these transport coefficients.

The first fits of $v(q)$ were made from the measured spin diffusion coefficient at saturated vapor pressure using Hone's result for the spin diffusion coefficient, based on the original Abrikosov–Khalatnikov approximate solution of the transport equation. A fit to transport coefficients can at best determine $v(q)$ to within an overall sign; the fact that $\varepsilon_{\text{int}} \sim -2x$ determines that $v(q)$ must be attractive for small q. The BBP fit was

$$v_{\text{BBP}}(q) = -v_0 \cos\frac{\beta q}{\hbar}, \tag{2.3.41}$$

where $v_0 = -0.0754\ m_4 s^2/n_4$ and $\beta = 3.16$ Å^{-1}.

The interactions derived by the Emery and BBP procedures give similar results for calculated properties of dilute solutions. It was found, though, that neither of these potentials yielded a consistent fit to the subsequently measured thermal conductivity [47] at $x = 1.3\%$ and 5%. This led Baym and Ebner [48] to reexamine the Abrikosov–Khalatnikov approximate solution, and to construct a fit to $v(q)$ using the variational solution of the kinetic equation, which provided a much better description of the experiments. Ebner [49] has also constructed a $v(q)$, using the exact solution for K and D_σ, in the form of a power series

$$v(q) = -\frac{m_4 s^2}{n_4} \sum_{n=0}^{4} a_n \left(\frac{q}{2p_5}\right)^{2n}, \tag{2.3.42}$$

where p_5 is the Fermi momentum of an $x = 5.0\%$ solution; the resulting best choice for the coefficients a_n is

$$a_0 = 0.0780,\ a_1 = -0.2627,\ a_2 = 0.4893,\ a_3 = -0.7329,\ a_4 = 0.4109. \tag{2.3.43}$$

The comparison with experiment of the transport coefficients as well as other Fermi liquid properties calculated from (2.3.42) are shown in Table 2.1.

TABLE 2.1. Comparison of experimental and theoretical values of the transport coefficients and Landau parameters of dilute solutions of ^{3}He in ^{4}He.

	1.3%		5.0%	
	Theory	Experiment	Theory	Experiment
DT^2(cm^2 K^2/sec)	18.6×10^{-6}	17.2×10^{-6}	80×10^{-6}	90×10^{-6}
KT(erg/sec cm)	9.6	11	27	24
$F_1^s/3$	0.0174		0.055	
m^*/m_3	2.38	2.38 ± 0.04	2.47	2.46 ± 0.04
$m/m_3 = 2.34$				
F_0^a	0.09	0.09 ± 0.03	0.04	0.08 ± 0.03
F_0^s	-0.20		-0.42	

The best fit (2.3.43) produces only 10% agreement between theory and experiment for K and D_σ; any adjustment of the parameters from (2.3.43) only worsens the overall agreement. This lack of agreement must arise from the approximation of the true scattering amplitude $t_{\mathbf{p},\mathbf{p}'}(\mathbf{q})$ by a velocity (and concentration) independent effective interaction $v(q)$; the concentration dependence should produce terms in $t_{\mathbf{pp}'}(\mathbf{q})$ on the order of $x^{1/3}$, and hence is capable of producing appreciable corrections to the transport coefficients.* Formulas and graphs of thermodynamic quantities calculated with (2.3.42) and (2.3.43), together with a complete discussion of the fitting problem, are given by Ebner and Edwards [2].

*H. H. Fu and C. J. Pethick [91] have evaluated concentration-dependent contributions to the effective interaction; they find that better agreement between theory and experiment can be obtained without increasing the number of adjustable parameters.

The Landau parameters are given in terms of the effective interaction, from (2.3.37), by

$$F_0^a = -N(0)\int_0^1 v(2p_f y)y\,dy$$

$$F_0^s = N(0)v(0) + F_0^a$$

$$F_1^s = -N(0)\int_0^1 v(2p_f y)y(1 - 2y^2)dy, \tag{2.3.44}$$

where $N(0) = m^* p_f/\pi^2\hbar^3$. Since the effective interaction has been assumed to depend only on the momentum transferred in the interaction, the effective mass m^* is given in terms of F_1^s by the result (1.1.56) for a Galilean invariant system. Note that F_0^a is slightly positive, implying a small reduction in the spin susceptibility from its value for a free Fermi gas. Since F_0^s is negative the ^{3}He is not expected to have a well-defined zero sound mode at zero pressure.

2.3.4. Microscopic Approaches to the Effective Interaction

Measured on the scale of characteristic interaction strengths, $m_4 s^2/n_4$, in liquid helium, the effective ^{3}He–^{3}He interaction (2.3.42, 2.3.43) is very weak. The coefficient $a_0 = 0.0780$ is surprisingly close to $\alpha^2 = 0.0812$ (for $\alpha = 0.285$ at $P = 0$), so that

$$v(q = 0) \simeq -\alpha^2 \frac{m_4 s^2}{n_4}. \tag{2.3.45}$$

One can understand physically the weakness of the effective interaction by asking how the ^{3}He atoms in solution become dynamically aware of each other. Since the force fields produced by ^{3}He atoms and ^{4}He atoms are identical, a ^{3}He atom cannot tell whether the potential it feels is produced by a ^{3}He or ^{4}He atom. It is only through the ways that the ^{3}He differ from their ^{4}He environment that an effective interaction is produced; only the differences in ^{3}He–^{4}He mass and statistics enable the ^{3}He atoms to identify each other.

Even though ^{3}He are fermions and ^{4}He are bosons, statistics play a minor role in determining the effective interaction. This is because the effective interaction is determined by the behavior of two opposite spin ^{3}He in the solution. The scattering of two ^{3}He particles of the same spin includes a further exchange term. But ^{3}He atoms of opposite spin do not obey the exclusion principle, and thus statistics alone offer a ^{3}He atom of one spin orientation no way of distinguishing a ^{3}He atom of opposite spin from a ^{4}He atom in the solution.

The important source of the interaction is thus the ^{3}He–^{4}He mass difference, which has the consequence that a ^{3}He atom occupies a volume in the solution $(1 + \alpha)$ times that occupied by a ^{4}He atom. Thus, in a sense, the effective interaction between two ^{3}He is equivalent to that between two "holes" of relative volume α in the liquid. Their interaction is expected to be of relative order α^2. (That the interaction is attractive does not, however, follow from so general a consideration.)

To discuss the theoretical basis for the result (2.3.45), we note that from (2.3.35) and (2.3.33)

$$v(q = 0) \;=\; t_{\mathbf{p},\mathbf{p}'}(0) \;=\; f_{\mathbf{p}\uparrow,\mathbf{p}'\downarrow} \;=\; \left(\frac{\partial \varepsilon_{\mathbf{p}\uparrow}}{\partial n_{\mathbf{p}'\downarrow}}\right)_{\mu_4} \qquad (2.3.46)$$

in the limit of zero concentration. In this limit (2.3.46) reduces to

$$v(q = 0) = \lim_{x\to 0}\left(\frac{\partial \varepsilon_{\uparrow}}{\partial n_{\downarrow}}\right)_{\mu_4} = \lim_{x\to 0}\left(\frac{\partial^2 E'}{\partial n_{\uparrow}\partial n_{\downarrow}}\right)_{\mu_4}, \qquad (2.3.47)$$

where $E' = E - \mu_4 n_4$, $n_\uparrow$ and $n_\downarrow$ denote the density of ^{3}He with up and down spins, and $\varepsilon_\uparrow$ is the energy of an up-spin ^{3}He quasiparticle of zero momentum. If we write the ground state energy in the form

$$E(n_4, n_\uparrow, n_\downarrow) = E_0(n) + n_3 E_1(n) + n_\uparrow n_\downarrow E_2(n) + \tfrac{1}{2}(n_\uparrow^2 + n_\uparrow^2)E_3(n) \qquad (2.3.48)$$

(where $n = n_3 + n_4$, $n_3 = n_\uparrow + n_\downarrow$), which (aside from kinetic energy terms of order $x^{5/3}$ irrelevant here) is perfectly general up to terms of order x^2, then to order x

$$\mu_4 = E_0' + n_3 E_1', \qquad (2.3.49)$$

while

$$\varepsilon_\uparrow = \mu_4 + E_1 + n_\downarrow E_2. \qquad (2.3.50)$$

$E_0(n)$ is the ground state energy of pure ^{4}He at density n, and in (2.3.49) the prime denotes differentiation. From (2.3.50) with (2.2.4) and (2.2.5) we have the exact results at $x = 0$,

$$\alpha = \frac{E_1'}{E_0''} \qquad (2.3.51)$$

and

$$v(0) = \left(\frac{\partial \varepsilon_\uparrow}{\partial n_\downarrow}\right)_{\mu_4} = -\frac{(E_1')^2}{E_0''} + E_2 = -\alpha^2\frac{m_4 s^2}{n_4} + E_2. \qquad (2.3.52)$$

We see then that the empirical result (2.3.45) is equivalent to the term E_2 in (2.3.48) being negligible.

The calculation below (2.2.7) is the simplest realization of the idea that ^{3}He (of opposite spin) become aware of each other only because their mass is different from m_4. The energy in that calculation (2.2.9) is essentially the first order perturbation due to the mass difference. In (2.2.9) the term E_2 vanishes, and the relation (2.3.45) holds exactly. One should also note that since in the hypothetical case $m_3 = m_4$ the exact ground state of N_4 ^{4}He plus two ^{3}He of opposite spin is the $N_4 + 2$ particle ^{4}He ground state, E_1 and E_2 both vanish, and thus $v(0) \equiv 0$. (We shall see below that in this case $E_3 \equiv 0$ as well.) Baym [92] has calculated E_2 to second order in the mass difference and finds to this order that $v(0) \approx -(\alpha^2 + 0.01)m_4s^2/n_4$.

Several microscopic calculations of the effective interaction have been based on the idea of a ^{3}He acting as a ^{4}He plus a hole of relative size α. Campbell [50] estimated from this point of view the modification of the bare ^{3}He–^{3}He interaction due to polarization of the ^{4}He background, using the experimental α as well as the average nearest neighbor distances in dilute solutions as input parameters. He found in his calculation that for small ^{3}He–^{3}He separation $r \lesssim 2.5$ Å the effective potential (as used by Emery) is the same as the bare potential, while for $r \gtrsim 4.8$ Å it is α^2 times the bare potential. The modification of the potential by the medium is qualitatively similar to that proposed by Emery [46]. The q dependent scattering amplitude he derived in this way is in reasonably good agreement up to $q \gtrsim p_5$ with the effective interaction (2.3.42, 2.3.43), except at small q where it is too attractive.

Woo, Tan, and Massey [41], and McMillan [51] have made considerable headway in deriving the ^{3}He–^{3}He scattering amplitude from first principles using the method of correlated basis functions. This work, which is reviewed by Woo in Reference [39], has so far yielded a qualitative effective interaction consistent with (2.3.42, 2.3.43). In McMillan's calculation the effective interaction is a sum of two terms, $-m_4s^2\alpha^2/n_4$, plus a dipolar interaction [cf. (2.4.32)].

We also mention the work of Østgaard [52], who has computed Landau parameters and the ^{3}He effective mass from Brueckner theory, and Saam [53] who has shown, in the framework of microscopic many-body Green's functions, the relation of the microscopic scattering amplitude to the effective interactions and Landau parameters.

The study of the dilute gas model of dilute solutions yields strong support for the relation (2.3.45). Saam [54] has computed the ground state energy to order x^2 in the dilute gas model and finds, in the notation of (2.3.48), the following results, for $n_3 \ll n_4$ and $m_3/m_4 = 3/4$:

$$E_0(n) = \left[\frac{v_0}{2} + \frac{4}{15}\lambda\right]n^2$$

$$E_1(n) = \left[\frac{v_0}{6} + \frac{2}{3}(0.1326)\lambda\right]n$$

$$E_2(n) = (0.0009)\lambda$$

$$E_3(n) = \frac{v_0}{36} + (0.0164)\lambda\,, \tag{2.3.53}$$

where

$$\lambda = 16\, v_0\left(\frac{na^3}{\pi}\right)^{1/2}$$

$$v_0 = \frac{4\pi a}{m_4}, \tag{2.3.54}$$

and a is the ^{4}He–^{4}He scattering length. The results (2.3.53) are valid to first order in the dilute gas parameter $(na^3)^{1/2}$. The numerical coefficients in (2.3.53) are functions only of the mass ratio m_3/m_4, given explicitly in Reference [54]. Then from (2.3.53) we have

$$\alpha = \frac{1}{6} - (0.0340)\frac{\lambda}{v_0}, \tag{2.3.55}$$

$$v(0) = -\left[\frac{v_0}{36} + (0.0173)\lambda\right], \tag{2.3.56}$$

while

$$\frac{m_4 s^2}{n_4}\alpha^2 = \left[\frac{v_0}{36} + (0.0164)\lambda\right] \simeq -v(0) + \frac{\lambda}{105}\left(\frac{m_4}{m_3} - 1\right)^2. \tag{2.3.57}$$

The latter is the leading term in an expansion in $(m_4 - m_3)/m_3$. Thus, in the dilute gas model (2.3.45) is exact to order $(\lambda/v_0)^0$, and it gives the term of order λ correctly to 5%.

The case $m_3 = m_4$ of the dilute gas model has the interesting result that $E_1 = E_2 = E_3 \equiv 0$ so that to order x^2

$$E(n_4, n_\uparrow, n_\downarrow) = E_0(n), \qquad m_3 = m_4, \tag{2.3.58}$$

where $E_0(n)$ is the same as in (2.3.53).* Then $\alpha = 0$ and $v(0) = 0$.

The result (2.2.58) is completely general, as we can see by the following argument. Imagine that we add just one ^{3}He of mass m_4 and spin up to pure ^{4}He. The ground state wave function of the $N_4 + 1$ particle state is the completely symmetric ^{4}He ground state, and (2.3.58) is the ground state energy. Now imagine that we add a second ^{3}He; if its spin is down, then the ground state wave function is again the completely symmetric ^{4}He ground state wave function for $N_4 + 2$ particles, the ^{3}He behave as marked ^{4}He, and (2.3.58) is valid. On the other hand, if the second ^{3}He has spin up, while the Hamiltonian is completely symmetric the ground state wave function must be antisymmetric with respect to interchange of the two ^{3}He—that is, it must be

*It should be pointed out that the derivation of (2.3.58) in the dilute gas model simply to order $(\lambda/v_0)^0$ is nontrivial, requiring examination of terms with explicit v_0^2 coefficients [54].

the lowest energy state belonging to the representation of the $N_4 + 2$ particle permutation group with Young tableau 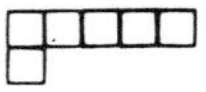 (the lowest energy state belonging to the representation 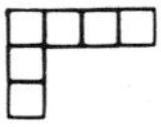 is expected to be of higher energy).

The crucial point now is that in a system with $N_4 + 1$ ^{4}He atoms and one ^{3}He, the phonon states must be completely symmetric—that is, belong to  —while the low-lying ^{3}He quasiparticle states are not possible excitations of pure ^{4}He and hence must belong to 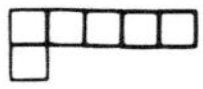 .

There is some evidence from Landau et al.'s [55] measurements of the osmotic pressure that the relation (2.3.45) becomes less valid at higher pressure [56]. Landau et al. fit a "speculative" ^{3}He–^{3}He effective interaction, and find $v(q = 0, P = 10 \text{ atm}) = 0.76v(q = 0, P = 0)$, $v(q = 0, P = 20 \text{ atm}) = 0.70v(q = 0, P = 0)$. They also find that $m_4 s^2 \alpha^2/n_4$ is 3% larger in magnitude than $|v(0)|$ at $P = 0$, 18% larger at $P = 10$ atm, and 21% larger at $P = 20$ atm. These differences are much smaller than those originally proposed by Eckstein et al. [56] in arguing against the validity of (2.3.45) at higher pressure. A reason for (2.3.45) becoming less valid with increasing pressure possibly lies in the fact that at zero pressure the change of volume occupied by the ^{4}He, due to the addition of ^{3}He, causes zero change in the ^{4}He energy, to first order.

2.4. INTERACTION BETWEEN THE ^{3}He AND ^{4}He

So far we have considered the purely Fermi liquid aspects of dilute solutions of ^{3}He in superfluid ^{4}He, assuming that the background ^{4}He plays the role of an inert ether. We have always assumed that the ^{4}He superfluid velocity $\mathbf{v}_s$ is fixed at zero, and neglected at low temperatures real phonon and roton excitations of the ^{4}He fluid. We consider now role played by the ^{3}He in modifying phenomena involving excitation of the ^{4}He fluid. We shall first determine the effects of superfluid flow on the ^{3}He quasiparticles, from which we can determine the ^{3}He contribution to the normal mass density of the system and the interaction of ^{3}He with long wavelength phonons. Then we shall describe, as a detailed illustrative example, the effects of the ^{3}He on first sound propagation in the system, and discuss their effects on other transport properties. We will not consider phenomena associated with the ^{3}He–roton interaction, which is not well understood theoretically. We refer the reader to References [57] and [58] for theoretical discussions of the ^{3}He–roton interaction, and Reference [59] for experimental results.

2.4.1. Effects of Superfluid Flow on ^{3}He Quasiparticles

In determining the ^{3}He quasiparticle properties so far we have only discussed situations in which the ^{4}He superfluid velocity $\mathbf{v}_s$ remains fixed at zero. For example, the quasiparticle energies (2.1.1) and the effective ^{3}He interactions are all computed at $\mathbf{v}_s = 0$. The superfluid velocity defines a preferred frame and destroys Galilean invariance of the ^{3}He subsystem, even when there are no ^{4}He excitations present. To compute the effect of superfluid flow on the ^{3}He quasiparticle properties, we begin by determining, through applying Galilean invariance arguments to the entire ^{3}He–^{4}He system, the dependence of the quasiparticle energy on $\mathbf{v}_s$.

In Section 1.1.3(d) we considered the general change in the energy of a quasiparticle due to a Galilean transformation to a frame with velocity $\mathbf{u}$. If the superfluid velocity vanishes in the initial (unprimed) frame, then in the primed frame $\mathbf{v}_s = -\mathbf{u}$. Hence from (1.1.61) we may write

$$\varepsilon'_{\mathbf{p}} = \varepsilon_{\mathbf{p}-m_3\mathbf{v}_s} + \mathbf{p}\cdot\mathbf{v}_s - \tfrac{1}{2}m_3v_s^2. \tag{2.4.1}$$

(Note that m_3 is the bare ^{3}He mass.) The quantity $\varepsilon'_{\mathbf{p}} = \varepsilon_{\mathbf{p}}\{n'_{\mathbf{p}'}, \mathbf{v}_s\}$ is the quasiparticle energy in the presence of a superfluid velocity $\mathbf{v}_s$, and a ^{3}He quasiparticle distribution (cf. 1.1.63)

$$n'_{\mathbf{p}} = n^0_{\mathbf{p}-m_3\mathbf{v}_s}. \tag{2.4.2}$$

Thus (with spin indices suppressed),

$$\varepsilon_{\mathbf{p}}\{n^0_{\mathbf{p}''-m_3\mathbf{v}_s}, \mathbf{v}_s\} = \varepsilon_{\mathbf{p}-m_3\mathbf{v}_s}\{n^0_{\mathbf{p}''}, \mathbf{v}_s = 0\} + \mathbf{p}\cdot\mathbf{v}_s - \tfrac{1}{2}m_3\mathbf{v}_s^2. \tag{2.4.3}$$

This equation is the fundamental consequence of Galilean invariance for the dependence of the quasiparticle energies on $\mathbf{v}_s$.

The generalization of (2.4.3) to include a background of ^{4}He excitations is straightforward; the point to bear in mind is that under a Galilean transformation the momentum $\mathbf{k}$ of an excitation is the *same* in the primed and unprimed frames, since creating a ^{4}He excitation does not change the number of ^{4}He atoms. Thus,

$$f'_{\mathbf{q}} = f_{\mathbf{q}}. \tag{2.4.4}$$

The excitation energy in the primed frame is given by

$$\omega'_{\mathbf{q}} = \omega_{\mathbf{q}} - \mathbf{q}\cdot\mathbf{u}. \tag{2.4.5}$$

[This is equivalent to (1.1.61) with $m = 0$.] We shall not, however, consider here effects of ^{4}He excitations on the dependence of the ^{3}He energies on $\mathbf{v}_s$.

For a single ^{3}He quasiparticle in ^{4}He we find from (2.4.3) and (2.1.1) that to order v_s^2

$$\varepsilon_{\mathbf{p}}\{\mathbf{v}_s\} = \varepsilon_0 + \frac{p^2}{2m} + \frac{\delta m}{m}\mathbf{p}\cdot\mathbf{v}_s - \frac{1}{2}\frac{\delta m}{m} m_3 \mathbf{v}_s^2, \tag{2.4.6}$$

where

$$\delta m = m - m_3. \tag{2.4.7}$$

As we shall see in the next section, this result is very useful in determining the interactions between long wavelength phonons and ^{3}He quasiparticles.

Expanding both sides of (2.4.3) to first order in $\mathbf{v}_s$ we derive the first variation of $\varepsilon_{\mathbf{p}}$ with $\mathbf{v}_s$:

$$\varepsilon_{\mathbf{p}}\{\mathbf{v}_s\} = \varepsilon_{\mathbf{p}}\{0\} + \left(1 - \frac{m_3}{m^*}\right)\mathbf{p}\cdot\mathbf{v}_s + m_3 \sum_{\mathbf{p}'} f_{\mathbf{pp}'}\mathbf{v}_s\cdot\nabla_p n^0_{p'}, \tag{2.4.8}$$

where we have used $\nabla_p \varepsilon_p = \mathbf{p}/m^*$. [Cf. (1.1.65).] This result enables us to compute the normal mass density carried by the ^{3}He quasiparticles in a hydrodynamic process. Consider a flow situation in which the superfluid velocity is nonzero and the ^{3}He Fermi sea is one of local equilibrium displaced by a momentum $\mathbf{p}_0$, that is,

$$n_{\mathbf{p}} = n^0_{\mathbf{p}-\mathbf{p}_0}. \tag{2.4.9}$$

Then the ^{3}He normal velocity $\mathbf{v}_3$ is defined in terms of the ^{3}He quasiparticle current (in unit volume) by

$$\mathbf{v}_3 n_3 \equiv \sum_{\mathbf{p}} (\nabla_p \varepsilon_{\mathbf{p}}) n_{\mathbf{p}}, \tag{2.4.10}$$

and the ^{3}He normal mass density ρ_{n3} is defined in terms of the ^{3}He momentum density (measured with respect to that in the frame in which the superfluid is at rest) by

$$\sum_{\mathbf{p}} (\mathbf{p} - m_3\mathbf{v}_3) n_{\mathbf{p}} \equiv \rho_{n3}(\mathbf{v}_3 - \mathbf{v}_s). \tag{2.4.11}$$

This normal mass density is measured in second sound experiments as well as in rotating bucket experiments. Using (2.4.9) in the left side of (2.4.11) we find

$$n_3(\mathbf{p}_0 - m_3\mathbf{v}_s) = \rho_{n3}(\mathbf{v}_3 - \mathbf{v}_s). \tag{2.4.12}$$

To compute ρ_{n3} we determine (2.4.10) to first order in $\mathbf{p}_0$ and $\mathbf{v}_s$; we have after a short calculation using (2.4.8) and (2.4.9):

$$\begin{aligned}\sum_{\mathbf{p}} (\nabla_p \varepsilon_{\mathbf{p}}) n_{\mathbf{p}} &= \sum_{\mathbf{p}} \left[\frac{\mathbf{p}}{m^*}\,\delta n_{\mathbf{p}} + \Big(\nabla_p \sum_{\mathbf{p}'} f_{\mathbf{pp}'}\Big)\delta n_{\mathbf{p}'} n^0_p + \nabla_p(\varepsilon_{\mathbf{p}}\{\mathbf{v}_s\} - \varepsilon_{\mathbf{p}}\{0\}) n^0_p\right] \\ &= \frac{n_3}{m^*}\mathbf{p}_0 + n_3\left(1 - \frac{m_3}{m^*}\right)\mathbf{v}_s + \sum_{\mathbf{pp}'} f_{\mathbf{pp}'}((\mathbf{p}_0 - m_3\mathbf{v}_s)\cdot\nabla n^0_p)\nabla n^0_{p'}.\end{aligned} \tag{2.4.13}$$

The final term is simply $(\mathbf{p}_0 - m_3\mathbf{v}_s)n_3F_1^s/3m^*$, so that

$$\mathbf{v}_3 - \mathbf{v}_s = \left(1 + \frac{F_1^s}{3}\right)\frac{\mathbf{p}_0 - m_3\mathbf{v}_s}{m^*}. \tag{2.4.14}$$

Substituting in (2.4.12) we see that the normal mass density is given by

$$\rho_{n3} = \frac{m^* n_3}{1 + F_1^s/3}. \tag{2.4.15}$$

From (1.3.121) (where for m read m_3), we see that

$$\rho_{n3} \geqslant m_3 n_3. \tag{2.4.16}$$

We may use (2.3.11) to write*

$$\frac{n_3}{\rho_{n3}} \equiv \frac{1}{m_n^*} = \frac{1}{m} + \frac{1}{6n_3}\sum_{\mathbf{pp}'} n_p^0 n_{p'}^0 (\nabla_p + \nabla_{p'})^2 f_{\mathbf{pp}'}. \tag{2.4.17}$$

As in pure ^{3}He, the effective mass m^* entering the specific heat differs from the mass m_n^* (2.4.17) entering flow experiments. The mass m_n^*, sometimes called the inertial mass, differs from the "bare" effective mass m by terms that reflect the lack of Galilean invariance of the ^{3}He–^{3}He interaction. It is clear from this form that if we assume Galilean invariance of the ^{3}He subsystem, so that $f_{\mathbf{pp}'}$ depends only on $\mathbf{p} - \mathbf{p}'$, then $m^*/m = 1 + F_1^s/3$ and

$$\rho_{n3} = m n_3. \tag{2.4.18}$$

After we have developed the full interaction between ^{3}He and phonons, we shall give (2.4.36) a first estimate of m_n^* assuming that the breaking of Galilean invariance arises purely through the phonon exchange interaction (Figure 2.4).

2.4.2. Interaction of ^{3}He Quasiparticles with Long Wavelength Phonons

(a) Effective Interaction Hamiltonian. The form of the coupling between ^{3}He quasiparticles and long wavelength phonons can be determined by a combination of thermodynamic and Galilean invariance arguments. In this derivation we shall neglect the ^{3}He concentration dependence of the ^{3}He-phonon interaction.

The long wavelength low-frequency deviations of the ^{4}He from equilibrium may be described by a ^{4}He density *variation* $\delta\rho_4(\mathbf{r}, t)$, and a local superfluid velocity $\mathbf{v}_s(\mathbf{r}, t)$ (given microscopically by the gradient of the phase of the ^{4}He condensate wave function). The interaction between a long wavelength phonon and a ^{3}He quasiparticle is determined by the dependence of the quasiparticle energy on the local ^{4}He density and velocity.

*This result was derived by G. Baym and A. Leggett (unpublished but see Reference [46]). Similar calculations are given by Khalatnikov [60], and Ebner and Edwards [2].

The dependence of $\varepsilon_{\mathbf{p}}$ on $\mathbf{v}_s$ is given by (2.4.6). The quasiparticle energy depends on $\delta\rho_4$ through the dependence of ε_0 and m on the ^{4}He density; the dependence of ε_0 on $\delta\rho_4$ can be written to second order as

$$\varepsilon_0\{\delta\rho_4\} = \varepsilon_0 + \frac{\partial\varepsilon_0}{\partial n_4}\delta\rho_4 + \frac{1}{2}\frac{\partial^2\varepsilon_0}{\partial n_4^2}(\delta\rho_4)^2, \tag{2.4.19}$$

with a similar result for $m\{\delta\rho_4\}$. The partial derivatives are at constant ^{3}He density. Then from (2.4.6) and (2.4.19) the variation of $\varepsilon_{\mathbf{p}}$ with $\delta\rho_4$ and $\mathbf{v}_s$ becomes, to second order,

$$\begin{aligned}\delta\varepsilon_{\mathbf{p}}(\mathbf{r}, t) = &\left(\frac{\partial\varepsilon_0}{\partial n_4} + \frac{p^2}{2}\frac{\partial m^{-1}}{\partial n_4}\right)\delta\rho_4(\mathbf{r}, t) + \frac{\delta m}{m}\,\mathbf{p}\cdot\mathbf{v}_s(\mathbf{r}, t)\\ &+ \frac{1}{2}\left[\left(\frac{\partial^2\varepsilon_0}{\partial n_4^2} + \frac{p^2}{2}\frac{\partial^2 m^{-1}}{\partial n_4^2}\right)(\delta\rho_4(\mathbf{r}, t))^2 - \frac{\delta m}{m}m_3\mathbf{v}_s(\mathbf{r}, t)^2\right]\\ &- m_3\frac{\partial m^{-1}}{\partial n_4}\,\mathbf{p}\cdot\mathbf{v}_s(\mathbf{r}, t)\delta\rho_4(\mathbf{r}, t),\end{aligned} \tag{2.4.20}$$

The final term is the variation with $\delta\rho_4$ of the $(\delta m/m)\mathbf{p}\cdot\mathbf{v}_s$ term in (4.6).

The point now is that to a quasiparticle, a long wavelength phonon looks like a uniform motion of the ^{4}He with a local velocity $\mathbf{v}_s$ and a local density $n_4 + \delta\rho_4$. Thus (2.4.20) is the effective interaction Hamiltonian between a quasiparticle and a long wavelength phonon, where $\delta\rho_4(\mathbf{r}, t)$ and $\mathbf{v}_s(\mathbf{r}, t)$ are the density variations and flow velocity associated with the phonon.

In order to calculate the amplitude for phonon-quasiparticle scattering we "quantize" the interaction (2.4.20) by treating $\mathbf{r}$ and $\mathbf{p}$ as the quasiparticle position and momentum *operators*, and by expanding $\delta\rho_4(\mathbf{r})$ and $\mathbf{v}_s(\mathbf{r})$ in terms of the operators $b_{\mathbf{q}}$ and $b_{\mathbf{q}}^\dagger$ that annihilate and create phonons of momentum $\mathbf{q}$. In a box of volume V these expansions are [20]

$$\delta\rho_4(\mathbf{r}) = \sum_{\mathbf{q}}\left(\frac{q^2 n_4}{2m_4\omega_q V}\right)^{1/2}(b_{\mathbf{q}}e^{i\mathbf{q}\cdot\mathbf{r}} + b_{\mathbf{q}}^\dagger e^{-i\mathbf{q}\cdot\mathbf{r}}) \tag{2.4.21}$$

and

$$\mathbf{v}_s(\mathbf{r}) = \sum_{\mathbf{q}}\left(\frac{\omega_{\mathbf{q}}}{2m_4 n_4 V}\right)^{1/2}\hat{\mathbf{q}}(b_{\mathbf{q}}e^{i\mathbf{q}\cdot\mathbf{r}} + b_{\mathbf{q}}^\dagger e^{-i\mathbf{q}\cdot\mathbf{r}}). \tag{2.4.22}$$

The operators $b_{\mathbf{q}}$ and $b_{\mathbf{q}}^\dagger$ obey the commutation relations

$$\begin{aligned}&[b_{\mathbf{q}}, b_{\mathbf{q}'}^\dagger] = \delta_{\mathbf{q}\mathbf{q}'}\\ &[b_{\mathbf{q}}, b_{\mathbf{q}'}] = [b_{\mathbf{q}}^\dagger, b_{\mathbf{q}'}^\dagger] = 0.\end{aligned} \tag{2.4.23}$$

The coefficients in the expansions are determined by (i) the particle conservation law for linearized free phonons [see (1.3.111)], and (ii) the commutation relation between the density and current operators. [See (1.3.116).]

Substituting (2.4.21) and (2.4.22) into (2.4.20) we find that the first terms in

(2.4.20), linear in $\delta\rho_4$ or $\mathbf{v}_s$, correspond to the absorption or emission of a phonon by a quasiparticle, as in Figure 2.3*a*. We denote these terms in (2.4.20) by V_1. In quantizing (2.4.20), the $\mathbf{p} \cdot \mathbf{v}_s(\mathbf{r})$ term must be symmetrized as $\frac{1}{2}[\mathbf{p} \cdot \mathbf{v}_s(\mathbf{r}) + \mathbf{v}_s(\mathbf{r}) \cdot \mathbf{p}]$, since the position and momentum of a quasiparticle fail to commute. This symmetrization derives from the fact that the momentum density operator for a collection of particles has the symmetrized form

$$\mathbf{g}(\mathbf{r}) = \frac{1}{2} \sum_i [\mathbf{p}_i \delta(\mathbf{r} - \mathbf{r}_i) + \delta(\mathbf{r} - \mathbf{r}_i)\mathbf{p}_i],$$

where $\mathbf{p}_i$ and $\mathbf{r}_i$ are the momentum and position operator of the *i*th particle. The terms $\sim \partial m^{-1}/\partial n_4$ in (2.4.20) must be similarly symmetrized; this symmetrization operation is not unique. However, we shall not need their explicitly symmetrized forms here. The terms quadratic in $\delta\rho_4$ and $\mathbf{v}_s$, which we denote collectively by V_2, correspond to the scattering of a phonon by a quasiparticle, as in Figure 2.3*b*; they lead in addition to two-phonon emission or absorption processes.

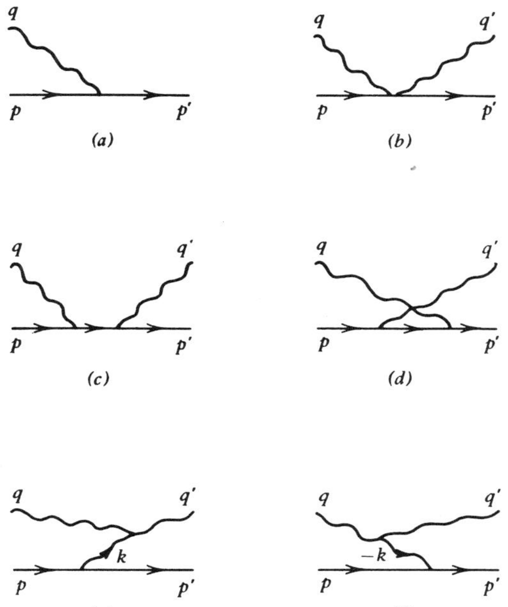

FIGURE 2.3. Interactions of phonons (wavy lines) with quasiparticles (solid lines). (*a*) Phonon absorption; (*b*), (*c*), and (*d*), three processes leading to scattering of phonons by quasiparticles; (*e*) and (*f*), "phonon-induced" scattering of phonons by quasiparticles.

The matrix element for a quasiparticle of momentum $\mathbf{p}$ (with $p \ll q_D$) to absorb a phonon of momentum $\mathbf{q}$, ending up with momentum $\mathbf{p}'$, as in Figure 2.3*a*, is then to leading order:

$$\langle \mathbf{p}' | V_1 | \mathbf{p}, \mathbf{q} \rangle = \delta_{\mathbf{p}', \mathbf{p}+\mathbf{q}} \left(\frac{q n_4}{2 m_4 s V} \right)^{1/2} \left[\frac{\partial \varepsilon_0}{\partial n_4} + \frac{\delta m\, s}{n_4 q} (\varepsilon_{\mathbf{p}'} - \varepsilon_{\mathbf{p}}) \right]. \quad (2.4.24)$$

We have omitted the contribution $\sim p^2 \partial m^{-1}/\partial n_4$, which is of order $(p/ms)^2$ compared with the $\partial\varepsilon_0/\partial n_4$ term. No Fermi liquid corrections are included in (2.4.24); thus $\varepsilon_{\mathbf{p}'} - \varepsilon_{\mathbf{p}} = \mathbf{q} \cdot (\mathbf{p} + \mathbf{p}')/2m$.

The matrix element for the direct scattering process illustrated in Figure 2.3*b* is

$$\langle \mathbf{p}'\mathbf{q}' | V_2 | \mathbf{p}\mathbf{q} \rangle = \delta_{\mathbf{p}+\mathbf{q}, \mathbf{p}'+\mathbf{q}'} \frac{\sqrt{q q'}\, n_4}{2 m_4 s V} \left[\frac{\partial^2 \varepsilon_0}{\partial n_4^2} - \frac{\delta m}{m} \frac{m_3 s^2}{n_4^2} \hat{\mathbf{q}} \cdot \hat{\mathbf{q}}' \right] \quad (2.4.25)$$

for $p, p' \ll q_D$; again we have neglected the terms $\sim \partial m^{-1}/\partial n_4$, since they vanish as p and $p' \to 0$.

The coefficients in (2.4.24) and (2.4.25) may be deduced from thermodynamic measurements. Equation 2.2.5 states

$$\frac{\partial \varepsilon_0}{\partial n_4} = \frac{m_4 s^2}{n_4} (1 + \alpha), \quad (2.2.5)$$

and thus

$$\frac{\partial^2 \varepsilon_0}{\partial n_4^2} = \frac{m_4 s^2}{n_4^2} \left[(1 + \alpha)(2u - 1) + n_4 \frac{\partial \alpha}{\partial n_4} \right], \quad (2.4.26)$$

where

$$u = \frac{n_4}{s} \frac{\partial s}{\partial n_4} \quad (2.4.27)$$

is the ^{4}He Grüneisen constant, ≈ 2.84 at vapor pressure at $T = 0.1$ K [61]. Since $n_4 \partial\alpha/\partial n_4 \simeq -1.03$ at zero pressure [62] the square bracket in (2.4.26) is $\simeq 4.98$.

This general procedure for relating matrix elements of processes involving long wavelength phonons to thermodynamic and hydrodynamic quantities is known as the Landau *quantum hydrodynamic* method. The spirit of the method is the same as that of the Landau Fermi-liquid theory. Its validity depends on the assumption that the scattering amplitudes have no singular structure as ω, the phonon frequency, and $\mathbf{q}$, the phonon wavenumber, both tend to zero. While this is an eminently reasonable property [53], the possibility of singular structure in the ^{3}He-phonon scattering vertices has not, to our knowledge, been definitely ruled out.

(b) Phonon-Exchange Interaction Between ^{3}He Quasiparticles. We have

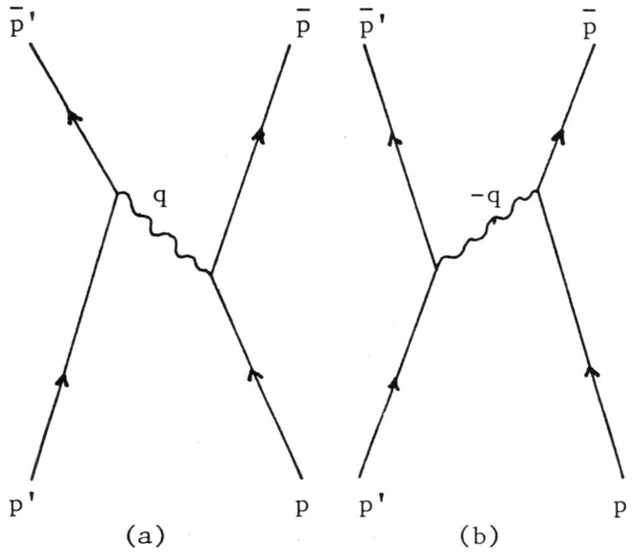

FIGURE 2.4. The two processes producing the phonon (wavy line) exchange scattering of two ^{3}He. Time increases upwards.

now developed sufficient machinery to write down the complete form of the phonon-exchange interaction between two ^{3}He quasiparticles. The complete process is the sum of the two terms shown in Figure 2.4. The first process (2.4*a*), in which quasiparticle **p** emits a phonon of momentum **q** which is absorbed by **p**′, has a matrix element given by second order perturbation theory as

$$\frac{\langle \bar{\mathbf{p}}' | V_1 | \mathbf{p}'\mathbf{q}\rangle\langle \bar{\mathbf{p}}\mathbf{q} | V_1 | \mathbf{p}\rangle}{\omega - sq} = \delta_{\mathbf{p}+\mathbf{p}',\, \bar{\mathbf{p}}+\bar{\mathbf{p}}'}\delta_{\mathbf{p},\, \bar{\mathbf{p}}+\mathbf{q}}\frac{n_4 q}{2m_4 sV}\frac{\left(\dfrac{\partial\varepsilon_0}{\partial n_4} - \dfrac{\delta m s\omega}{n_4 q}\right)^2}{\omega - sq}, \tag{2.4.28}$$

where we have assumed energy conservation between the initial and final states, and have written

$$\omega = \varepsilon_{\mathbf{p}} - \varepsilon_{\bar{\mathbf{p}}} = \varepsilon_{\bar{\mathbf{p}}'} - \varepsilon_{\mathbf{p}'}. \tag{2.4.29}$$

The second process has a matrix element

$$\frac{\langle \bar{\mathbf{p}} | V_1 | \mathbf{p},\, -\mathbf{q}\rangle\langle \bar{\mathbf{p}}',\, -\mathbf{q} | V_1 | \mathbf{p}'\rangle}{-\omega - sq} = -\delta_{\mathbf{p}+\mathbf{p}',\, \bar{\mathbf{p}}+\bar{\mathbf{p}}'}\delta_{\mathbf{p},\, \bar{\mathbf{p}}+\mathbf{q}}\frac{n_4 q}{2m_4 sV}\frac{\left(\dfrac{\partial\varepsilon_0}{\partial n_4} + \dfrac{\delta m s\omega}{n_4 q}\right)^2}{\omega + sq}. \tag{2.4.30}$$

We write the total matrix element for the phonon exchange scattering, the sum of (2.4.28) and (2.4.30), as $\delta_{\mathbf{p}+\mathbf{p}',\,\bar{\mathbf{p}}+\bar{\mathbf{p}}'}\delta_{\mathbf{p},\,\bar{\mathbf{p}}+\mathbf{q}}\langle\bar{\mathbf{p}}\bar{\mathbf{p}}'\,|\,V_{\text{ph-ex}}\,|\,\mathbf{p}\mathbf{p}'\rangle/V$, where

$$\langle\bar{\mathbf{p}}\bar{\mathbf{p}}'\,|\,V_{\text{ph-ex}}\,|\,\mathbf{p}\mathbf{p}'\rangle = -\frac{m_4 s^2}{n_4}\left[(1+\alpha)^2 - \frac{\omega^2}{\omega^2 - s^2q^2}(1+\alpha+\delta m/m_4)^2\right]. \tag{2.4.31}$$

For scattering of quasiparticles with $\omega = 0$, (2.4.31) reduces to the final term of (2.3.5).

The second term of (2.4.31) is effectively a dipolar interaction between the quasiparticles, which depends on their momenta. For $q \ll p,\ p' \ll ms$ the second term is

$$\langle\bar{\mathbf{p}}\bar{\mathbf{p}}'\,|\,V_{\text{dip}}\,|\,\mathbf{p}\mathbf{p}'\rangle = -\left(\frac{\mathbf{p}\cdot\hat{\mathbf{q}}}{m}\right)\left(\frac{\mathbf{p}'\cdot\hat{\mathbf{q}}}{m}\right)\left(1+\alpha+\frac{\delta m}{m_4}\right)^2\frac{m_4}{n_4}. \tag{2.4.32}$$

This interaction is the exact quantum analogue of the classical dipolar interaction of two hard spheres at large separation in a classical fluid. It is of order $(v_f/s)^2$ smaller than the first term in (2.4.31).

[For a classical sphere of volume v_3, $\delta m = m_4 n_4 v_3/2$, while $n_4 v_3 = 1 + \alpha$; for this case (2.4.32) reduces to

$$V_{\text{dip}} = -\left(\frac{3v_3}{2}\right)^2 m_4 n_4\left(\frac{\mathbf{p}\cdot\hat{\mathbf{q}}}{m}\right)\left(\frac{\mathbf{p}'\cdot\hat{\mathbf{q}}}{m}\right). \tag{2.4.33}$$

Because of the momentum dependence of this interaction, the velocity of the particle with momentum $\mathbf{p}$ is

$$\mathbf{v}_{\mathbf{p}} = \frac{\mathbf{p}}{m} + \nabla_p V_{\text{dip}},$$

with a similar expression for $\mathbf{v}_{\mathbf{p}'}$. The total classical energy $p^2/2m + p'^2/2m + V_{\text{dip}}$ to lowest order in the dipole interaction is then

$$\frac{m}{2}(\mathbf{v}_{\mathbf{p}}^2 + \mathbf{v}_{\mathbf{p}'}^2) + \left(\frac{3v_3}{2}\right)^2 m_4 n_4(\mathbf{v}_{\mathbf{p}}\cdot\hat{\mathbf{q}})(\mathbf{v}_{\mathbf{p}'}\cdot\hat{\mathbf{q}}). \tag{2.4.34}$$

This is the form in which the classical dipolar interaction is usually given [63].]

The phonon-exchange interaction contributes directly to the scattering amplitude $t_{\mathbf{p}\mathbf{p}'}(\bar{\mathbf{p}} - \mathbf{p})$ [in (2.3.31)]. The total scattering amplitude must include the exchange term $\langle\bar{\mathbf{p}}'\bar{\mathbf{p}}\,|\,V_{\text{ph-ex}}\,|\,\mathbf{p}\mathbf{p}'\rangle$ as well. When one computes the Landau parameters from (2.3.32), the direct term $\langle\bar{\mathbf{p}}\bar{\mathbf{p}}'\,|\,V_{\text{ph-ex}}\,|\,\mathbf{p}\mathbf{p}'\rangle$ leads to the final term in (2.3.5); one should note though that the exchange term is included in the first part, $f^{\text{dir}}_{\mathbf{p}\sigma,\,\mathbf{p}'\sigma'}$, the Landau parameter defined at constant n_4. The total contribution of the phonon-exchange interaction to $f_{\mathbf{p}\sigma,\,\mathbf{p}'\sigma'}$ is

$$-\frac{m_4 s^2}{n_4}\left[(1+\alpha)^2(1-\delta_{\sigma\sigma'}) - \delta_{\sigma\sigma'}\frac{\omega^2}{s^2q^2}\left(1+\alpha+\frac{\delta m}{m_4}\right)^2\right] \tag{2.4.35}$$

where $\omega = \varepsilon_p - \varepsilon_{p'}$, $\mathbf{q} = \mathbf{p} - \mathbf{p}'$, and we have neglected the ω^2 in the denominator in (2.4.31).

We may make a first estimate of the effect of ^{3}He–^{3}He interactions on the ^{3}He normal mass density by assuming that the lack of the Galilean invariance of the quasiparticle interactions is due entirely to the phonon exchange term (2.4.35). Substituting (2.4.35) into (2.4.17) we find

$$\frac{1}{m_n^*} = \frac{1}{m}\left[1 + \frac{1}{6}\frac{n_3}{n_4}\frac{m_4}{m}\left(1 + \alpha + \frac{\delta m}{m_4}\right)^2\right]$$
$$\simeq \frac{1 + 0.5x}{m}, \tag{2.4.36}$$

an estimate in qualitatively good agreement with the concentration dependence of m_n^* as determined by Brubaker et al. [44]. A detailed comparison of theory with experiment is given in Reference [2]. Disatnik and Brucker [64] have also given a computation of the normal mass density taking into account the lack of Galilean invariance in the ^{3}He effective interaction.

2.4.3. Scattering of Phonons by ^{3}He Quasiparticles

At temperatures below ~0.6 K, where thermally excited phonons and rotons are few in number, transport processes in the ^{4}He, such as thermal conduction and momentum transport (viscosity) by phonons, are limited by scattering and absorption of phonons against ^{3}He quasiparticles, as well as by boundary scattering of phonons for temperatures well below 100 mK. In this section we evaluate the amplitude for the scattering of a long wavelength phonon against a ^{3}He quasiparticle.

The complete scattering amplitude is a sum of the amplitudes for the five processes shown in Figure 2.3*b*, *c*, *d*, *e*, *f*. The first three processes are analogous to the nonrelativistic scattering of photons by electrons. The latter two processes involve phonon–phonon scattering. The amplitude for the process *b* is given by (2.4.25). The amplitude for processes *c* and *d* is easily computed using (2.4.24) in the standard second order perturbation theory expressions, with the result

$$\delta_{\mathbf{p}+\mathbf{q},\,\mathbf{p}'+\mathbf{q}'}\frac{m_4 s\sqrt{qq'}}{2n_4 V}\left[-\left(\frac{\delta m}{m_4}\right)^2\frac{\hat{\mathbf{q}}\cdot\hat{\mathbf{q}}'}{m}\right.$$
$$\left.+\left(1 + \alpha + \frac{\delta m}{m_4}\right)^2 s^2\left(\frac{1}{\varepsilon_p + sq - \varepsilon_{\mathbf{p}+\mathbf{q}}} + \frac{1}{\varepsilon_{p'} - sq - \varepsilon_{\mathbf{p}'-\mathbf{q}}}\right)\right]. \tag{2.4.37}$$

In deriving (2.4.37) we have neglected the exclusion principle in the intermediate states; in dilute solutions exclusion of intermediate states reduces the scattering rate by terms of relative order x. Since the ^{3}He Fermi velocity v_f is $\ll s$ it is sufficient to expand the energy denominators in (2.4.37) to terms proportional to $(sq)^{-2}$. Furthermore, in the temperature regime where ^{3}He-phonon scattering is important, the relevant phonon energies are much smaller than m_4s^2. This implies that quasiparticle-phonon scattering on energy shell is essentially elastic—that is, $q = q'$ to terms of relative order $\kappa T/ms^2$ or v_f/s. Thus, the total amplitude for processes b, c, and d, which we call the direct quasiparticle-phonon scattering, reduces to

$$\langle \mathbf{p}'\mathbf{q}'|T_{\text{dir}}|\mathbf{pq}\rangle \simeq \delta_{\mathbf{p}+\mathbf{q},\,\mathbf{p}'+\mathbf{q}'}\frac{sq}{2n_4V}\left\{\left[\left(1+\alpha+\frac{\delta m}{m_4}\right)^2\frac{m_4}{m}-\frac{\delta m}{m_4}\right]\cos\theta + \frac{n_4^2}{m_4s^2}\frac{\partial^2\varepsilon_0}{\partial n_4^2}\right\}, \quad (2.4.38)$$

where θ is the angle between $\mathbf{q}$ and $\mathbf{q}'$.

The two processes in Figure 2.3e and f are the scattering of a phonon by a quasiparticle through the exchange of a virtual phonon, analogous to the phonon-exchange interaction between two ^{3}He considered in the previous section. In order to calculate their contribution, denoted by $\langle \mathbf{p}'\mathbf{q}'|T_{\text{ph}}|\mathbf{pq}\rangle$, to the total scattering amplitude, we need to find the three-phonon vertex as it occurs in e and f. This may be found for long wavelength phonons by the quantum hydrodynamic method by expanding the total energy of the system to third order in the density variation $\delta\rho_4(\mathbf{r})$ and flow velocity $\mathbf{v}_s(\mathbf{r})$ associated with the phonons; the third order term in the energy is

$$V_3 = \int d^3r\left[\frac{1}{2}\,m_4\mathbf{v}_s(\mathbf{r})\delta\rho_4(\mathbf{r})\cdot\mathbf{v}_s(\mathbf{r}) + \frac{1}{6}\frac{\partial}{\partial n_4}\left(\frac{m_4s^2}{n_4}\right)(\delta\rho_4(\mathbf{r}))^3\right]. \quad (2.4.39)$$

The first term is from the kinetic energy of flow, while the coefficient in the latter term is $(\partial^3E_0/\partial n_4^3)/3!$, where $E_0(n_4)$ is the ground state energy density of pure ^{4}He. With the expansions (2.4.21) and (2.4.22) we see that the three-phonon matrix element in Figure 2.3e is

$$\langle \mathbf{q}'|V_3|\mathbf{q},\mathbf{k}\rangle = \delta_{\mathbf{q}+\mathbf{k},\,\mathbf{q}'}\left(\frac{qq'ks}{8m_4n_4V}\right)^{1/2}[2u - 1 + \hat{\mathbf{q}}\cdot\hat{\mathbf{q}}' + \hat{\mathbf{q}}\cdot\hat{\mathbf{k}} + \hat{\mathbf{q}}'\cdot\hat{\mathbf{k}}], \quad (2.4.40)$$

while that in Figure 2.3f is $\langle \mathbf{q}',-\mathbf{k}|V_3|\mathbf{q}\rangle = \langle \mathbf{q}|V_3|\mathbf{q}',-\mathbf{k}\rangle$.

The scattering amplitude due to phonon exchange $\langle \mathbf{p}'\mathbf{q}'|T_{\text{ph}}|\mathbf{pq}\rangle$ is simply evaluated on the energy shell when the scattering is elastic; then the δm term in (2.4.24) vanishes, as does $(\hat{\mathbf{q}} + \hat{\mathbf{q}}')\cdot\hat{\mathbf{k}}$. The result is

$$\langle \mathbf{p}'\mathbf{q}'|T_{\text{ph}}|\mathbf{pq}\rangle = -\delta_{\mathbf{p}+\mathbf{q},\,\mathbf{p}'+\mathbf{q}'}\frac{sq}{2n_4V}(1+\alpha)(2u - 1 + \cos\theta). \quad (2.4.41)$$

Adding (2.4.41) and (2.4.38), and using (2.4.26), we see that the total phonon-quasiparticle scattering amplitude is

$$\langle \mathbf{p}'\mathbf{q}'|T_{\text{qp-ph}}|\mathbf{pq}\rangle$$

$$= \delta_{\mathbf{p}+\mathbf{q},\,\mathbf{p}'+\mathbf{q}'}\frac{sq}{2n_4V}\left[n_4\frac{\partial\alpha}{\partial n_4} + \left(1+\alpha+\frac{\delta m}{m_4}\right)\left(1+\alpha-\frac{m_3}{m_4}\right)\frac{m_4}{m}\cos\theta\right]. \quad (2.4.42)$$

This result is valid for q, q', p, $p' \ll ms$.

The rate at which phonons are scattered from $\mathbf{q}$ to $\mathbf{q}'$ by the ^{3}He is given by

$$2 \sum_{\mathbf{pp}'} 2\pi\delta(\varepsilon_{\mathbf{p}} + sq - \varepsilon_{\mathbf{p}'} - sq')\,|\langle \mathbf{p}'\mathbf{q}'|T_{\text{qp-ph}}|\mathbf{pq}\rangle|^2\,[1 + f_{\mathbf{q}'}]n_{\mathbf{p}}(1 - n_{\mathbf{p}'})\,; \quad (2.4.43)$$

the initial factor of 2 accounts for the ^{3}He spin. If we assume the ^{3}He to be in thermal equilibrium and neglect the Pauli principle corrections, we find from (2.4.43), by multiplying by $\delta(\cos\theta - \hat{\mathbf{q}}\cdot\hat{\mathbf{q}}')$ and summing over all $\mathbf{q}'$, that the rate at which phonons of momentum $\mathbf{q}$ are scattered by angle θ, divided by $1 + f_{\mathbf{q}'}$, is

$$\begin{aligned}\Gamma(q, \theta) &= \frac{n_3 s q^4}{8\pi n_4^2}\left[n_4 \frac{\partial \alpha}{\partial n_4} + \left(1 + \alpha + \frac{\delta m}{m_4}\right)\left(1 + \alpha - \frac{m_3}{m_4}\right)\frac{m_4}{m}\cos\theta\right]^2 \\ &\simeq \frac{x s q^4}{8\pi n_4}(-1.03 + 0.70\cos\theta)^2. \end{aligned} \quad (2.4.44)$$

This scattering rate has the q^4 dependence characteristic of Rayleigh scattering. Consequently, its importance in limiting phonon transport phenomena increases rapidly with temperature. For very long wavelengths, however, and consequently at very low temperatures, the absorption of phonons by ^{3}He viscosity (as in ultrasonic attenuation) plays a more dominant role in limiting rates of phonon transport. In Section 2.4.6 we summarize calculations of transport phenomena involving phonon-quasiparticle scattering, after we develop the theory of first sound propagation in dilute solutions.

We indicate the relation of the calculations of this section to the Landau parameters describing the phonon-quasiparticle interaction. In the processes shown in Figure 2.3 b, c and d, one never encounters vanishing energy denominators. We may therefore identify the forward phonon-quasiparticle scattering amplitude (2.4.38) with the Landau parameter:

$$\begin{aligned}\left(\frac{\delta \omega_{\mathbf{q}}}{\delta n_{\mathbf{p}}}\right)_{f_{\mathbf{q}'},\, n_{\mathbf{p}'},\, n_4} &= V\langle \mathbf{pq}|T_{\text{dir}}|\mathbf{pq}\rangle \\ &= \frac{sq}{2n_4}\left[\left(1 + \alpha + \frac{\delta m}{m_4}\right)^2 \frac{m_4}{m} - \frac{\delta m}{m_4}\right. \\ &\quad \left. + (1 + \alpha)(2u - 1) + n_4 \frac{\partial \alpha}{\partial n_4}\right]. \end{aligned} \quad (2.4.45)$$

The addition of the phonon-exchange scattering $V\langle \mathbf{pq}|T_{\text{ph}}|\mathbf{pq}\rangle$ to this result converts it from the Landau parameter defined at fixed n_4 to that defined at fixed μ_4 [cf. (2.3.3) and (2.3.4)]:

$$\begin{aligned}\left(\frac{\delta \omega_{\mathbf{q}}}{\delta n_{\mathbf{p}}}\right)_{f_{\mathbf{q}'},\, n_{\mathbf{p}'},\, \mu_4} &= V\langle \mathbf{pq}|T_{\text{qp-ph}}|\mathbf{pq}\rangle \\ &= \frac{sq}{2n_4}\left[n_4 \frac{\partial \alpha}{\partial n_4} + \left(1 + \alpha + \frac{\delta m}{m_4}\right)\left(1 + \alpha - \frac{m_3}{m_4}\right)\frac{m_4}{m}\right]. \end{aligned} \quad (2.4.46)$$

The shift in the $T = 0$ phonon frequency produced by the addition of ^{3}He to pure ^{4}He, at constant μ_4, is then, to lowest order in the ^{3}He concentration x,

$$\delta\omega_{\mathbf{q}} = \sum_{\mathbf{p}} \langle \mathbf{pq} | T_{\text{qp-ph}} | \mathbf{pq} \rangle n_{\mathbf{p}} \,. \tag{2.4.47}$$

Since (2.4.46) is proportional to $\mathbf{q}$, we can write (2.4.47) as a shift in the $T = 0$ sound velocity:

$$\begin{aligned} s(x, \mu_4) &= s(0, \mu_4)\left\{1 + \frac{x}{2}\left[n_4 \frac{\partial \alpha}{\partial n_4} + \left(1 + \alpha + \frac{\delta m}{m_4}\right)\left(1 + \alpha - \frac{m_3}{m_4}\right)\frac{m_4}{m}\right]\right\} \\ &\simeq s(0, \mu_4)\,(1 - 0.17x)\,. \end{aligned} \tag{2.4.48}$$

The latter result is at $P = 0$.

We turn now to the full theory of first sound propagation.

2.4.4. First Sound in Dilute Solutions

The addition of ^{3}He atoms to superfluid ^{4}He modifies the first sound velocity, and, by providing additional degrees of freedom for energy absorption, increases the attenuation of first sound over its value in pure ^{4}He. For temperatures below about 0.3 K for ^{3}He concentrations of a few percent, the primary mechanism for sound attenuation is absorption by the ^{3}He (via the ^{3}He viscosity), a process which is kinematically possible because of ^{3}He–^{3}He collisions. Rayleigh scattering of sound by ^{3}He quasiparticles gives negligible attenuation. In the temperature range from 0.3 K to about 0.6–0.7 K, attenuation due to thermal phonons becomes important. This latter process depends critically on the scattering of thermal phonons by the ^{3}He. For temperatures above 0.7 K rotons and phonon-phonon scattering become important; however, we shall in this section assume temperatures below this and neglect thermal rotons.

The propagation of first sound in dilute solutions is described by the kinetic equations for the ^{3}He quasiparticles and phonons, together with the superfluid acceleration equation. The ^{3}He kinetic equation is [cf. (1.2.12a)]

$$\frac{\partial n_{\mathbf{p}}}{\partial t} + \nabla_p \varepsilon_{\mathbf{p}} \cdot \nabla_r n_{\mathbf{p}} - \nabla_r \varepsilon_{\mathbf{p}} \cdot \nabla_p n_{\mathbf{p}} = I_{3-3} + I_{3-\text{ph}} \,, \tag{2.4.49}$$

where I_{3-3} is the ^{3}He–^{3}He collision integral, and $I_{3-\text{ph}}$ describes collisions between ^{3}He quasiparticles and phonons. The phonon kinetic equation is similarly

$$\frac{\partial f_{\mathbf{q}}}{\partial t} + \nabla_q \omega_{\mathbf{q}} \cdot \nabla_r f_{\mathbf{q}} - \nabla_r \omega_{\mathbf{q}} \cdot \nabla_q f_{\mathbf{q}} = I_{\text{ph}-3} + I_{\text{ph}-\text{ph}} \,, \tag{2.4.50}$$

where $I_{\text{ph}-3}$ describes phonon-^{3}He collisions and absorption of phonons by the ^{3}He, and $I_{\text{ph}-\text{ph}}$ describes phonon-phonon scattering. We discuss the form of the phonon-^{3}He collision term in Section 2.4.6(a).

The linearized equation of motion for superfluid flow is

$$m_4 \frac{\partial \mathbf{v}_s}{\partial t} = -\nabla \mu_4 , \tag{2.4.51}$$

where m_4 is the bare ^{4}He mass, and μ_4 the ^{4}He chemical potential. In computing the propagation of first sound we shall neglect Fermi liquid effects due to interactions between excitations; at low ^{3}He concentrations and low temperatures these are small. Then, in the presence of superfluid flow, $\varepsilon_{\mathbf{p}}$ is given by (2.4.1), while

$$\omega_{\mathbf{q}}\{\mathbf{v}_s\} = sq + \mathbf{q}\cdot\mathbf{v}_s . \tag{2.4.52}$$

(a) Conservation Laws. We now write the linearized conservation laws for ^{3}He number, overall mass, energy, and momentum. Summing (2.4.49) over $\mathbf{p}$, using (2.4.1), and noting that collisions conserve ^{3}He atoms, we have

$$\frac{\partial n_3}{\partial t} + n_3 \nabla\cdot\mathbf{v}_3 = 0; \tag{2.4.53}$$

n_3 is the ^{3}He number density and the ^{3}He velocity $\mathbf{v}_3$ is defined by (2.4.10).

The equation expressing conservation of mass is

$$m_4 \frac{\partial n_4}{\partial t} + m_3 \frac{\partial n_3}{\partial t} + \nabla\cdot\mathbf{g} = 0 , \tag{2.4.54}$$

where the total momentum density is given (in unit volume) by

$$\mathbf{g} = \sum_{\mathbf{p}} \mathbf{p} n_{\mathbf{p}} + \sum_{\mathbf{q}} \mathbf{q} f_{\mathbf{q}} + m_4 n_4 \mathbf{v}_s . \tag{2.4.55}$$

In linear order

$$\sum_{\mathbf{p}} \mathbf{p} n_{\mathbf{p}} = n_3(m\mathbf{v}_3 - \delta m \mathbf{v}_s) = n_3[m_3\mathbf{v}_3 + \delta m(\mathbf{v}_3 - \mathbf{v}_s)] \tag{2.4.56}$$

[compare with (2.4.11) and (2.4.18)]. Similarly

$$\sum \mathbf{q} f_{\mathbf{q}} = \rho_{\mathrm{ph}}(\mathbf{v}_{\mathrm{ph}} - \mathbf{v}_s) , \tag{2.4.57}$$

thus defining the phonon normal velocity $\mathbf{v}_{\mathrm{ph}}$; the phonon normal mass density is given by

$$\rho_{\mathrm{ph}} = -\int \frac{d^3q}{(2\pi)^3} \frac{q^2}{3} \frac{\partial f_q^0}{\partial \omega_q} = \frac{2\pi^2}{45} \frac{(\kappa T)^4}{s^5 \hbar^3} . \tag{2.4.58}$$

The momentum density is then

$$\mathbf{g} = \rho_s \mathbf{v}_s + n_3 m \mathbf{v}_3 + \rho_{\mathrm{ph}} \mathbf{v}_{\mathrm{ph}} , \tag{2.4.59}$$

and the resulting equation of mass conservation is

$$m_4\frac{\partial n_4}{\partial t} + m_3\frac{\partial n_3}{\partial t} + \rho_s\nabla\cdot\mathbf{v}_s + mn_3\nabla\cdot\mathbf{v}_3 + \rho_{\text{ph}}\nabla\cdot\mathbf{v}_{\text{ph}} = 0, \quad (2.4.60)$$

where

$$\rho_s = m_4 n_4 - \delta m\, n_3 - \rho_{\text{ph}} \quad (2.4.60)$$

is the superfluid mass density. Note that the effective mass addition $\delta m\, n_3$ of the ^{3}He quasiparticles (which is actually ^{4}He mass) is carried with the ^{3}He quasiparticles in low frequency phenomena, and hence is not counted as superfluid mass density.

The equation for conservation of momentum may be derived by multiplying (2.4.49) by **p**, summing over all **p**, multiplying (2.4.50) by **q**, summing over all **q**, using (2.4.55), (2.4.51), (2.4.52), and (2.4.1), and noting that total momentum is conserved in all collisions; one finds in linear order [cf. (1.2.31)]

$$\frac{\partial g_i}{\partial t} + \nabla_j T_{ij}$$
$$= -n_4\nabla_i\mu_4 - n_3\nabla_i\varepsilon_0(n_4) + \frac{5}{2}\frac{P_f}{m}\nabla_i m(n_4) - \rho_{\text{ph}} s\nabla_i s(n_4), \quad (2.4.62)$$

where

$$T_{ij} = \sum_{\mathbf{p}}\frac{p_i p_j}{m}n_{\mathbf{p}} + \sum_{\mathbf{q}}\frac{sq_iq_j}{q}f_{\mathbf{q}}, \quad (2.4.63)$$

and P_f is the pressure of a free Fermi gas of mass m at temperature T and density n_3. The right side of (2.4.62) can, from (2.1.7), be written as

$$\nabla_i\Big(E - n_4\mu_4 - \sum_{\mathbf{p}}\varepsilon_{\mathbf{p}}n_{\mathbf{p}} - \sum_{\mathbf{q}}\omega_{\mathbf{q}}n_{\mathbf{q}}\Big),$$

where E is the ground state energy density.

The equation for energy conservation can be similarly derived by taking moments of the kinetic equations, using conservation of total excitation energies in collisions; in linear order

$$\sum_{\mathbf{p}}\varepsilon_p\Big(\frac{\partial}{\partial t} + \frac{\mathbf{p}}{m}\cdot\nabla_r\Big)n_{\mathbf{p}} + \sum_{\mathbf{q}}\omega_q\Big(\frac{\partial}{\partial t} + s\hat{\mathbf{q}}\cdot\nabla_r\Big)f_{\mathbf{q}}$$
$$= \Big[\Big(\varepsilon_0 n_3 + \frac{5}{2}P_f\Big)\frac{\delta m}{m} + s^2\rho_{\text{ph}}\Big]\nabla\cdot\mathbf{v}_s. \quad (2.4.64)$$

If we neglect phonon and ^{3}He thermal conductivity, which is valid in considering first sound propagation above $\sim$20 to 30 mK for ^{3}He concentrations $\gtrsim 1\%$, $\nabla_r n_{\mathbf{p}}$ and $\nabla_r f_{\mathbf{q}}$ can be replaced by their local equilibrium values, and (2.4.64) becomes

$$\sum_{\mathbf{p}}\varepsilon_p\frac{\partial n_{\mathbf{p}}}{\partial t} + \sum_{\mathbf{q}}\omega_q\frac{\partial f_{\mathbf{q}}}{\partial t} + \Big(\varepsilon_0 n_3 + \frac{5}{2}P_f\Big)\nabla\cdot\mathbf{v}_3 + s^2\rho_{\text{ph}}\nabla\cdot\mathbf{v}_{\text{ph}} = 0. \quad (2.4.65)$$

(b) ^{3}He Kinetic Equation in Relaxation Time Approximation. In order to use the conservation laws to describe the propagation of sound it is necessary to solve the ^{3}He and phonon kinetic equations in the presence of a sound wave. We shall carry out the solution here for the ^{3}He kinetic equation assuming a relaxation time approximation (1.2.72) for ^{3}He–^{3}He collisions, and neglecting thermal phonons and Fermi-liquid effects. This is adequate to describe sound propagation below 0.3 K.

The primary problem is to determine $T_{ij} = \Sigma(p_i p_j/m)\,\delta n_{\mathbf{p}}$ in the presence of a sound wave. This can be done by solving the ^{3}He kinetic equation in the presence of an external potential $Ue^{i\mathbf{q}\cdot\mathbf{r}-i\omega t}$. In the relaxation time approximation this equation takes the form

$$(\omega - \mathbf{q}\cdot\mathbf{v}_{\mathbf{p}})\,\delta n_{\mathbf{p}} + U\mathbf{q}\cdot\mathbf{v}_{\mathbf{p}}\frac{\partial n_p^0}{\partial \varepsilon_p} = -\frac{i}{\tau}\,\delta \bar{n}_{\mathbf{p}}^{\text{l.e.}}, \tag{2.4.66}$$

where $\delta\bar{n}^{\text{l.e.}}$, defined by (1.2.61), is given by

$$\delta\bar{n}_{\mathbf{p}}^{\text{l.e.}} = \delta n_{\mathbf{p}} - \delta n_{\mathbf{p}}^0 + m\mathbf{v}_3\cdot\nabla_p n_p^0\,; \tag{2.4.67}$$

$n_p^0 + \delta n_p^0$ is a local equilibrium distribution spherically symmetric about the origin, with the same quasiparticle number and energy density as $n_{\mathbf{p}}$, and $\mathbf{v}_3$ is the local ^{3}He velocity.

From (2.4.66) and (2.4.67) we readily find

$$\delta n_{\mathbf{p}} = \frac{\delta n_{\mathbf{p}}^0 - (m\mathbf{v}_3 - i\tau U\mathbf{q})\cdot\mathbf{v}_{\mathbf{p}}\partial n_p^0/\partial\varepsilon_p}{1 - i\tau\omega + i\tau\mathbf{q}\cdot\mathbf{v}_{\mathbf{p}}}, \tag{2.4.68}$$

together with the particle and momentum conservation laws

$$\omega\delta n_3 = n_3\mathbf{q}\cdot\mathbf{v}_3 \tag{2.4.69}$$

$$\omega m n_3\mathbf{v}_3 - \sum_{\mathbf{p}}(\mathbf{q}\cdot\mathbf{v}_{\mathbf{p}})\,\mathbf{p}\delta n_{\mathbf{p}} = n_3\mathbf{q}U, \tag{2.4.70}$$

where δn_3 is the variation in the ^{3}He density. The first sound velocity is large compared with typical ^{3}He velocities, and thus we are interested in $\omega \gg qv_p$, where v_p is a typical quasiparticle velocity. Taking moments of (2.4.68) with respect to $\mathbf{v}_{\mathbf{p}}^2$ and $(\mathbf{v}_{\mathbf{p}}\cdot\hat{\mathbf{q}})^2$, expanding in qv_p/ω, keeping terms of order $\mathbf{q}^0\delta n_p^0$, $\mathbf{q}\cdot\mathbf{v}_3$ and q^2U, and using the conservation laws (2.4.69) and (2.4.70), we discover

$$(1 - i\tau\omega)\sum_{\mathbf{p}}\mathbf{v}_{\mathbf{p}}^2\delta n_{\mathbf{p}} = \sum_{\mathbf{p}}\mathbf{v}_{\mathbf{p}}^2\delta n_p^0 - \frac{5i\tau\omega}{mn_3}P_f\delta n_3 \tag{2.4.71}$$

and

$$(1 - i\tau\omega)\sum_{\mathbf{p}}(\mathbf{v}_{\mathbf{p}}\cdot\hat{\mathbf{q}})^2\delta n_{\mathbf{p}} = \frac{1}{3}\sum_{\mathbf{p}}\mathbf{v}_{\mathbf{p}}^2\delta n_p^0 - \frac{3i\tau\omega}{mn_3}P_f\delta n_3\,. \tag{2.4.72}$$

Since

$$\sum \mathbf{v}_\mathbf{p}^2 (\delta n_\mathbf{p}^0 - \delta n_\mathbf{p}) = 0\,, \tag{2.4.73}$$

we find

$$\sum \mathbf{v}_\mathbf{p}^2 \delta n_\mathbf{p} = \frac{5P_f}{mn_3}\,\delta n_3 \tag{2.4.74}$$

$$\begin{aligned}\sum (\mathbf{v}_\mathbf{p}\cdot\hat{\mathbf{q}})^2 \delta n_\mathbf{p} &= \frac{P_f}{3m}\left(\frac{5\delta n_3}{n_3} - \frac{4i\tau}{1 - i\tau\omega}\,\mathbf{q}\cdot\mathbf{v}_3\right)\\ &= \left(1 - \frac{4}{5}\,\frac{i\tau\omega}{1 - i\tau\omega}\right)\frac{5}{3}\,\frac{P_f}{mn_3}\,\delta n_3\,.\end{aligned} \tag{2.4.75}$$

Equation 2.4.74 implies that for $\omega \gg qv_p$ the variation of the quasiparticle energy $\Sigma\varepsilon_p\delta n_\mathbf{p}$ is given by the adiabatic result

$$\sum \varepsilon_p \delta n_\mathbf{p} = \left(\varepsilon_0 + \frac{5}{2}\,\frac{P_f}{n_3}\right)\delta n_3\,; \tag{2.4.76}$$

the coefficient $\varepsilon_0 + 5p_f/2n_3$ is the derivative of the quasiparticle energy $\Sigma\varepsilon_p n_p^0$ with respect to n_3 at fixed quasiparticle entropy density s_3. From (2.4.75) we find that the variation in the stress tensor is given by

$$\sum \frac{p_i p_j}{m}\,\delta n_\mathbf{p} = \left(\frac{\partial P_f}{\partial n_3}\right)_{s_3}\delta n_3 \delta_{ij} - \frac{\tau P_f}{1 - i\tau\omega}\left(\nabla_i v_{3j} + \nabla_j v_{3i} - \frac{2}{3}\,\delta_{ij}\boldsymbol{\nabla}\cdot\mathbf{v}_3\right). \tag{2.4.77}$$

Equation 2.4.77 is essentially the result for T_{ij} in the hydrodynamic limit [cf. (1.2.123)] only with the first viscosity η replaced by $\tau P_f/\,(1 - i\tau\omega)$. The relaxation time τ can thus be identified with that appearing in the first viscosity:

$$\tau = \frac{\eta}{P_f}\,. \tag{2.4.78}$$

One may verify by application of the conservation laws that the ^{3}He density-density correlation function $\chi_{33}(q, \omega)$ (1.3.106) is given in the limit $\omega \gg qv_p$ by*

$$\chi_{33}(q, \omega) = \frac{n_3 q^2}{m\omega^2 - (5P_f/3n)_3 q^2\left(1 - \dfrac{4}{5}\,\dfrac{i\tau\omega}{1 - i\tau\omega}\right)}\,. \tag{2.4.79}$$

It is also straightforward to verify that in the presence of a sound wave, which

*This result is only valid for frequencies sufficiently small that the ^{3}He carry their screening clouds δm with them. At very high frequencies χ_{33} must approach $n_3q^2/m_3\omega^2$. (Cf. Chapter 1, Section 1.3.3.)

couples to the ^{3}He via (2.4.20), equations (2.4.76) and (2.4.77) still correctly describe the linear response of the ^{3}He when $\omega \gg qv_p$.

(c) Dispersion Relation for First Sound. We may now use the conservation laws to find the relation of ω to q in a first sound wave. The first step is to Fourier transform the conservation laws (2.4.53), (2.4.54), (2.4.60), and (2.4.62), neglecting the terms due to thermal phonons. Substituting **g** in (2.4.54) from (2.4.59) and using (2.4.51) and (2.4.53) to eliminate $\mathbf{v}_s$ and $\mathbf{v}_3$ we find

$$m_4\omega^2\delta n_4 - \frac{\rho_s q^2}{m_4}\,\delta\mu_4 = \delta m\omega^2\delta n_3 \tag{2.4.80}$$

where δn_3, δn_4, and $\delta\mu_4$ are the local variations in the ^{3}He density and ^{4}He density and chemical potential caused by the sound wave. The variation of μ_4 can be written with the aid of (2.1.7) as

$$\begin{aligned}\delta\mu_4 &= \delta\,\frac{\partial}{\partial n_4}\,E(\{n_\mathbf{p}\}, n_4) = \left(\frac{\partial^2 E}{\partial n_4^2}\right)_{n_\mathbf{p}}\delta n_4 + \sum_\mathbf{p}\frac{\partial^2 E}{\partial n_4\partial n_\mathbf{p}}\,\delta n_\mathbf{p}\\ &= \left(\frac{\partial\mu_4}{\partial n_4}\right)_{s_3, n_3}\delta n_4 + \sum_\mathbf{p}\frac{\partial\varepsilon_p}{\partial n_4}\,\delta n_\mathbf{p};\end{aligned} \tag{2.4.81}$$

the n_4 derivatives are carried out with the ^{3}He distribution function, or equivalently the ^{3}He number density and entropy density s_3 held fixed. We evaluate the sum using (2.4.74) for the term $\sim p^2\delta n_\mathbf{p}$, and find

$$\sum_\mathbf{p}\frac{\partial\varepsilon_p}{\partial n_4}\,\delta n_\mathbf{p} = \left(\frac{\partial\varepsilon_0}{\partial n_4} - \frac{5}{2}\,\frac{P_f}{n_3}\,\frac{w}{n_4}\right)\delta n_3 = \left(\frac{\partial\mu_3}{\partial n_4}\right)_{s_3, n_3}\delta n_3\,, \tag{2.4.82}$$

where

$$w = \frac{n_4}{m}\,\frac{\partial m}{\partial n_4}. \tag{2.4.83}$$

The derivative $(\partial\mu_3/\partial n_4)_{s_3, n_3}$ acts as an effective temperature dependent coupling of first sound to the ^{3}He density. Substituting (2.4.81) and (2.4.82) into (2.4.80) we have

$$\left[m_4\omega^2 - \frac{\rho_s}{m_4}\left(\frac{\partial\mu_4}{\partial n_4}\right)_{s_3, n_3} q^2\right]\delta n_4 = \left[\delta m\omega^2 + \frac{\rho_s}{m_4}\left(\frac{\partial\mu_3}{\partial n_4}\right)_{s_3, n_3} q^2\right]\delta n_3. \tag{2.4.84}$$

To lowest order in x, ρ_s on the right side may be replaced by $m_4 n_4$.

It remains to compute δn_3 in terms of δn_4. We substitute (2.4.59) into the momentum conservation law (2.4.62), use (2.4.77) for T_{ij}, eliminate $\mathbf{v}_s$ and $\mathbf{v}_3$ in terms of $\delta\mu_4$ and δn_3, and find

$$\delta n_3 = \chi_{33}(q, \omega)\left[\frac{\delta m}{m_4}\,\delta\mu_4 + \left(\frac{\partial\mu_3}{\partial n_4}\right)_{s_3, n_3}\delta n_4\right] \tag{2.4.85}$$

with χ_{33} given by (2.4.79). For small concentrations $\delta\mu_4$ here may be replaced by $(m_4 s^2/n_4)\delta n_4$. Then from (2.4.84) and (2.4.85) we find the dispersion relation

$$\omega^2 = \frac{q^2}{m_4}\left[\frac{\rho_s}{m_4}\left(\frac{\partial\mu_4}{\partial n_4}\right)_{s_3,n_3} + \chi_{33}(q,\omega)\left(\frac{\delta m s^2}{n_4} + \left(\frac{\partial\mu_3}{\partial n_4}\right)_{s_3,n_3}\right)\left(\delta m\frac{\omega^2}{q^2} + n_4\left(\frac{\partial\mu_3}{\partial n_4}\right)_{s_3,n_3}\right)\right]. \quad (2.4.86)$$

As $n_3 \to 0$, this reduces to the dispersion relation $\omega^2 = s^2 q^2$.

(d) Sound Attenuation. The amplitude attenuation α_1 of first sound is defined by writing the complex solution of (2.4.86) for real ω in the form

$$q(\omega) = \frac{\omega}{s} + i\alpha_1. \quad (2.4.87)$$

Then to leading order in the ^{3}He concentration we find from (2.4.86)

$$\alpha_1 = -\frac{\omega}{2m_4 s^3}\left(\delta m s^2 + \frac{\partial\mu_4}{\partial n_4}\right)^2 \operatorname{Im}\chi_{33}(q = \omega/s, \omega). \quad (2.4.88)$$

We may replace $\partial\mu_3/\partial n_4$ to lowest order by $\partial\varepsilon_0/\partial n_4$, and expand Im χ to first order in P_f, with the result

$$\alpha_1 = \frac{2}{3}\frac{P_f(n_3,T)}{n_4 m s^2}\frac{m_4}{ms}\left(1 + \alpha + \frac{\delta m}{m_4}\right)^2 \frac{\omega^2\tau}{1 + \omega^2\tau^2}. \quad (2.4.89)$$

The scattering time τ is that for viscosity, and is expected to have the form at low temperatures

$$\tau = \frac{A}{T^2}\left[1 + B\left(\frac{T}{T_f}\right)^2\right], \quad (2.4.90)$$

where T_f is the ^{3}He Fermi temperature. (According to the discussion in Section 1.4.3(b) the term in $\tau \propto T^{-1}$ is small.) Baym and Ebner [65] determined A and B empirically for an $x = 0.050$ solution by requiring α_1 to have a maximum at the observed temperatures [66] $T \simeq 0.055$ K for $\omega/2\pi = 20$ MHz, and $T \simeq 0.121$ K for $\omega/2\pi = 60$ MHz. This yields the values for $x = 0.050$, $A \simeq 20 \times 10^{-12}$ sec K^2, $B \simeq 2$, while a theoretical calculation of A from the BBP effective ^{3}He–^{3}He interaction (normalized to fit the spin diffusion data) gives $A = 17.8 \times 10^{-12}$ sec K^2. A more complete estimate of τ up to temperatures $T \gg T_f$ is given in Reference [66]. The comparison of (2.4.89) with data of Abraham et al. [67] on a 5% solution at $P = 0$ is shown in Figure 2.5. The extension of the theory of sound attenuation in dilute solutions to include attenuation by thermal phonons (whose mean free path is limited by ^{3}He scattering) is given in Reference [67]; the results

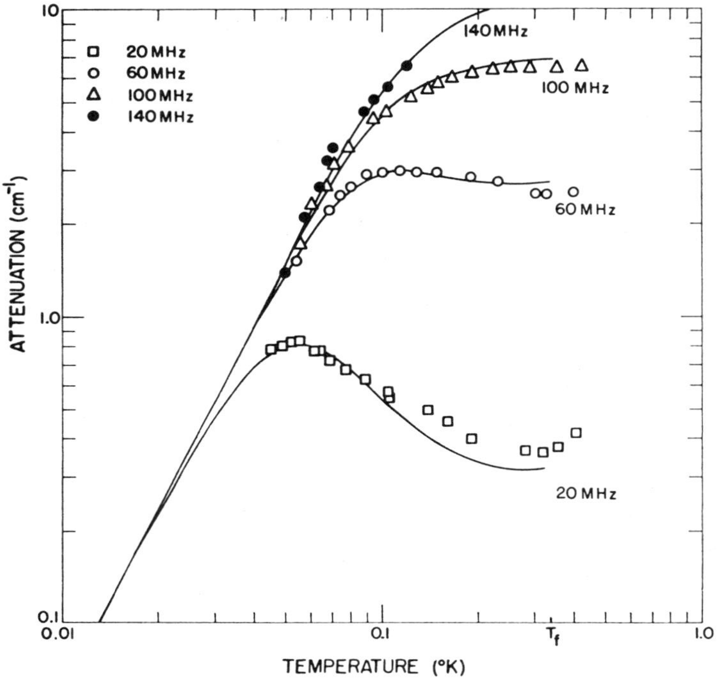

FIGURE 2.5. Calculated attenuation of first sound a 5% solution of ^{3}He in ^{4}He. The solid lines are the calculated attenuation at 20, 60, 100, and 140 MHz/sec. The data points are those of Abraham et al. [66].

are in very good agreement with experiment. The theory of first sound attenuation is further discussed in References [68–70].

(e) Temperature Dependence of the Sound Velocity. From the dispersion relation (2.4.86), together with Re χ_{33} evaluated from (2.4.79) to first order in P_f, we find that the sound velocity has the form

$$s(n_3, n_4, T) = s(n_4) + \delta s(n_3, n_4, T), \tag{2.4.91}$$

where $s(n_4)$ denotes the sound velocity in pure ^{4}He at density n_4 and

$$\delta s = \frac{1}{2m_4 s}\left\{\delta\left[\frac{\rho_s}{m_4}\left(\frac{\partial\mu_4}{\partial n_4}\right)_{s_3, n_3}\right] + \frac{n_3 n_4}{ms^2}\left(\frac{\delta ms^2}{n_4} + \left(\frac{\partial\mu_3}{\partial n_4}\right)_{s_3, n_3}\right)^2\left[1 + \frac{P_f}{n_3 ms^2}\left(3 - \frac{4/3}{1 + \omega^2\tau^2}\right)\right]\right\} \tag{2.4.92}$$

is the small shift in the sound velocity. The first term, the variation of $(\rho_s/m_4)(\partial\mu_4/\partial n_4)$ from its value in pure ^{4}He, is evaluated by writing

$$\left(\frac{\partial \mu_4}{\partial n_4}\right)_{s_3, n_3} = \frac{\partial^2}{\partial n_4^2}\left[E_0(n_4) + \sum_{\mathbf{p}} \varepsilon_p n_p\right] = \frac{m_4 s^2}{n_4} + \sum_{\mathbf{p}} \frac{\partial^3 \varepsilon_p}{\partial n_4^2} n_p$$

$$= \frac{m_4 s^2}{n_4} + \frac{\partial^2 \varepsilon_0}{\partial n_4^2} n_3 + \frac{3}{2} P_f m \frac{\partial^2 m^{-1}}{\partial n_4^2}, \qquad (2.4.93)$$

and using $\rho_s = m_4 n_4 - \delta m n_3$. To compare (2.4.92) with experiments carried out at fixed pressure, it is necessary to include the modification of the ^{4}He density in $s(n_4)$ due to the presence of the ^{3}He. This can be done by using (2.3.26) and (2.3.29) to write

$$s(n_4) = s_0 + \frac{\partial s}{\partial n_4}\delta n_4$$

$$= s_0\left[1 - u\left((1 + \alpha_0)x + \frac{P_f}{m_4 s^2 n_4}\left(1 - \frac{3}{2} w\right)\right)\right], \qquad (2.4.94)$$

where α_0 and s_0 refer to zero ^{3}He concentration values in pure ^{4}He at the given pressure, and $u = (n_4/s)\partial s/\partial n_4$. The temperature dependence of P_f in (2.4.94) includes effects of thermal expansion on s. The net result for the shift in the sound velocity, at fixed pressure, is to first order in x and P_f:

$$\frac{\delta s(x, T, \omega)}{s_0} = \frac{x}{2}\left[n_4 \frac{\partial \alpha_0}{\partial n_4} + \lambda(\lambda - 1)\frac{m}{m_4}\right]$$
$$+ \frac{P_f}{2 m_4 s^2 n_4}\left[3(w - \lambda)^2 + w\lambda + (3w - 2)u\right.$$
$$\left. - \frac{3}{2}\frac{n_4^2}{m}\frac{\partial^2 m}{\partial n_4^2} - \frac{4\lambda^2/3}{1 + \omega^2\tau^2}\right]; \qquad (2.4.95)$$

in deriving (2.4.95) we have used (2.4.82), and written

$$\lambda = 1 + \alpha_0 + \frac{\delta m}{m_4}.$$

The temperature dependent part of this result is compared with the experiments of Abraham et al. [71] on a 5.5% solution at $P = 0$ in Figure 2.6. Note that s decreases with temperature and increases with frequency. In the curves in this figure the values $w = 1.1$, $(n_4^2/m)\partial^2 m/\partial n_4^2 = 3.2$ were used.

The first term $\sim x$ in (2.4.95) is that already found in (2.4.48) from the point of view of quantum mechanical perturbation theory. In fact, the complete expressions (2.4.89) for the attenuation and (2.4.95) for the temperature dependence of s can be equivalently derived in this manner; see References [65] and [72] for details. Measurements of the low temperature shift in s with ^{3}He concentration provide a determination of $n_4\partial\alpha/\partial n_4$ at zero concentration: Abraham et al. [62] find

$$n_4 \frac{\partial \alpha}{\partial n_4} = -1.03. \qquad (2.4.96)$$

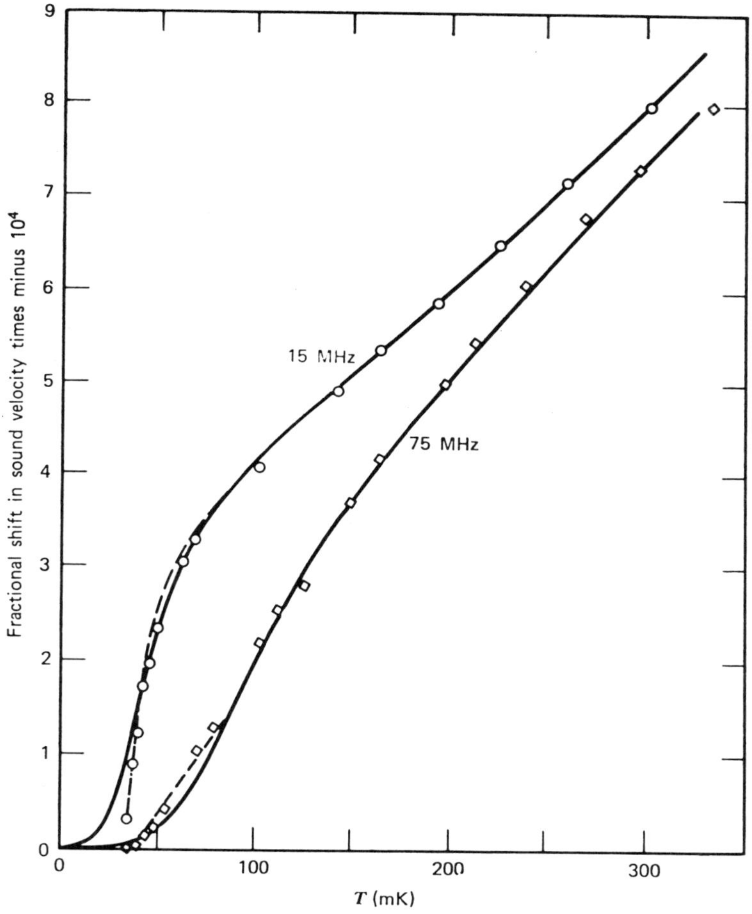

FIGURE 2.6. Fractional shift of the sound velocity of a 5.5% solution as a function of temperature at two frequencies. The data points are from Reference [71].

2.4.5. Second Sound

Second sound in dilute solutions is a hydrodynamic mode in which there is a counter-oscillation of the normal and superfluid mass densities. The normal mass density is the sum of that due to ^{4}He phonons (2.4.85) and rotons, plus that due to the ^{3}He, ρ_{n3} (2.4.15). Unlike in pure ^{4}He, the normal mass density remains finite at $T = 0$.

The detailed calculation of first sound propagation illustrates the general methods by which one can derive the hydrodynamic properties of dilute solutions. In the case of first sound, one is interested in situations where the frequency of the sound varies from values much less than the collision frequency (the hydrodynamic regime) to values much greater than the collision frequency (the collisionless regime). On the other hand, second sound exists only in the hydrodynamic regime, and its properties, which may be determined quite simply from the hydrodynamic equations, involve only thermodynamic derivatives, which may be found without having to solve the transport equations for the excitations. The basic theory of second sound was derived from the hydrodynamic equations by Khalatnikov [20]. Guyer et al. [73] and Throop and van Leeuwen [74] determined the effects of ^{3}He–^{3}He interactions on the second sound velocity. Additional aspects of second sound in dilute solutions are discussed in References [68–70], [75], and [76]. Brubaker et al. [44] have done extensive measurements of the second sound velocity, as a function of ^{3}He concentration, temperature, and pressure, and from these measurements have deduced the ^{3}He inertial mass $m_n^* = \rho_{n3}/n_3$. See Reference [77] for further measurements of ρ_{n3} by the oscillating disk method.

Fourth sound, in which the normal mass density remains stationary, is discussed in References [70] and [78].

2.4.6. Transport Properties

(a) Phonon Kinetic Equation. The presence of ^{3}He impurities in superfluid ^{4}He changes the transport properties of the superfluid in two ways. First, the ^{3}He quasiparticles, carrying energy, momentum, and spin, provide a new mechanism for transport. At extremely low temperature, where the thermal excitations of the ^{4}He are negligible, the contributions of the ^{3}He dominate the transport coefficients. Second, transport by phonons and rotons is limited by their interactions with the ^{3}He, either by scattering or by absorption or emission by the ^{3}He. In the temperature range up to $\sim$ 0.6 K for a few percent ^{3}He concentration, where rotons can be neglected, phonon mean free paths are determined entirely by the ^{3}He-phonon interaction (except at very low temperatures where boundary scattering becomes important [65]).

The phonon transport properties are determined by the phonon kinetic equation (2.4.50). The phonon-^{3}He collision term is a sum of a scattering term, I_{scatt}, which can be determined from the phonon-^{3}He scattering theory developed in Section 2.4.3, and an absorption term, I_{abs}, which can be determined from the theory of sound attenuation in Section 2.4.4(d). Because ^{3}He–^{3}He scattering rates are so relatively rapid, these collision terms may be evaluated under the assumption that the ^{3}He quasiparticles maintain local equilibrium among themselves. Then in the frame in which $\mathbf{v}_s = 0$, denoted by a bar,

$$\bar{I}_{\text{scatt}} = -2\sum_{\mathbf{pp'q}} 2\pi\delta(\bar{\varepsilon}_{\mathbf{p}} + \bar{\omega}_{\mathbf{q}} - \bar{\varepsilon}_{\mathbf{p}'} - \bar{\omega}_{\mathbf{q}'})\left|\langle \mathbf{p'q'}|T_{\text{qp-ph}}|\mathbf{pq}\rangle\right|^2 \times [\bar{f}_{\mathbf{q}}(1+\bar{f}_{\mathbf{q}'})\bar{n}_{\mathbf{p}}(1-\bar{n}_{\mathbf{p}'}) - (1+\bar{f}_{\mathbf{q}})\bar{f}_{\mathbf{q}'}(1-\bar{n}_{\mathbf{p}})\bar{n}_{\mathbf{p}'}], \quad (2.4.97)$$

where $\langle|T_{\text{qp-ph}}|\rangle$ is the ^{3}He phonon scattering amplitude; the factor of 2 is the spin sum.

To compute phonon transport properties as well as sound attention one is interested only in small deviations from global equilibrium with $\mathbf{v}_s = 0$. Linearizing (2.4.97). and using the fact that to lowest order ^{3}He-phonon scattering is elastic ($q = q'$, $p = p'$), we have

$$\bar{I}_{\text{scatt}} = -2\sum_{\mathbf{pp'q}} 2\pi\delta(\varepsilon_p + \omega_q - \varepsilon_{p'} - \omega_{q'})\left|\langle \mathbf{p'q'}|T_{\text{qp-ph}}|\mathbf{pq}\rangle\right|^2 \times \{n_p^0(1-n_p^0)[\delta\bar{f}_{\mathbf{q}} - \delta\bar{f}_{\mathbf{q}'}] + f_q^0(1+f_q^0)[\delta\bar{n}_{\mathbf{p}} - \delta\bar{n}_{\mathbf{p}'}]\}. \quad (2.4.98)$$

The transformation of (2.4.98) back to the (unbarred) lab frame is trivial. Since the phonon momentum $\mathbf{q}$ is independent of the frame in which it is measured, we have $\bar{f}_{\mathbf{q}} = f_{\mathbf{q}}$; furthermore, the rate of change of this distribution is the same in both frames so that

$$I_{\text{scatt}} = \bar{I}_{\text{scatt}}. \quad (2.4.99)$$

In addition, $\bar{n}_{\mathbf{p}} = n_{\mathbf{p}+m_3\mathbf{v}_3}$ [cf. (2.4.2)] so that in linear order for the ^{3}He in local equilibrium

$$\delta\bar{n}_{\mathbf{p}} = \mathbf{p}\cdot(\mathbf{v}_s - \mathbf{v}_3)\frac{\partial n_p^0}{\partial\varepsilon_p}, \quad (2.4.100)$$

where $\mathbf{v}_3$ is the ^{3}He flow velocity. Thus

$$I_{\text{scatt}} = -2\pi\sum_{\mathbf{pp'q'}} 2\pi\delta(\varepsilon_p + \omega_q - \varepsilon_{p'} - \omega_{q'})\left|\langle \mathbf{p'q'}|T_{\text{qh-ph}}|\mathbf{pq}\rangle\right|^2 \times n_p^0(1-n_p^0)\left[\delta f_{\mathbf{q}} - \delta f_{\mathbf{q}'} - \frac{\partial f_q^0}{\partial\omega_q}(\mathbf{q}-\mathbf{q}')\cdot(\mathbf{v}_s - \mathbf{v}_3)\right]. \quad (2.4.101)$$

To lowest order (2.4.42) for $T_{\text{qp-ph}}$ implies that the scattering contains only s, p, and d waves. The detailed further reduction of the collision term (2.4.101) into spherical harmonic components is given in Reference [79].

The rate at which phonons are absorbed by ^{3}He is given in terms of the amplitude attenuation α_1 by

$$\frac{1}{\tau_{\text{abs}}(q)} = 2s\alpha_1. \quad (2.4.102)$$

To evaluate α_1 for thermal phonons we use (2.4.89) with the ^{3}He–^{3}He scattering rate τ^{-1} replaced by the more general quantum result [80] $\tau^{-1}[1 + (\hbar\omega/2\pi\kappa T)^2]$; thus

$$\frac{1}{\tau_{\rm abs}(q)} = \frac{4}{3}\frac{m_4 P_f(T)}{m^2 n_4}\frac{(1+\alpha+\delta m/m_4)^2 q^2 \tau[1+(\hbar s q/2\pi\kappa T)^2]}{[1+(\hbar s q/2\pi\kappa T)^2]^2+(sq\tau)^2}. \quad (2.4.103)$$

When the ^{3}He are in local equilibrium with temperature $T_3(\mathbf{r}, t)$ and velocity $\mathbf{v}_3(\mathbf{r}, t)$, the absorption process causes the phonon distribution to relax to local temperature T_3 and velocity $\mathbf{v}_3$. The net rate of absorption $I_{\rm abs}(q)$—the difference of the absorption and emission rates—then becomes

$$I_{\rm abs} = -\frac{1}{\tau_{\rm abs}}\left\{\delta f_{\mathbf{q}} - \frac{\partial f_q^0}{\partial \omega_q}\left[qT\delta\left(\frac{s}{T_3}\right) + \mathbf{q}\cdot(\mathbf{v}_s - \mathbf{v}_3)\right]\right\}, \quad (2.4.104)$$

where δs and δT_3 are the first order deviations from global equilibrium. The phonon kinetic equation with (2.4.101) and (2.4.104) is the starting point for calculating the phonon contributions to transport in dilute solutions. The application to sound attenuation is described in Reference [67], to viscosity in Reference [79], and to thermal conductivity in Reference [65]. We now summarize the main results for the viscosity and thermal conductivity.

(b) First Viscosity. The first viscosity is the sum of a ^{3}He contribution η_3 and a phonon and roton contribution η_4. The latter contribution is very unlike that in pure ^{4}He because of scattering of ^{4}He excitations by the ^{3}He quasiparticles.

For temperatures below ~ 0.1 K for $x \gtrsim 1\%$, η_4 is completely negligible (except for boundary scattering effects) and $\eta \simeq \eta_3$. For $T \ll T_f$, the ^{3}He Fermi temperature, η_3 is $\propto T^{-2}$, and is given exactly in terms of the ^{3}He–^{3}He scattering amplitude on the Fermi surface by (1.2.129). At higher temperatures one can compute η_3 in the relaxation time approximation [cf. (2.4.77)] as

$$\eta_3 = P_f(T)\tau. \quad (2.4.105)$$

The phonon viscosity $\eta_{\rm ph}$ which begins to become important above ~ 0.1 K may be computed by solving the phonon kinetic equation, and one finds [79]

$$\eta_{\rm ph} = -\sum_{\mathbf{q}} \frac{s^2 q^2}{15}\frac{\partial f_q^0}{\partial \omega_q}\tau_v(q), \quad (2.4.106)$$

where

$$\tau_v^{-1}(q) = \tau_{\rm abs}^{-1}(q) + \tau_2^{-1}(q) \quad (2.4.107)$$

and τ_2^{-1} is the phonon-^{3}He scattering rate for $\delta f_{\mathbf{q}} \propto P_2(\cos\theta_{\mathbf{q}})$, computed from (2.4.101). Note that $\eta_{\rm ph}$ diverges if the absorption term is neglected.

There has been some discrepancy between measurements [81, 82] and theory of the viscosity, but the recent experiments of Kuenhold, Crum, and Sarwinski [83] are in good agreement with the theory described here. Kuenhold and Ebner have used these viscosity measurements, together with the

^{3}He thermal conductivity and spin diffusion measurements, to construct an improved ^{3}He–^{3}He effective interaction, valid for higher q as well [84, 85].

The viscosity in the high temperature regime, including roton effects as well as phonon–phonon scattering has been calculated by Zharkov [86]. The theory of the three second-viscosity coefficients in dilute solutions has been discussed by Yaniv and Disatnik [68].

(c) Thermal Conductivity. The structure of the thermal conductivity, a sum of the ^{3}He plus phonon and roton thermal conductivities, is quite similar to that of the first viscosity. For $T \lesssim 0.1$ K, ^{3}He thermal conductivity ($\propto T^{-1}$ for $T \ll T_f$) is dominant, except at very low T where boundary scattering is important. The phonon thermal conductivity is computed analogously to the phonon viscosity; one finds [65]

$$K_{\text{ph}} = -\sum_{\mathbf{q}} \frac{s^2 q}{3} \frac{\partial f_q^0}{\partial T} \tau_{\text{K}}(q), \tag{2.4.108}$$

where

$$\tau_{\text{K}}^{-1}(q) = \tau_{\text{abs}}^{-1}(q) + \tau_{\text{I}}^{-1}(q), \tag{2.4.109}$$

and τ_{I}^{-1} is the phonon-^{3}He scattering rate for $\delta f_{\mathbf{q}} \propto P_1(\cos\theta_{\mathbf{q}})$.

Measurements of the thermal conductivity at $P = 0$ were done by Abel et al. [87]. More recently Rosenbaum et al. [88] have studied the thermal conductivity over a range of pressures; they also discuss the boundary scattering problem. The experiments are in qualitative agreement with theory; however, it appears that (2.4.42), with parameters taken from experiment, underestimates the scattering amplitude by possibly as much as 40%. The reason for this discrepancy is not understood at present.* The discrepancy is not as apparent in the viscosity measurements, since below 0.6 K the phonon viscosity is small compared with the ^{3}He viscosity. The theory of thermal conductivity at high temperatures is discussed by Khalatnikov and Zharkov [89].

REFERENCES

1. D. O. Edwards, D. F. Brewer, P. Seligmann, M. Skertic, and M. Yaqub, *Phys. Rev. Lett.* **15,** 773 (1965).
2. C. Ebner and D. O. Edwards, *Phys. Reports* **2C,** 77 (1970).
3. A. T. Berestov, A. V. Voronel' and M. Sh. Giterman, *Zh. Eksp. Teor. Fiz. Pis. Red.* **15,** 273 (1972) [English transl.: *Sov. Phys.—JETP* **15,** 190 (1972)].

*Berker [90] has examined the corrections to (2.4.108) that result if one does not approximate the ^{3}He-phonon scattering as elastic. These corrections reduce the theoretical thermal conductivity by ~10 to 20%.

4. G. Ahlers, in *The Physics of Liquid and Solid Helium*, Part 1, edited by K. H. Bennemann and J. B. Ketterson (Wiley, New York, 1976), p. 85.
5. J. M. J. van Leeuwen and E. G. D. Cohen, in *Proceedings of the Eighth International Conference on Low Temperature Physics*, London 1962, edited by R. O. Davies (Butterworth, London, 1963), p. 43.
6. J. Bardeen, G. Baym, and D. Pines, *Phys. Rev. Lett.* **17**, 372 (1966); *Phys. Rev.* **156**, 207 (1967).
7. J. Landau, J. T. Tough, N. R. Brubaker, and D. O. Edwards, *Phys. Rev. A* **2**, 2472 (1970).
8. M. B. Hoffberg, *Phys. Rev. A* **5**, 1963 (1972).
9. W. Hsu and D. Pines, *J. Stat. Phys.* **38**, 273 (1985).
10. H. M. Guo, D. O. Edwards, R. E. Sarwinski, and J. T. Tough, *Phys. Rev. Lett.* **27**, 1259 (1971); W. F. Saam, *Phys. Rev. A* **5**, 335 (1972).
11. C. J. Palin, W. F. Vinen, E. R. Pike, and J. M. Vaughn, *J. Phys. C* **4**, L225 (1971).
12. B. N. Ganguly and A. Griffin, *Can. J. Phys.* **46**, 1895 (1968).
13. R. L. Woerner, D. A. Rockwell, and T. J. Greytak, *Phys. Rev. Lett.* **30**, 1114 (1973).
14. C. M. Surko and R. E. Slusher, *Phys. Rev. Lett.* **30**, 1111 (1973).
15. J. M. Rowe, D. L. Price, and G. E. Ostrowski, *Phys. Rev. Lett.* **31**, 510 (1973).
16. W. F. Saam and J. P. Laheurte, *Phys. Rev. A* **4**, 1170 (1971).
17. H. Dandache, J. P. Laheurte, and M. Papoular, *J. de Phys.* **34**, 643 (1973).
18. H. Dandache and J. P. Laheurte, *J. de Phys.* **34**, 635 (1973).
19. J. P. Laheurte, *J. Low Temp. Phys.* **12**, 127 (1973).
20. I. M. Khalatnikov, *An Introduction to the Theory of Superfluidity* (Benjamin, New York, 1965).
21. J. Wilks, *The Properties of Liquid and Solid Helium* (Clarendon Press, Oxford, 1967).
22. W. E. Keller, *Helium-3 and Helium-4* (Plenum Press, New York, 1969).
23. K. W. Taconis and R. de Bruyn Ouboter, *Progress in Low Temperature Physics, Vol. IV*, edited by C. J. Gorter, (North-Holland Publishing, Amsterdam, 1970), p. 38.
24. J. C. Wheatley, *Am. J. Phys.* **36**, 181 (1968); *Progress in Low Temperature Physics, Vol. VI*, edited by C. J. Gorter, (North-Holland Publishing, Amsterdam, 1970), p. 77.
25. D. O. Edwards, in *Proceedings of the Eleventh International Conference on Low Temperature Physics*, St. Andrews, Scotland, 1968, edited by J. F. Allen, D. M. Finlayson, and D. M. McCall, (St. Andrews University, 1969), p. 352.
26. R. Radebaugh, National Bureau of Standards Report No. 362 (U.S. Government Printing Office, Washington, D. C., 1967).
27. L. D. Landau and I. Pomeranchuk, *Dokl. Akad. Nauk SSSR* **59**, 669 (1948); I. Pomeranchuk, *Zh. Eksp. Teor. Fiz.* **19**, 42 (1949).
28. P. Seligmann, D. O. Edwards, R. E. Sarwinski, and J. T. Tough, *Phys. Rev.* **181**, 415 (1969).
29. A. C. Anderson, W. R. Roach, R. E. Sarwinski, and J. C. Wheatley, *Phys. Rev. Lett.* **16**, 263 (1966).
30. M. J. Stephen and L. Mittag, *Phys. Rev. Lett.* **31**, 923 (1973).
31. C. M. Varma, *Phys. Lett.* **45A**, 301 (1973).
32. C. Ebner, *Phys. Rev. A* **3**, 1201 (1971).

33. G. E. Watson, J. D. Reppy, and R. C. Richardson, *Phys. Rev.* **188**, 384 (1969); B. M. Abraham, O. G. Brandt, and Y. Eckstein, in *Proceedings of the Twelfth International Conference on Low Temperature Physics*, Kyoto, Japan, 1970, edited by E. Kanda (Keigaku, Tokyo, 1971); C. Boghosian and H. Meyer, *Phys. Lett.* **25A**, 352 (1967).

34. B. M. Abraham, O. G. Brandt, Y. Eckstein, J. Munarin, and G. Baym, *Phys. Rev.* **188**, 309 (1969).

35. G. Baym, *Phys. Rev. Lett.* **17**, 952 (1966).

36. W. L. McMillan, *Phys. Rev.* **138**, A442 (1965).

37. R. P. Feynman, *Phys. Rev.* **94**, 362 (1954).

38. W. E. Massey and C. W. Woo, *Phys. Rev. Lett.* **19**, 301 (1967).

39. C. W. Woo, in *The Physics of Liquid and Solid Helium*, Part 1, edited by K. H. Bennemann and J. B. Ketterson (Wiley, New York, 1976), p. 349.

40. T. B. Davison and E. Feenberg, *Phys. Rev.* **178**, 306 (1969).

41. C. W. Woo, H. T. Tan, and W. E. Massey, *Phys. Rev. Lett.* **22**, 278 (1969); and *Phys. Rev.* **185**, 287 (1969).

42. R. P. Feynman and M. Cohen, *Phys. Rev.* **102**, 1189 (1956).

43. V. R. Pandharipande and N. Itoh, *Phys. Rev. A* **8**, 2564 (1973).

44. N. R. Brubaker, D. O. Edwards, R. E. Sarwinski, P. Seligmann, and R. A. Sherlock, *Phys. Rev. Lett.* **25**, 715 (1970); *J. Low Temp. Phys.* **3**, 619 (1970).

45. A. J. Leggett, *Phys. Rev. Lett.* **31**, 352 (1973).

46. V. J. Emery, *Phys. Rev.* **148**, 138 (1966); **161**, 194 (1967).

47. W. R. Abel, R. T. Johnson, J. C. Wheatley, and W. Zimmermann, Jr., *Phys. Rev. Lett.* **18**, 737 (1967).

48. G. Baym and C. Ebner, *Phys. Rev.* **170**, 346 (1968).

49. C. Ebner, *Phys. Rev.* **185**, 392 (1969).

50. L. J. Campbell, *Phys. Rev. Lett.* **19**, 156 (1967).

51. W. L. McMillan, *Phys. Rev.* **175**, 266 (1968); **182**, 299 (1969); also, C. S. Hsu, *Phys. Rev. A* **8**, 376 (1973).

52. E. Østgaard, *Phys. Rev. A* **1**, 1048 (1970).

53. W. F. Saam, *Ann. Phys. (N.Y.)* **53**, 219 (1969); H. Winter, *Phys. Lett.* **31A**, 489 (1970).

54. W. F. Saam, *Ann. Phys. (N.Y.)* **53**, 239 (1969).

55. J. Landau, J. T. Tough, N. R. Brubaker, and D. O. Edwards, *Phys. Rev. A* **2**, 2472 (1970).

56. S. G. Eckstein, Y. Eckstein, C. G. Kuper, and A. Ron, *Phys. Rev. Lett.* **25**, 97 (1970).

57. H. T. Tan and C. W. Woo, *Phys. Rev. Lett.* **30**, 365 (1973).

58. A. Bagchi and J. Ruvalds, *Phys. Rev. A* **8**, 1973 (1973).

59. G. A. Herzlinger and J. G. King, *Phys. Lett.* **40A**, 1 (1972).

60. I. M. Khalatnikov. *Zh. Eksp. Teor. Fiz. Pis. Red.* **5**, 288 (1967) [English transl.: *JETP Lett.* **5**, 235 (1967)]; *Zh. Eksp. Teor. Fiz.* **55**, 1919 (1968) [English transl.: *Sov. Phys.—JETP* **28**, 1014 (1969)].

61. B. M. Abraham, Y. Eckstein, J. B. Ketterson, M. Kuchnir, and P. R. Roach, *Phys. Rev. A* **1**, 250 (1970).

62. B. M. Abraham, Y. Eckstein, J. B. Ketterson, M. Kuchnir, and J. H. Vignos, *Phys. Rev.* **181**, 347 (1969).

63. L. D. Landau and E. M. Lifshitz, *Fluid Mechanics* (Addison-Wesley, Reading, Mass,

64. Y. Disatnik and H. Brucker, *J. Low Temp. Phys* **7,** 501 (1972); Y. Disatnik and A. Yaniv, *J. Low Temp. Phys.* **10,** 595 (1973).

65. G. Baym and C. Ebner, *Phys. Rev.* **164,** 235 (1967).

66. B. M. Abraham, Y. Eckstein, J. B. Ketterson, and J. Vignos, *Phys. Rev. Lett.* **17,** 1254 (1966).

67. G. Baym, W. F. Saam, and C. Ebner, *Phys. Rev.* **173,** 306 (1968).

68. A. Yaniv and Y. Disatnik, *J. Low Temp. Phys.* **10,** 793 (1973).

69. J. Seiden and A. Ghozlan, *C. R. Acad. Sci. B***269,** 160 (1969).

70. T. A. Karchava and D. G. Sanikidze, *Zh. Eksp. Teor. Fiz.* **57,** 1349 (1969) [English transl.: *Sov. Phys.*-JETP **30,** 731 (1970)].

71. B. M. Abraham, Y. Eckstein, J. B. Ketterson, and M. Kuchnir, *Phys. Rev. Lett.* **20,** 251 (1968).

72. G. Baym, in *Proceedings of the Eleventh International Conference on Low Temperature Physics, op. cit.,* Vol. I, p. 385.

73. R. A. Guyer, P. V. E. McClintock, and K. H. Mueller, *Phys. Lett.* **27A,** 611 (1968).

74. G. J. Throop and J. M. J. van Leeuwen, *Phys. Rev.* **188,** 468 (1969).

75. A. Ghozlan, *J. de Phys.* **33,** 415 (1972).

76. Z. M. Galasiewicz, *Phys. Lett.* **34A,** 7 (1971).

77. T. I. Sobolev and B. N. Esel'son, *Zh. Eksp. Teor. Fiz.* **60,** 240 (1971) [English transl.: *Sov. Phys.—JETP* **33,** 136 (1971)].

78. D. G. Sanikidze and A. N. Shaanova, *Zh. Eksp. Teor. Fiz. Pis. Red.* **10,** 482 (1969) [English transl.: *JETP Lett.* **10,** 310 (1969); N. E. Dyumin, B. N. Esel'son, E. Ya. Rudavskii, and I. A. Serbin, *Zh. Eksp. Teor. Fiz.* **56,** 747 (1969) [English transl.: *Sov. Phys.—JETP* **29,** 406 (1969)].

79. G. Baym and W. F. Saam, *Phys. Rev.* **171,** 172 (1968).

80. A. A. Abrikosov and I. M. Khalatnikov. *Rept. Progr. Phys.* **22,** 329 (1959).

81. M. P. Bertinat, D. S. Betts, D. F. Brewer, and G. J. Butterworth, *Phys. Rev. Lett.* **28,** 472 (1972).

82. R.W.H. Webeler and G. Allen, *Phys. Lett.* **29A,** 93 (1969).

83. K. A. Kuenhold, D. B. Crum, and R. E. Sarwinski, *Phys. Lett.* **41A,** 1 (1972).

84. K. A. Kuenhold and C. Ebner, *Phys. Rev. A***9,** 2724 (1974).

85. C. Ebner, *Phys. Rev.* **156,** 222 (1967).

86. V. N. Zharkov, *Zh. Eksp. Teor. Fiz.* **33,** 929 (1957) [English transl.: *Sov. Phys.—JETP* **6,** 714 (1958)].

87. W. R. Abel and J. Wheatley, *Phys. Rev. Lett* **21,** 1231 (1968).

88. R. L. Rosenbaum, J. Landau, and Y. Eckstein, *J. Low Temp. Phys.* **16,** 131 (1974).

89. I. M. Khalatnikov and V. N. Zharkov, *Zh. Eksp. Teor. Fiz.* **32,** 1108 (1957). [English transl.: *Sov. Phys.—JETP* **5,** 905 (1957)].

90. A. N. Berker (private communication).
91. H. H. Fu and C. J. Pethick, *Phys. Rev. B* **14,** 3837 (1976).
92. G. Baym, *J. Low Temp. Phys.* **18,** 335 (1975).

3

FURTHER DEVELOPMENTS

INTRODUCTION

The concepts of the Landau theory of Fermi liquids form a vital part of condensed matter physics. As further experiments and theoretical advances have deepened our understanding of the helium liquids, the techniques of the theory have found application in study of new systems. In this chapter we give a brief (intentionally not exhaustive) guide to important lines of development in the field of Fermi liquids since the original writing of the first two chapters of this book. Inasmuch as these developments have been primarily in applications to particular systems, rather than in the general form of the theory, this survey is organized according to the physical systems in question. We also call the reader's attention to other reviews of the physics of Fermi liquids [1–4].

3.1. LIQUID ^{3}He

3.1.1. Quasiparticle Spectrum and Thermodynamic Properties

One of the major developments has concerned measurements of the specific heat of normal liquid ^{3}He and the interpretation of these measurements in terms of the quasiparticle spectrum. Experiments have been carried out at a number of different laboratories including Helsinki [5], Cornell [6, 7], Grenoble [8], and Bell Labs [9, 10]. Above $T = 30$ mK the measurements are consistent with one another to better than $\pm 5\%$, while below 30 mK discrepancies are as large as 20%. The measurements of Greywall [9, 10] extend over the largest range of pressures (from the vapor pressure to the melting line) and temperatures (from 7 mK to 2.5 K), and in addition, satisfy a number of thermodynamic consistency checks.

The values of m^* determined by Greywall (see Chapter 1: Appendix C) are about 10% lower than those obtained earlier by Mota et al. [Chap. 1: 59], Abel et al. [11], and Anderson et al. [12], as well as those of Hebral et al. [8], and are about 20% higher than those of References [5] and [6], but they agree rather well with the determination of m^* at the melting pressure by Halperin et al. [7]. An attempt to understand differences between these measurements in terms of temperature scales is given by Greywall [10]. Up to $T \sim 200$ mK at vapor pressure and 100 mK at the melting pressure the specific heat can be fitted very well by the low temperature linear term (1.1.41) plus a term $n\kappa\Gamma T^3 \ln T$, (1.4.79), a form which, as we argued in Section 1.4.4., is expected to hold quite generally for a normal Fermi liquid. Combining the empirical coefficient $\Gamma = (36.8 + 5.69 P_{\text{bar}})\text{K}^{-3}$ with the results of Reference [Chap. 1: 90], one arrives at the estimates of F_1^a included in Appendix C of Chapter 1.

The effective mass of quasiparticle excitations close to the Fermi surface is significantly enhanced owing to the quasiparticle's dressing cloud of spin fluctuations. The behavior of the measured specific heat at higher temperatures, from the initial onset of the $T^3 \ln T$ correction and higher, reflects the fact that with increasing energy away from the Fermi surface, the cloud, as described in Section 1.4.4., is shaken off and the quasiparticle effective mass returns to a value closer to that for a free Fermi gas. We note that a similar effect occurs in the behavior of quasiparticles near the Fermi surface in finite nuclei (Section 3.4.1. below).

To calculate the finite temperature corrections to the specific heat and estimate quantitatively the characteristic energy range over which the effective mass falls requires including spin fluctuations at finite wavenumbers, to which the Landau theory is not applicable. Calculations of the specific heat in this spirit are described in References [13–15]. The calculation in Reference [15] considers quantitatively the difference between the dynamical and statistical quasiparticle energy contributions to the specific heat. These calculations provide an understanding of why the $T^3 \log T$ behavior extends to higher temperatures than is predicted by the paramagnon model, as discussed below equation (1.4.112). Examples of attempts to understand the quasiparticle spectrum from a more microscopic point of view are found in References [16–18].

3.1.2. Measurements of Transport Properties

Developments in determining the transport properties of normal ^{3}He include improved measurements of transport coefficients, as well as measurements on the superfluid phases close to T_c, discussed below, from which one can extract information about lifetimes in the normal phase. Among the former are improved measurements of the thermal conductivity in the normal phase by Greywall [19], measurements of characteristic lifetimes from spin diffusion by Masuhara et al. [20], viscosity measurements by Parpia et al. [21], and determination of relaxation times, including τ_η, equation (1.2.129a), from measurements of sound propagation [22]; the Landau parameter F_2^s has also been inferred from sound propagation [23].

3.1.3. Density and Spin Fluctuations

(a) Long-Wavelength Regime. One striking prediction of the Landau theory is the existence of spin waves in the collisionless regime [Chap. 1: 41], in which the spin current precesses about the magnetization. The theory of these excitations is discussed in Section 1.3.2(b) [cf. equation (1.3.73)], and as discussed in Section 1.3.2(d), their existence was shown in the observation of the Leggett–Rice effect. Direct observation of transverse spin waves by the Ohio State group [20, 24] experimentally confirms Leggett's original predictions. Transverse spin waves have also been seen in dilute solutions of ^{3}He in superfluid ^{4}He [25, 26].

The Landau theory also predicts the existence of a transverse zero sound mode, which has been detected in measurements of the transverse acoustic impedance by Roach and Ketterson [Chap. 1: 116], and has been further studied by Milliken et al. [27]; calculations of the transverse acoustic impedance, as well as further references, are given in Reference [28].

It is tempting to try to describe, in analogy with hydrodynamics, the collective modes in Fermi liquids as those of a viscoelastic medium [29]. Such a description provides a qualitative picture of the modes, but it works quantitatively only when a few spherical harmonic components of the distribution function are excited, for example, in certain limits such as that of strong coupling, $F_0^s \gg 1$, for longitudinal zero or first sound [30].

(b) Short Wavelengths. Significant advances have been made in the study of the density and spin fluctuation spectrum of liquid ^{3}He via neutron scattering, which explores momenta beyond the range of applicability of the normal Landau theory. Carrying out neutron scattering from liquid ^{3}He is particularly difficult because the neutron absorption cross section of the ^{3}He nucleus is some thousand times larger than the scattering cross section [31]. The first neutron scattering measurements of liquid ^{3}He in the millikelvin range, by Sköld et al. [32], showed both the zero sound density mode, and for the same wavenumber, the broadened

"paramagnon" spin mode at lower frequencies. Later experiments are described in References [33] and [34].

The physics of the higher-momentum excitation regime is complicated by the presence of excitations of multiple particle–hole pairs, as well as quasiparticle lifetime effects away from the Fermi surface. A simple first extension of the Landau theory to higher-momenta excitations which preserves the concept of the effective field is to allow the Landau parameters to depend on the momentum in the particle–hole channel, while ignoring lifetime effects. This "polarization potential" approach has been developed by Pines and co-workers [35]. Phenomenological approaches to Fermi liquids beyond the Landau theory are reviewed by Levin and Valls [36].

3.1.4. Calculations of Scattering Amplitudes

The Landau theory requires, in addition to the Landau parameters, knowledge of the quasiparticle scattering amplitudes at the Fermi surface. While one can gain information about certain aspects of the scattering amplitudes from the Landau parameters [see Section 1.4.1, in particular, equations (1.4.27), (1.4.34), and the $s-p$ approximation, in equation (1.4.38)], one must generally turn to theoretical calculations for fuller evaluations. The problem of calculating the scattering amplitudes, or even the Landau parameters, is nontrivial for several reasons: first, liquid ^{3}He is a strongly coupled system, and perturbative techniques are not valid; and second, the scattering amplitudes must be antisymmetric under exchange of the two particles, so that the two particle–hole channels, as discussed in Section 1.4.1 [see Figure 1.6, as well as the discussion in the neighborhood of equation (1.4.32)], must be treated on the same footing. The basic approach employed is to build up the scattering amplitudes from a "bare" interaction, which has no particle–hole correlations, by adding in the effects of single particle–hole intermediate states in the two particle–hole channels.

The first attempt to carry out a self-consistent calculation was that by Bäckman et al. [37] for the Landau parameters in nuclear matter, in which they took a bare scattering kernel based on the Brueckner G-matrix. The same approach was then applied to the Landau parameters of liquid ^{3}He by Babu and Brown [38]. Later, Ainsworth et al. [39] improved the Babu–Brown calculation by using a properly antisymmetrized bare input kernel, and were able to account quantitatively for the parameters F_0^s, F_0^a, and F_1^s of liquid ^{3}He; the model was extended to arbitrarily polarized systems in Reference [40]. Pfitzner and Wölfle [41] have carried out similar calculations using Krotschek's microscopic calculation [42] of the bare scattering kernel.

Calculations of the scattering amplitudes for general scattering angles based on use of pseudopotentials have been given by Levin and Valls [43, 36], and Sauls and Serene [44]. Bedell and Pines [45] have taken an approach based on the Aldrich–Pines polarization potential [35] as the bare input in generating the scattering amplitudes. The fits to the transport coefficients that emerge are quite satisfactory. Pfitzner and Wölfe [46], in order to incorporate the antisymmetry of the amplitudes at a more fundamental level than in the Bedell–Pines calculation, included the properties of the exchange operator on the Fermi surface to generate fully antisymmetric scattering amplitudes from the Aldrich–Pines polarization potential.

3.1.5. Superfluid ^{3}He and the Landau Theory of Fermi Liquids

The discovery of the superfluidity of liquid ^{3}He [Chap. 1: 9, 122, 123] has opened up an important new area of application of the Landau theory of Fermi liquids. Because the pairing correlation energies are so small—and thus the superfluid transition temperature T_c is $\sim 10^{-3}T_f$—the basic quasiparticles of the superfluid states are essentially those of the normal state with pairing correlations included, that is, Bogoliubov transformed linear superpositions of normal state quasiparticles and quasiholes. In addition, the Fermi liquid interactions between the quasiparticles in the superfluid are closely related to those in the normal state, which are described by the usual Landau theory of the normal state. Not only is the theory useful for describing equilibrium and nonequilibrium phenomena in the superfluid states, but information gathered from the superfluids (primarily in the neighborhood of the transition temperature) has been useful in deducing properties of the normal state as well. In the case of anisotropic superfluids, thermodynamic properties can depend on Landau parameters other than those that determine the corresponding normal state property. For example, the magnetic susceptibility of ^{3}He–B, which is in the Balian–Werthamer state, depends on F_2^a as well as F_0^a and m^* [47].

Equilibrium properties of the superfluid readily described in terms of the Landau theory include the specific heat, compressibility, and magnetic susceptibilities, as in the normal state, as well as the normal mass density [48] [Chap. 1: 48, 106, 113]. One effect that can be understood qualitatively in terms of Fermi liquid theory is the reduction of the effect of an electric field on orienting the anisotropy direction of the orbital order parameter (the **d** vector) in ^{3}He–A by a factor $\sim 1/(1 + F_0^s)^2$, discussed theoretically in Reference [49] and experimentally in Reference [50]. The basic physical effect is that the electric dipoles induced by the field are screened by a factor, which is $\sim 1/(1 + F_0^s)$ at long wavelengths; in making detailed estimates, however, it is important to take into account the finite-wavelength behavior of the screening.

The superfluid exhibits the familiar nonequilibrium phenomena of the normal state. In addition, further transport and relaxation processes and collective modes associated with the additional degrees of freedom of the superfluid are present, for example, relaxation of the quasiparticle distribution after a change in the superfluid order parameters (orbital relaxation), and relaxation of magnetic modes (such as the wall-pinned mode). The spin and orbital dynamical modes are reviewed experimentally in Reference [51] and theoretically in Reference [52]. To generalize the Landau Fermi-liquid description of nonequilibrium processes to superfluids, one starts with the kinetic equation for the distribution function, which has not only components of the form (1.2.3) but anomalous components, for example, $\sim \langle aa \rangle$ and $\langle a^\dagger a^\dagger \rangle$ as well. In the long-wavelength, low-frequency limit, the kinetic equation reduces to a scalar equation for the Bogoliubov transformed superfluid state quasiparticles [Chap. 1: 106]. In the absence of strong coupling effects, the scattering amplitudes in the collision term are simply Bogoliubov transformed normal-state scattering amplitudes (e.g., Reference [53]).

Measurements of transport processes in the superfluid provide further information on scattering processes in the normal phase at the Fermi surface, in particular the relaxation time τ, equation (1.2.91). The reason is that here scattering of the quasiparticles involves different linear superpositions of scattering amplitudes than one has in the normal state, due to mixing of normal state quasiparticles and holes in the superfluid. Close to T_c, where the superfluid gap parameter Δ tends to zero, the transport coefficients and relaxation rates depend only on the same quantities, τ_0 and the λ's, as in the normal state, but in somewhat different combinations. For instance, measurements of spin relaxation [54], which causes among other effects the damping of the wall-pinned mode, and of orbital relaxation [55] directly enable one to deduce τ [56, 57], and from viscosity measurements [58], together with the behavior of the gap near T_c, one can deduce the quantity λ_η [53].

Below T_c the transport coefficients are not in general simple functions of the normal state quantities. In the low temperature limit the kinetic equation can be solved analytically [59], and in the B-phase, where the magnitude of the gap is isotropic, it has been solved numerically by Einzel for spin-independent transport coefficients [60]. In addition transport coefficients have been derived using simplifying approximations for the collision term [61]. Properties of ^3He in the ultracold limit, in which mean free paths become very long, are reviewed in Reference [62], where references to earlier work can be found.

The collective modes in the superfluid are driven by the mean fields generated by the quasiparticle interactions, as in the normal state, together with the variations of the gap arising from its dependence on the dynamical state of the system. This latter dependence is included when employing the full matrix kinetic equation. Examples of calculations taking these effects into account are those of sound propagation by Wölfle [63] and Serene [64].

Fermi liquid effects produce qualitative changes in the temperature dependence of physical quantities such as the magnetic susceptibility and the normal mass density. Such changes arise through the dependence of the densities of states on energy, and hence on temperature, in the Fermi liquid interaction corrections, for example, the analogs of the factors $1 + F_0^a$ in the susceptibility and $1 + F_1^s/3$ in the effective mass in the normal state [Chap. 1: 113].

Investigation of magnetic field dependence in the superfluid phases has also been a useful source of information on quasiparticle interactions: for example, in ^{3}He–B measurements of the splitting of the real squashing mode [65, 66], of the pair-breaking edge [67], and of longitudinal magnetic resonance [68] have been used to estimate F_2^a. Similar Fermi liquid effects also occur in superconductors; there one sees the temperature dependence of the normal mass density in measurements of the magnetic penetration depth, from which one can, in clean Type II superconductors, infer F_1^s, a quantity not otherwise accessible in a nontranslationally invariant system.

3.2. DILUTE SOLUTIONS OF ^{3}He IN SUPERFLUID ^{4}He

3.2.1. Equilibrium and Transport Properties

Primary developments in the study of dilute solutions in bulk have been more refined measurements of the equilibrium and transport properties, including at elevated pressures, which have led to better characterization of the Landau parameters, more precise tests of the effective interactions between excitations, and better understanding of the ^{3}He quasiparticle spectrum at higher momenta.

Among the equilibrium measurements have been that of the specific heat of solutions up to 1% concentration in the temperature range 0.07 to 1 K, by Greywall [69], and up to 2.6% by Owers-Bradley et al. [70]. The latter investigators find a somewhat smaller zero concentration effective mass ($m = 2.23 \pm 0.02$) than previously determined. Measurements of the relative magnetic susceptibility at 5% and 1.3% have been carried out by

Ahonen et al. [71], from which they infer that $F_0^a = 0.03$ for 5% solutions, compared with the earlier value 0.08. The osmotic pressure has been remeasured by Corruccini [72] at pressures of 0, 10, and 20 bar; the values for F_0^s deduced from these measurements are in good agreement with a BBP-like effective interaction fitted to the measured spin diffusion coefficient at these pressures [73].

Direct measurements of collective spin waves in mixtures, by nuclear magnetic resonance, were first carried out by Owers–Bradley et al. [25] (analyzed in [74]), and later by Ishimoto et al. [26]; also [75]. The Landau parameters F_0^a and F_1^a inferred from the shift in the resonance frequency are consistent with earlier theoretical estimates. New features of the transverse spin modes at the special point, between 3% and 5% ^{3}He concentration, at which $F_0^a = F_1^a/3$ [26] have been noted by Bedell [76].

One interesting approach to determining the properties of dilute solutions has been to study directly the interactions of phonons with ^{3}He quasiparticles by excitation of heat pulses, from the second sound regime down to the ballistic phonon regime [77, 78]. Related measurements have been made of the attenuation of second sound by Murdock and Corruccini [79], and Church et al. [80]. While the results of these experiments are generally in good agreement with the theory discussed in Chapter 2, certain quantitative discrepancies remain in need of further investigation. Similarly, measurements at higher temperatures and very low concentrations by the Duke group [81, 82] indicate departures from the expected leading concentration dependence of the effective thermal conductivity and thermal diffusivity. This group has also determined the roton–^{3}He scattering cross section, and at the same time observed departures from the expected thermal relaxation times [82].

Experimental searches in dilute solutions to discover whether the ^{3}He itself becomes a superfluid have failed to find evidence of a superfluid transition down to 0.2 to 0.5 mK (depending on concentration and pressure) [25, 83, 84], a result not unexpected, given current theoretical estimates for the maximum T_c as a function of concentration, $\sim 10^{-8}$ K at low pressure, and $\sim 2 \times 10^{-8}$ K at 20 bar [85].

3.2.2. Higher-Momentum Excitations

While the Landau–Pomeranchuk quadratic ^{3}He excitation spectrum is valid at low momenta, deviations at higher momenta, which can be explored in a variety of experiments, become important, as does the coupling of the ^{3}He excitation modes to those of the ^{4}He. The first hints of the inadequacy of the Landau–Pomeranchuk spectrum were found in experiments on second sound, Reference [Chap. 2: 44]. These measurements led Pitaevskii [86], Varma [87], and Stephen and Mittag [88] to suggest that hybridization of the ^{3}He and roton spectra, where they cross near the roton minimum, could lead to a roton-like minimum in the ^{3}He spectrum itself.

In the first neutron scattering experiments on dilute solutions, Rowe et al. [89] were able to establish the dependence on ^{3}He concentration of the ^{4}He spectrum in the neighborhood of the roton minimum. Later, Hilton et al. [90] measured both the ^{4}He and ^{3}He branches of the excitation spectrum by neutron scattering, and found negative deviations from the ^{3}He quadratic law, but not a minimum. Subsequently, Greywall, studying the specific heat [69] and the normal mass density deduced from the velocity of second sound [91], was able to establish that the ^{3}He spectrum is inconsistent with a simple parabola corrected by either a quartic term or a term with a roton-like minimum. His deduced spectrum is slightly above measurements of Hilton et al. [90]. In addition, Eselson et al. [92] inferred the form of the spectrum from normal mass density measurements, finding a spectrum somewhat lower at $k > 1.4$ Å^{-1} than Hilton et al.

In order to understand the higher-momentum spectrum, it is necessary to take into account the density dependence of the ^{4}He spectrum as ^{3}He is added to the system, as well as the interactions at finite temperature between the ^{3}He and ^{4}He-like excitations [93]. Detailed calculations have been performed by Bhatt [94], Götze et al. [95], and most recently by Hsu et al. [85], who find a spectrum in good agreement with those extracted from experiment.

3.3. SPIN-POLARIZED SYSTEMS

3.3.1. Spin-Polarized ^{3}He

Spin-polarized liquid ^{3}He provides a further testing ground for Fermi liquid theory. While one can create the spin-polarized liquid by imposing large magnetic fields, of order 10^5 gauss to create 1% polarization, much larger polarizations can be created by the elegant technique, proposed by Castaing and Nozières [96], of rapidly melting polarized solid ^{3}He; the paramagnetic solid is more easily polarized since the characteristic energies to flip a spin is of order millikelvin, compared with the Fermi energy in the liquid, of order kelvin. This latter experiment was first carried out by Chapellier, Frossati, and Rasmussen [97]; also [98, 99]. The present state of the technique is reviewed in Reference [100]. In such experiments, where one typically reaches polarizations of at least 10%, the longitudinal spin relaxation time plays a crucial role in determining how long one can perform measurements on the polarized liquid. Production of polarized liquid ^{3}He by rapid cooling of polarized ^{3}He gas has been carried out [101]; polarization by means of interactions with an electron-spin-polarized substrate has also been proposed [102] and observed [103].

The dominant spin relaxation in present experiments is at the surface of the container [96, 104]; the spin can also relax in the bulk, via the nuclear dipole–dipole interaction, as first described by Ipatova and Eliashberg [105]. The bulk relaxation rate depends strongly on Fermi liquid interactions (F_0^a) through renormalization of the nuclear dipoles in the medium [106, 107].

In a spin-polarized Fermi liquid, the invariance of the quasiparticle interactions under rotation of the quasiparticle spins is broken. As a consequence of this loss of symmetry, the Fermi liquid interactions no longer assume the simple form (1.1.26); the Fermi liquid theory for polarized systems is given in Reference [40]. Through the correlations between spin and quasiparticle velocities, the density and spin collective and transport modes become coupled. Furthermore, *s*-wave quasiparticle interactions, which must take place between particles of opposite spin, become fewer in number in a polarized medium; with increasing polarization the dominant quasiparticle interactions become those between particles of the same spin, which take place through *p*-wave and higher channels, in which the interactions are generally weaker. These effects tend to increase transport coefficients with polarization, as observed in viscosity [108]. The theory of the transport coefficients is discussed in References [109–111]; exact transport coefficients in a multicomponent system are derived from the Boltzmann equation in Reference [112], and applied to arbitrarily polarized liquid ^{3}He in Reference [113]. (See also Reference [114] on the perils of the relaxation time approximation for the kinetic equation of polarized Fermi liquids.) In the superfluid phases of ^{3}He, rather modest magnetic fields can produce large effects, as in the A_1 phase, in which only one spin population is paired. A careful discussion of Fermi liquid effects on supercurrents in this phase has been given by Muzikar [115].

3.3.2. Dilute Solutions of ^{3}He in ^{4}He

The ^{3}He quasiparticles in dilute solutions of ^{3}He in ^{4}He can be substantially spin polarized by application of magnetic fields, since the Fermi energy is much smaller than in pure liquid ^{3}He. The general theory is reviewed in Reference [116], where references to earlier work can be found. Among the first measurements in polarized dilute solutions were those of second sound velocity by Greywall and Paalanen [117]. As in spin-polarized liquid ^{3}He, with increasing polarization interactions between quasiparticles become dominantly between those with spin parallel to the polarization axis; in the fully polarized system the *s*-wave interaction is removed entirely. The residual interactions are weaker than between opposite-spin quasiparticles; in particular, in dilute solutions the *p*-wave amplitudes, of order p_f^2, are proportional to $x^{2/3}$, where x is the ^{3}He concentration. Consequently, transport coefficients increase with polarization. Such an increase in viscosity has been deduced from measurements of damping of second sound [118], confirming earlier indications of such an effect [117].

3.3.3. Other Systems

Normally, liquid deuterium is in the form of D_2 molecules. However, as pointed out by Miller, Nosanow, and Parish [119], spin polarization in deuterium inhibits formation of molecules, and a sufficiently polarized liquid would be composed of atomic deuterium, a new Fermi liquid, $D_\downarrow$. For further theoretical work see References [120] and [121], and on transport properties in particular see References [110, 111, 122, 123]. Experimental attempts to produce liquid $D_\downarrow$ are described in References [124] and [125].

Similarly, polarization also prevents formation of molecular hydrogen. Spin-polarized hydrogen thus forms a Bose system, $H_\downarrow$, with collective properties of the spins analogous to those of liquid (and gaseous) ^{3}He [126–130].

3.4. NUCLEAR APPLICATIONS

The nuclear shell model, the description of the nucleus in terms of quasiparticle excitations, is precisely a Landau Fermi-liquid theory of the nucleus. The reason that the shell model works is the same as in bulk Fermi liquids, that scattering of particles near the Fermi surface is inhibited by the Pauli principle (an effect first noted in studies of the scattering of nucleons from nuclei [131] and recognized as the basis of the resolution of the apparent contradiction between the shell model and the strongly interacting compound nucleus by Weisskopf [132]), enabling the existence of long-lived quasiparticle excitations. As in bulk systems, the quasiparticles are single-particle excitations—although no longer in plane-wave states—carrying a dressing cloud. For a review of the shell model from a many-body point of view, see Reference [133].

The Landau theory provides the conceptual framework for understanding the quasiparticles and their interactions in nuclei and extended nuclear and neutron star matter, even though the straightforward few-parameter description must be extended to apply to these systems; the basic theory is given by Migdal [Chap. 1: 107]. Nuclear systems have several complications not present in single-component bulk Fermi liquids such as ^{3}He. On the one hand, being composed of neutrons and protons, such systems have an additional degree of freedom, isospin. (Nuclei with different neutron and proton numbers are the nuclear isospin analog of spin-polarized ^{3}He.) Furthermore, spin and space degrees of freedom are coupled by both spin–orbit and tensor forces, removing the independence of space and spin rotational symmetries as in ^{3}He. In addition, the nuclear surface plays a strong role in determining the properties of finite nuclei.

3.4.1. Particles and Quasiparticles

Nuclei, unlike bulk systems, offer one the clear possibility of studying individual quasiparticle and quasihole states. These states can be probed by comparison of the ground states of nuclei that differ only by a single proton or neutron. For example, one may regard ^{207}Pb as a single-neutron quasihole in ^{208}Pb, and ^{209}Bi a single-proton quasiparticle in ^{208}Pb. Furthermore, probes such as electron scattering excite quasiparticle–quasihole states, among others, and provide measures of ground state charge and magnetization distributions, and matrix elements of these quantities between the ground state and excited states [134].

Although the Landau theory is quite generally formulated in terms of quasiparticles, the amplitude Z_p (in a bulk system) for a single quasiparticle of momentum **p** actually to be in a single-particle state of momentum **p** —the *single-particle wave function renormalization constant*—is less than unity. The quantity Z_{p_f} is equal to the discontinuity in the true single-particle distribution at the Fermi surface [Chap. 1: 8]. Such wave function renormalizations of the single-particle excitations occur as well in nuclei. Nuclear wave function renormalization has been studied in a very interesting series of electron scattering measurements in the Pb region [134, 135]. For instance, by comparison of ^{206}Pb with ^{205}Tl, which contains in the shell model one fewer $3s_{1/2}$ proton, one finds that this proton quasihole is composed of $\sim 65\%$ single proton and of $\sim 35\%$ higher multiparticle–hole excitations, occupation probabilities characteristic of quasiparticle and hole excitations in the Pb region [136].

3.4.2. Quasiparticle Interactions in Nuclei and Nuclear Matter

(a) Landau Parameters in Nuclear Matter. In the presence of isospin degrees of freedom and tensor forces, which couple space and spin degrees of freedom, the Landau effective interaction assumes the form [137, 138]:*

$$f_{\mathbf{p}\sigma\tau,\mathbf{p}'\sigma'\tau'} = f + f'\tau\cdot\tau' + g\sigma\cdot\sigma' + g'\sigma\cdot\sigma'\tau\cdot\tau'$$
$$+ (q^2/p_f^2)S_{12}(\mathbf{q})[h + h'\tau\cdot\tau'], \qquad (3.4.1)$$

where $\mathbf{q} = \mathbf{p} - \mathbf{p}'$, and S_{12} is the tensor operator:

$$S_{12}(\mathbf{q}) = 3\sigma\cdot\hat{\mathbf{q}}\sigma'\cdot\hat{\mathbf{q}} - \sigma\cdot\sigma', \qquad (3.4.2)$$

where σ and τ on the right side are Pauli matrices. The Landau parameters f and g here are the same as the f^s and f^a earlier. The parameters f, f', g, g', h, and h' are functions of the quasiparticle momenta $\mathbf{p}$ and $\mathbf{p}'$, and can be expanded on the Fermi surface, as usual [e.g., equation (1.1.30)], in Legendre polynomials. The dimensionless Landau parameters are defined in symmetric nuclear matter by, for example, $F_l^s = N(0)f_l^s$, where $N(0) = 2p_f m^*/\pi^2$ is the total density of states at the Fermi surface for neutrons and protons. As in single component Fermi liquids, the compressibility is simply related to the parameter F_0; in nuclear matter, the parameter F_0' is similarly related to the symmetry energy associated with changing a proton into a neutron.

As Friman and Dhar [138] showed, the Landau parameters obey two forward scattering sum rules, corresponding to the two total isospin channels of the pair of quasiparticles. These sum rules are generalizations of equation (1.4.36); with the tensor contributions neglected for simplicity, they may be written in the forms

$$\sum_l \left[\frac{F_l}{1 + F_l/(2l+1)} + 3\frac{G_l'}{1 + G_l'/(2l+1)}\right] = 0$$
$$\sum_l \left[\frac{2}{3}\frac{F_l}{1 + F_l/(2l+1)} + \frac{F_l'}{1 + F_L'/(2l+1)} + \frac{G_l}{1 + G_l/(2l+1)}\right] = 0. \qquad (3.4.3)$$

The tensor contributions, given in [138], are of order 10% in the first sum rule, but are in fact substantial in the second. The stability conditions on the Landau parameters, in the presence of the tensor force, are derived in Reference [137].

*Note that unlike in Chapter 1, for example, equation (1.1.26), we here denote the spin by σ and the isospin by τ.

(b) Collective Modes. The collective modes of bulk nuclear matter include, in addition to zero sound and spin modes as in liquid ^{3}He, modes involving isospin excitations. In nuclei the monopole, or breathing mode [139], is an example of a zero sound mode, while the magnetic dipole ($M1$) modes correspond to the spin modes [140]. Similarly, nuclear giant dipole resonances are isospin zero sound, while the Gamow–Teller giant resonances are spin–isospin zero sound [141]. Attempts to deduce bulk Landau parameters of nuclear matter, particularly F_0^s, from the excitation energies of collective modes are generally complicated by the presence of the nuclear surface. The growth of density instabilities in nuclear matter at subnuclear densities where $F_0^s < -1$, a mechanism of possible importance in fragmentation processes in laboratory heavy ion collisions, has been studied in Reference [142].

The onset of the conjectured pion condensed state of bulk nuclear matter [143] is signalled by an instability of the spin–isospin collective mode. Such instabilities are strongly controlled by the Landau parameter g'; determinations of g'_0 [144] indicate that pion condensation should not set in until the density is at least twice that of normal nuclear matter [143, 145].

(c) Transport Properties and Relaxation Processes. The dynamical behavior of neutron stars, such as their oscillations, and thermal and magnetic history, are dependent on transport coefficients such as viscosities and electrical and thermal conductivities. Since neutron star interiors are at temperatures well below the nucleon and electron Fermi temperatures, transport coefficients can be calculated by standard Fermi liquid techniques [146, 112]. The matter in laboratory heavy ion collisions is also a Fermi liquid; while the dynamics of collisions cannot be strictly described in hydrodynamic terms, transport theory calculations of finite temperature nuclear matter are useful for understanding characteristic relaxation times in collisions [147, 148].

(d) Relativistic Systems. The Landau theory also describes relativistic systems such as the Dirac treatment of nuclei [149, 150], nuclear matter under extreme conditions, quark matter, and relativistic plasmas; the basic framework of the Landau theory of relativistic Fermi liquids is given by Baym and Chin [151]. Long-range magnetic forces, which are unscreened, become important in electromagnetic plasmas when particle velocities are close to the speed of light (see below), as do color-magnetic forces associated with transverse gluon exchange in a quark-gluon plasma. Such long-range interactions lead to divergent Landau parameters in the Landau theory in its simplest form; to construct a consistent theory free of

divergences, the renormalization of the quasiparticle structure by the magnetic interactions must be properly taken into account [152].

3.5. ELECTRONS IN METALS

Just as for the nuclear shell model, the Landau Fermi-liquid theory provides the underpinning of the one-electron picture of metals. Again the small density of low-lying states allows the existence of long-lived quasiparticle states, even though interactions of an electron with the ionic lattice and other electrons are strong. Complicating features in metals are the long range of the Coulomb interaction, the lack of translational invariance, which leads to the band structure, the interaction of electrons with lattice vibrations, and the spin–orbit force.

The long-range part of the Coulomb interaction can be taken into account in the Landau theory by writing the quasiparticle interactions as the sum of a long-range Coulombic term plus a short-range part, as in equation (1.2.8); for detailed application see Pines and Nozières [Chap. 1: 6]. While the electron density and current can be tracked in terms of the quasiparticles, the lack of translational invariance implies that the momentum density cannot be calculated simply as a sum of quasiparticle crystal momenta alone, nor can the electron effective masses be written as band masses corrected by factors $(1 + F_1^s)^{-1}$ alone (cf. Section 2.3.1. for the dilute helium solutions). Furthermore, coupling of electrons to the phonons introduces important frequency dependence, on the scale of the Debye energy, ω_D. At frequencies well above ω_D the phonon dressing clouds are shaken off, while at frequencies well below ω_D the effective mass of quasiparticles near the Fermi surface is markedly increased by coupling to the phonons. In simple metals, where the spin–orbit force is small and can be taken into account as a perturbation, the Landau theory is well developed; see, for example, References [153–155] [Chap. 1: 46, 102–104], as well as Reference [156], a two-component Fermi-liquid theory for liquid metallic hydrogen. Effects of spin–orbit and tensor interactions are discussed by Quader and co-workers [157].

Fermi liquid concepts have proven fruitful in studies of disordered metals and the metal–insulator transition. McMillan [158] constructed a phenomenological theory for very dirty metals in which the basic single quasiparticle states are not plane waves, as in a translationally invariant system, but rather are states of electrons in a strong spatially random potential resulting from the disorder. In an apparently different approach, Finkel'stein mapped the disorder problem onto a nonlinear σ-model [159] with results that resembled those of a Fermi liquid theory. Subsequently, in References [160] the precise connection was made between Finkel'stein's work and a Landau Fermi-liquid theory with scale-dependent Landau parameters.

The Landau theory has also been usefully applied to a number of strongly correlated systems. Nozières has made an ingenious use of the theory to account for the low temperature properties of a Kondo impurity in a metal [161]. In this problem, equilibrium properties such as the heat capacity and magnetic susceptibility, and transport properties such as the electrical resistivity can be evaluated in terms of the Landau Fermi-liquid interaction between conduction electrons induced by the presence of the impurity. The application of Fermi liquid theory to the two-band Anderson model of localization is given by Yip [162]. The theory is also helpful in understanding the properties of heavy electron ("heavy fermion") compounds. In UPt_3, for instance, the specific heat in the normal state is linear in T at temperatures on the order of a few kelvin, while the temperature-dependent part of the resistivity varies as T^2—behaviors characteristic of a normal Fermi liquid with electron–electron scattering (in the nontranslationally invariant medium) as the dominant source of the electrical resistance. Simplified Fermi liquid models have been used to discuss the properties of these compounds [163–165], but a rigorous formulation of Landau theory applicable to strongly correlated systems in which spin–orbit coupling and band structure play important roles remains a subject for future investigation.

One of the fundamental assumptions of Landau Fermi-liquid theory is that the states of the interacting Fermi system can be classified in a one-to-one correspondence with those of a free Fermi gas. Microscopically, this condition amounts to the statement that there must be a nonzero overlap between the quantum states of quasiparticles in the interacting system and those in the free system, or equivalently that the residue at the pole of the single-particle propagator—the single-particle wave function renormalization constant, Z_p— must be nonzero. During the past several decades a number of systems, known as *marginal Fermi liquids*, in which this condition is violated at the Fermi surface, have been investigated. Although the vanishing of Z_{p_f} implies that the particle number density does not have a discontinuity at the Fermi surface, marginal Fermi liquids have properties resembling those of normal Fermi liquids.

A simple example is the electron gas interacting by exchange of transverse photons, in addition to the usual Coulomb forces. Effects of transverse photon exchange are usually neglected because they are formally of order $(v_f/c)^2$, compared with those of the Coulomb interaction, where c is the speed of light. However, Holstein, Norton, and Pincus [166] (see also

[152, 167]) demonstrated that close to the Fermi surface Z_p, calculated on-energy-shell, behaves as $[A \ln(p_f/|p - p_f|)]^{-1}$, vanishing at p_f, and therefore Landau's basic assumption fails for excitations right at the Fermi surface. The constant A equals $(2\alpha/3\pi)(v_f/c)^2$, where $\alpha = e^2/\hbar c$ is the fine structure constant. Since Z_p does not vanish away from the Fermi surface, the concepts of Fermi liquid theory still provide a useful framework for describing the properties of the system. Holstein et al. [166] also found that the effective mass behaves as $mA \ln(p_f/|p - p_f|) \sim 1/Z_p$, where m is the bare electron mass; consequently, the specific heat behaves as $T \ln T$, rather than as T as for a normal Fermi liquid. For an electron gas at metallic densities, the divergences become important only at unobservably low temperatures, because $\alpha(v_f/c)^2 \sim 10^{-7}$. As mentioned in Section 3.4.2(d), analogous effects occur in quark matter through exchanges of transverse gluons; these become very important because v_f/c and the dimensionless coupling parameter $g^2/4\pi\hbar c$, where g is the qcd coupling constant, are both of order unity.

Similar effects occur in one-dimensional conductors with a short-range interaction between electrons. In the Born approximation one finds that the imaginary part of the electron self-energy at the Fermi surface is proportional to $|\epsilon - \mu|$, where ϵ is the quasiparticle energy and μ the chemical potential; hence the real part of the self-energy varies as $(\epsilon - \mu)\ln|\epsilon - \mu|$, and Z_p is $\sim 1/\ln(p_f/|p - p_f|)$. Higher-order effects modify this result, but Z_p still vanishes at the Fermi surface: the basic physics is that the important low-energy excitations are density or spin fluctuations, which behave as bosons rather than fermions. For a review, see Reference [168].

Marginal Fermi liquids have recently attracted attention because they are able to explain many normal state properties of high temperature superconductors [169]. For example, if the imaginary part of the self-energy is $\sim |\epsilon - \mu|$, the thermally averaged quasiparticle lifetime varies as $1/T$ and the electrical resistivity as T, in agreement with experiment. In the model of Varma et al. [169], the $|\epsilon - \mu|$ behavior arises from coupling of fermions to low-frequency spin and density fluctuations. While further work is required to determine the extent to which such a model applies to high transition temperature superconductivity, such materials provide in general a fruitful area for future application of the Landau Fermi-liquid theory.

REFERENCES

1. P. Wölfle, *Rep. Prog. Phys.* **42**, 270 (1979).
2. P. Wölfle, in *Progress in Low Temperature Physics*, Vol. 7A, edited by D. F. Brewer, (North-Holland, Amsterdam, 1978), p. 191.
3. D. Vollhardt, *Rev. Mod. Phys.* **56**, 99 (1984).
4. D. Vollhardt and P. Wölfle, *The Superfluid Phases of ^3He* (Taylor and Francis, London, 1989).
5. T. A. Alvesalo, T. Haavasoja, M. T. Manninen, and A. T. Soinne, *Phys. Rev. Lett.* **44**, 1076 (1980); T. A. Alvesalo, T. Haavasoja, and M. T. Manninen, *J. Low Temp. Phys.* **45**, 373 (1981).
6. E. K. Zeise, J. Saunders, A. I. Ahonen, C. N. Archie, and R. C. Richardson, *Physica* **108B**, 1213 (1981).
7. W. P. Halperin, F. B. Rasmussen, C. N. Archie, and R. C. Richardson, *J. Low Temp. Phys.* **31**, 617 (1978).
8. B. Hebral, G. Frossati, H. Godfrin, and D. Thoulouze, *Phys. Lett.* **85A**, 290 (1981).
9. D. S. Greywall, *Phys. Rev. B* **27**, 2747 (1983).
10. D. S. Greywall, *Phys. Rev. B* **31**, 2675 (1985); *B* **33**, 7520 (1986).
11. W. R. Abel, A. C. Anderson, W. C. Black, and J. C. Wheatley, *Phys. Rev.* **147**, 111 (1966).
12. A. C. Anderson, W. Reese, and J. C. Wheatley, *Phys. Rev.* **130**, 495 (1963).
13. G. E. Brown, C. J. Pethick, and A. Zaringhalam, *J. Low Temp. Phys.* **48**, 349 (1982).
14. V. K. Mishra, G. E. Brown, and C. J. Pethick, *J. Low Temp. Phys.* **52**, 379 (1983).
15. D. Coffey and C. J. Pethick, *Phys. Rev. B* **37**, 1647 (1988).
16. B. L. Friman and E. Krotschek, *Phys. Rev. Lett.* **49**, 1705 (1982).
17. E. Krotschek and R. A. Smith, *Phys. Rev. B* **27**, 4222 (1983).
18. S. Fantoni, V. R. Pandharipande, and K. E. Schmidt, *Phys. Rev. Lett.* **48**, 878 (1982).
19. D. S. Greywall, *Phys. Rev. B* **29**, 4933 (1984).
20. N. Masuhara, D. Candela, D. O. Edwards, R. F. Hoyt, H. N. Scholz, D. S. Sherrill, and R. Combescot, *Phys. Rev. Lett.* **53**, 1168 (1984).
21. J. M. Parpia, D. J. Sandiford, J. E. Berthold, and J. D. Reppy, *Phys. Rev. Lett.* **40**, 565 (1978).
22. J. B. Ketterson, P. R. Roach, B. M. Abraham, and P. D. Roach, in *Quantum Statistics and the Many-Body Problem*, edited by S. B. Trickey, W. P. Kirk, and J. W. Dufty (Plenum, New York, 1975), p. 35.
23. B. N. Engel and G. G. Ihas, *Phys. Rev. Lett.* **55**, 955 (1985).
24. D. Candela, N. Masuhara, D. S. Sherrill, and D. O. Edwards, *J. Low Temp. Phys.* **63**, 331 (1986).
25. J. R. Owers-Bradley, H. Chocholacs, R. M. Mueller, C. Buchal, M. Kubota, and F. Pobell, *Phys. Rev. Lett.* **51**, 2120 (1983).
26. H. Ishimoto, H. Fukuyama, N. Nishida, Y. Miura, Y. Takano, T. Fukuda, T. Tazaki, and S. Ogawa, *Phys. Rev. Lett.* **59**, 904 (1987).
27. F. P. Milliken, R. W. Richardson, and S. J. Williamson, *Physica* **108B**, 1201 (1981).

28. E. G. Flowers and R. W. Richardson, *Phys. Rev. B* **17**, 1238 (1978); D. Einzel, H. Højgaard Jensen, H. Smith, and P. Wölfle, *J. Low Temp. Phys.* **53**, 695 (1983).

29. R. E. Nettleton, *J. Low Temp. Phys.* **22**, 407 (1976); **24**, 275 (1976); **26**, 277 (1977); G. F. Bertsch, in *Nuclear Physics with Heavy Ions and Mesons, Les Houches, Session XXX*, edited by R. Balian, M. Rho, and G. Ripka (North-Holland, Amsterdam, 1978), p. 175; I. Rudnick, *J. Low Temp. Phys.* **40**, 287 (1980).

30. K. Bedell and C. J. Pethick, *J. Low Temp. Phys.* **49**, 213 (1982).

31. D. L. Price, in *The Physics of Liquid and Solid Helium*, Part 2, edited by K. H. Bennemann and J. B. Ketterson (Wiley, New York, 1978), p. 675.

32. K. Sköld, C. A. Pelizzari, R. Kleb, and G. E. Ostrowski, *Phys. Rev. Lett.* **37**, 842 (1976).

33. P. A. Hilton, R. A. Cowley, R. Scherm, and W. G. Stirling, *J. Phys. C* **13**, L295 (1980).

34. K. Sköld and C. A. Pelizzari, *Philos. Trans. R. Soc. London Ser. B* **290**, 605 (1980).

35. C. H. Aldrich, C. J. Pethick, and D. Pines, *Phys. Rev. Lett.* **37**, 845 (1976); C. H. Aldrich and D. Pines, *J. Low Temp. Phys.* **25**, 673 (1976); C. H. Aldrich and D. Pines, *J. Low Temp. Phys.* **32**, 689 (1978); K. S. Bedell and K. F. Quader, *Phys. Lett.* **96A**, 91 (1983).

36. K. Levin and O. Valls, *Phys. Rept.* **98**, 1 (1983).

37. S. O. Bäckman, C.-G. Källman, and O. Sjöberg, *Phys. Lett.* **43B**, 263 (1973); O. Sjöberg, *Ann. Phys. (N.Y.)* **78**, 39 (1973).

38. S. Babu and G. E. Brown, *Ann. Phys. (N.Y.)* **78**, 1 (1973).

39. T. Ainsworth, K. S. Bedell, G. E. Brown, and K. F. Quader, *J. Low Temp. Phys.* **50**, 319 (1983).

40. K. F. Quader and K. S. Bedell, *J. Low Temp. Phys.* **58**, 89 (1985).

41. M. Pfitzner and P. Wölfle, *Physica* **109** & **110B**, 1253 (1984); *Phys. Rev. B* **35**, 4699 (1987).

42. E. Krotschek, in *Quantum Fluids and Solids—1983*, edited by E. D. Adams and G. G. Ihas (AIP, New York, 1983), p. 132.

43. K. Levin and O. Valls, *Phys. Rev. B* **20**, 105, 120 (1979).

44. J. Sauls and J. W. Serene, *Phys. Rev. B* **24**, 1983 (1981).

45. K. Bedell and D. Pines, *Phys. Rev. Lett.* **45**, 39 (1980).

46. M. Pfitzner and P. Wölfle, *J. Low Temp. Phys.* **51**, 535 (1983).

47. J. Czerwonko, *Acta Phys. Polonica* **32**, 335 (1967).

48. A. J. Leggett, *Phys. Rev.* **147**, 119 (1966).

49. I. A. Fomin, C. J. Pethick, and J. W. Serene, *Phys. Rev. Lett.* **40**, 1144 (1978).

50. G. W. Swift, J. P. Eisenstein, and R. E. Packard, *Phys. Rev. Lett.* **45**, 1955 (1980).

51. J. C. Wheatley, in *Progress in Low Temperature Physics*, Vol. 7A, edited by D. F. Brewer (North-Holland, Amsterdam, 1978), p. 1.

52. W. F. Brinkman and M. C. Cross, in *Progress in Low Temperature Physics*, Vol. 7A, edited by D. F. Brewer (North-Holland, Amsterdam, 1978), p. 105.

53. P. Bhattacharyya, C. J. Pethick, and H. Smith, *Phys. Rev. B* **15**, 3367 (1977).

54. R. A. Webb, R. E. Sager, and J. C. Wheatley, *Phys. Rev. Lett.* **35**, 1164 (1975).

55. D. N. Paulson, M. Krusius, and J. C. Wheatley, *Phys. Rev. Lett.* **36**, 1322 (1976).

56. P. Bhattacharyya, C. J. Pethick, and H. Smith, *Phys. Rev. Lett.* **35**, 473 (1975).

57. C. J. Pethick and H. Smith, *Phys. Rev. Lett.* **37**, 226 (1976).

58. T. A. Alvesalo, H. K. Collan, M. T. Loponen, O. V. Lounasmaa, and M. C. Veuro, *J. Low Temp. Phys.* **19**, 1 (1975).
59. C. J. Pethick, H. Smith, and P. Bhattacharyya, *Phys. Rev. B* **15**, 3384 (1977).
60. D. Einzel, *J. Low Temp. Phys.* **54**, 427 (1984).
61. D. Einzel and P. Wölfle, *J. Low Temp. Phys.* **32**, 19, 39 (1978).
62. H. Smith, in *Progress in Low Temperature Physics*, Vol. 11, edited by D. F. Brewer (North-Holland, Amsterdam, 1987), p. 75.
63. P. Wölfle, *Phys. Rev. B* **14**, 89 (1976).
64. J. W. Serene, Ph.D. thesis, Cornell University, 1974 (unpublished).
65. O. Avenel, E. Varaquaux, and H. Ebisawa, *Phys. Rev. Lett.* **45**, 1952 (1980).
66. R. S. Fishman and J. A. Sauls, *Phys. Rev. B* **38**, 2526 (1988).
67. R. Movshovich, N. Kim, and D. M. Lee, *Phys. Rev. Lett.* **64**, 431 (1990).
68. D. Candela, D. O. Edwards, A. Heff, N. Masuhara, Y. Oda, and D. S. Sherrill, *Phys. Rev. Lett.* **61**, 420 (1988).
69. D. S. Greywall, *Phys. Rev. Lett.* **41**, 177 (1978).
70. J. R. Owers-Bradley, P. C. Main, G. J. Batey, R. J. Church, and R. M. Bowley, *J. Low Temp. Phys.* **72**, 201 (1988).
71. A. I. Ahonen, M. A. Paalanen, R. C. Richardson, and Y. Takano, *J. Low Temp. Phys.* **25**, 733 (1976).
72. L. Corruccini, *Phys. Rev. B* **30**, 3735 (1984).
73. E. S. Murdock, K. R. Mountfield, and L. R. Corruccini, *J. Low Temp. Phys.* **31**, 581 (1978).
74. R. M. Bowley and J. R. Owers-Bradley, *J. Low Temp. Phys.* **63**, 331 (1986).
75. H. Akimoto, T. Kawae, G.-H. Oh, O. Ishikawa, T. Hata, T. Kodama, and T. Shigi, *Jpn. J. Appl. Phys.* **26**, suppl. 26-3, 65 (1987).
76. K. S. Bedell, *Phys. Rev. Lett.* **62**, 167 (1989).
77. W. J. P. de Voogt and H. C. Kramers, *Physica* **84B**, 328 (1976).
78. L. P. J. Husson, C. E. D. Ouwerkerk, A. L. Reesink, and R. de Bruyn Ouboter, *Physica* **122B**, 8, 183 (1983); L. P. J. Husson and R. de Bruyn Ouboter, *Physica* **122B**, 201 (1983).
79. E. S. Murdock and L. R. Corruccini, *J. Low Temp. Phys.* **46**, 219 (1982).
80. R. J. Church, J. R. Owers-Bradley, P. C. Main, G. McHale, and R. M. Bowley, *Jpn. J. Appl. Phys.* **26**, suppl. 26-3, 21 (1987).
81. M. Dingus, F. Zhong, J. Tuttle, and H. Meyer, *J. Low Temp. Phys.* **65**, 213 (1986); F. Zhong, J. Tuttle, and H. Meyer, *J. Low Temp. Phys.*, **79**, 9 (1990).
82. J. Tuttle, F. Zhong, and H. Meyer, *J. Low Temp. Phys.* **82**, 15 (1991).
83. H. Chocholacs, R. M. Mueller, J. R. Owers-Bradley, C. Buchal, M. Kubota, and F. Pobell, *Physica* **126B + C**, 1247 (1984).
84. H. Ishimoto, H. Fukuyama, N. Nishida, Y. Miura, Y. Takano, T. Fukuda, T. Tazaki, and S. Ogawa, *Jpn. J. Appl. Phys.* **26**, suppl. 26-3, 67 (1987).
85. W. Hsu and D. Pines, *J. Stat. Phys.* **38**, 273 (1985); W. Hsu, D. Pines, and C. H. Aldrich, *Phys. Rev. B* **32**, 7179 (1985).
86. L. P. Pitaevskii, *U.S.–U.S.S.R. Symposium on Condensed Matter Physics*, Berkeley, 1973 (unpublished).
87. C. Varma, *Phys. Lett.* **45A**, 301 (1973).

88. M. J. Stephen and L. Mittag, *Phys. Rev. Lett.* **31**, 923 (1973).

89. J. M. Rowe, D. L. Price, and G. E. Ostrowski, *Phys. Rev. Lett.* **31**, 510 (1973).

90. P. A. Hilton, R. Scherm, and W. G. Stirling, *J. Low Temp. Phys.* **27**, 851 (1977).

91. D. S. Greywall, *Phys. Rev. B* **20**, 2643 (1979).

92. B. N. Eselson, V. A. Slyusaren, V. I. Sobolev, and M. A. Strzhemechnyi, *Zh. Eksp. Teor. Fiz. Pis. Red.* **21**, 253 (1975) [English transl.: *JETP Lett.* **21**, 115 (1975)].

93. A. Bagchi and J. Ruvalds, *Phys. Rev. A* **8**, 1973 (1973); J. Ruvalds, J. Slinkman, A. K. Rajagopal, and A. Bagchi, *Phys. Rev. B* **16**, 2047 (1977).

94. R. Bhatt, *Phys. Rev. B* **18**, 2108 (1978).

95. W. Götze, M. Lücke, and A. Szprynger, *Phys. Rev. B* **19**, 206 (1979).

96. B. Castaing and P. Nozières, *J. de Phys.* **40**, 257 (1979).

97. M. Chapellier, G. Frossati, and F. B. Rasmussen, *Phys. Rev. Lett.* **42**, 904 (1979).

98. G. Schumacher, D. Thoulouze, B. Castaing, Y. Chabre, P. Segransan, and J. Joffrin, *J. de Phys.* **40**, L143 (1979).

99. G. Frossati, *J. de Phys.* **41**, C7-95 (1980).

100. G. Frossati, *Jpn. J. Appl. Phys.* **26**, suppl. 26-3, 1833 (1987).

101. P.-J. Nacher, M. Leduc, G. Tastevin, L. Wiesenfeld, and F. Laloë, *Jpn. J. Appl. Phys.* **26**, suppl. 26-3, 205 (1987).

102. M. Chapellier, *J. de Phys. Lett.* **43**, L609 (1982); S. A. Langer, K. DeConde, and D. L. Stein, *J. Low Temp. Phys.* **57**, 249 (1984).

103. L. W. Engel and K. DeConde, *Phys. Rev. B* **33**, 2035 (1986).

104. H. Godfrin, G. Frossati, B. Hebral, and D. Thoulouze, *J. de Phys.* **41**, C7-275 (1981).

105. I. P. Ipatova and G. M. Eliashberg, *Zh. Eksp. Teor. Fiz.* **43**, 1795 (1962) [English transl.: *Sov. Phys.-JETP* **16**, 1269 (1963)].

106. D. Vollhardt and P. Wölfle, *Phys. Rev. Lett.* **47**, 190 (1981).

107. K. S. Bedell and D. E. Meltzer, *J. Low Temp. Phys.* **63**, 215 (1986).

108. G. A. Vermeulen, A. Schuhl, F. B. Rasmussen, J. Joffrin, G. Frossati, and M. Chapellier, *Phys. Rev. Lett.* **60**, 2315 (1988).

109. C. Lhuillier and F. Laloë, *J. de Phys.* **40**, 239 (1979); **41**, C7-51 (1980).

110. C. Lhuillier and F. Laloë, *J. de Phys.* **43**, 197, 225 (1982).

111. C. Lhuillier, *J. de Phys.* **44**, 1 (1983).

112. R. A. Anderson, C. J. Pethick, and K. F. Quader, *Phys. Rev. B* **35**, 1620 (1987).

113. D. W. Hess and K. F. Quader, *Phys. Rev. B* **36**, 756 (1987).

114. S. M. Troian and N. D. Mermin, *J. Low Temp. Phys.* **59**, 115 (1985).

115. P. Muzikar, *J. Low Temp. Phys.* **46**, 533 (1982).

116. A. E. Meyerovich, in *Progress in Low Temperature Physics*, Vol. 11, edited by D. F. Brewer (North-Holland, Amsterdam, 1987), p. 1.

117. D. S. Greywall and M. A. Paalanen, *Phys. Rev. Lett.* **46**, 1292 (1981); *Physica* **109** & **110B**, 1575 (1982).

118. J. R. Owers-Bradley, P. C. Main, R. J. Church, T. M. M. Hampson, G. McHale, and R. M. Bowley, *Phys. Rev. Lett.* **61**, 1619 (1988).

119. M. D. Miller, L. H. Nosanow, and L. J. Parish, *Phys. Rev. B* **15**, 214 (1976).

120. R. M. Panoff, J. W. Clark, M. A. Lee, K. E. Schmidt, M. H. Kalos, and G. V. Chester, *Phys. Rev. Lett.* **48**, 1675 (1982); R. M. Panoff and J. W. Clark, *Phys. Rev. B* **36**, 5527 (1987).
121. A. J. Leggett, *J. de Phys.* **41**, C7-19 (1980); A. G. K. Modawi, Ph.D. thesis, University of Sussex, 1981 (unpublished).
122. W. J. Mullin and K. Miyake, *J. Low Temp. Phys.* **53**, 313 (1983).
123. E. P. Bashkin and A. E. Meyerovich, *Adv. Phys.* **30**, 1 (1981).
124. I. F. Silvera and J. T. M. Walraven, *Phys. Rev. Lett.* **45**, 1268 (1980); and in *Progress in Low Temperature Physics*, Vol. 10, edited by D. F. Brewer (North-Holland, Amsterdam, 1986), p. 139.
125. I. Shinkoda, M. W. Reynolds, R. W. Cline, and W. N. Hardy, *Phys. Rev. Lett.* **57**, 1243 (1986); *Jpn. J. Appl. Phys.* **26**, suppl. 26-3, 243 (1987).
126. I. F. Silvera, *Physica* **109** & **110B**, 1499 (1982).
127. B. R. Johnson, J. S. Denker, N. Bigelow, L. P. Levy, J. H. Freed, and D. M. Lee, *Phys. Rev. Lett.* **52**, 1508 (1984).
128. L. P. Levy and A. E. Ruckenstein, *Phys. Rev. Lett.* **52**, 1512 (1984).
129. J. T. M. Walraven, *Physica* **126B** + **C**, 176 (1984).
130. D. M. Lee, *Jpn. J. Appl. Phys.* **26**, suppl. 26-3, 1841 (1987).
131. R. Serber, *Phys. Rev.* **72**, 1114 (1947).
132. V. Weisskopf, *Science* **113**, 101 (1951).
133. C. Mahaux, P. F. Bortignon, R. A. Broglia, and C. H. Dasso, *Phys. Rept.* **120**, 1 (1985).
134. B. Frois and C. N. Papanicolas, *Ann. Rev. Nucl. Part. Sci.* **37**, 133 (1987).
135. C. N. Papanicolas et al., *Phys. Rev. Lett.* **45**, 106 (1980); J. M. Cavedon et al., *Phys. Rev. Lett.* **49**, 978 (1982), *Nucl. Phys. A* **396**, 409c (1983); J. Heisenberg et al., *Phys. Rev. C* **25**, 2292 (1982); C. N. Papanicolas et al., *Phys. Rev. Lett.* **58**, 2296 (1987); E. Quint et al., *Phys. Rev. Lett.* **58**, 1727 (1987).
136. V. R. Pandharipande. C. N. Papanicolas, and J. Wambach, *Phys. Rev. Lett.* **53**, 1133 (1984).
137. J. Dąbrowski and P. Haensel, *Ann. Phys. (N.Y.)* **97**, 452 (1976).
138. B. L. Friman and A. K. Dhar, *Phys. Lett.* **85B**, 1 (1979).
139. J.-P. Blaizot, *Phys. Rept.* **64**, 171 (1980).
140. C. N. Papanicolas et al., *Phys. Rev. Lett.* **58**, 2296 (1987); R. M. Laszewski and J. Wambach, *Comments Part. Nucl. Phys.* **14**, 321 (1985).
141. K. Goeke and J. Speth, *Ann. Rev. Nucl. Part. Sci.* **32**, 65 (1982).
142. C. J. Pethick and D. G. Ravenhall, *Ann. Phys. (N.Y.)* **183**, 131 (1988).
143. G. Baym and D. K. Campbell, in *Mesons in Nuclei*, Vol. 3, edited by M. Rho and D. Wilkinson (North-Holland, Amsterdam, 1979), p. 1031; G. E. Brown and W. Weise, *Phys. Rept.* **27**, 1 (1976).
144. J. Speth, E. Werner, and W. Wild, *Phys. Rev. C* **33**, 127 (1977).

145. S.-O. Bäckman and W. Weise, in *Mesons in Nuclei*, Vol. 3, edited by M. Rho and D. Wilkinson (North-Holland, Amsterdam, 1979), p. 1095.

146. E. G. Flowers and N. Itoh, *Astrophys. J.* **206**, 218 (1976); **230**, 847 (1979); **250**, 750 (1981).

147. P. Danielewicz, *Phys. Lett.* **146B**, 168 (1984).

148. C. J. Pethick and D. G. Ravenhall, *Nucl. Phys. A* **471**, 19c (1987).

149. B. D. Serot and J. D. Walecka, *Adv. Nucl. Phys.* Vol. 16, edited by J. W. Negele and E. Vogt (Plenum, New York, 1985).

150. G. Brown, W. Weise, J. Speth, and G. Baym, *Comments Part. Nucl. Phys.* **17**, 39 (1987).

151. G. Baym and S. A. Chin, *Nucl. Phys. A* **262**, 527 (1976).

152. C. J. Pethick, G. Baym, and H. Monien, *Nucl. Phys. A* **498**, 313c (1989); G. Baym, H. Monien, C. J. Pethick, and D. G. Ravenhall, *Phys. Rev. Lett.* **64**, 1867 (1990).

153. J. M. Luttinger, *Phys. Rev.* **119**, 1153 (1960).

154. B. L. Jones and J. W. McClure, *Phys. Rev.* **143**, 133 (1966).

155. J. W. Wilkins, *Observable Many-Body Effects in Metals* (Nordita, Copenhagen, 1968).

156. J. Oliva and N. W. Ashcroft, *Phys. Rev. B* **23**, 6399 (1981); *B* **25**, 223 (1982).

157. K. F. Quader and S. Yip, *Ann. Phys. (N.Y.)* **195**, 1 (1989); T. Fujita and K. F. Quader, *Phys. Rev. B* **36**, 5152 (1987).

158. W. L. McMillan, *Phys. Rev. B* **31**, 2750 (1985).

159. A. M. Finkel'stein, *Z. Phys. B—Cond. Matter* **56**, 189 (1984); A. M. Finkel'stein, *Zh. Eksp. Teor. Fiz.* **86**, 367 (1984) [English transl.: *Sov. Phys.-JETP* **59**, 212 (1984); *Zh. Eksp. Teor. Fiz. Pis. Red.* **40**, 63 (1984) [English transl.: *JETP Lett.* **40**, 796 (1984)].

160. C. Castellani, G. Kotliar, and P. A. Lee, *Phys. Rev. Lett.* **59**, 323 (1987); C. Castellani, C. Di Castro, G. Kotliar, P. A. Lee, and G. Strinati, *Phys. Rev. B* **37**, 9046 (1988).

161. P. Nozières, *J. Low Temp. Phys.* **17**, 31 (1974).

162. S. Yip, *Phys. Rev. B* **38**, 8785 (1988).

163. O. Valls and Z. Tešanović, *Phys. Rev. Lett.* **53**, 1497 (1984).

164. C. J. Pethick, D. Pines, K. F. Quader, K. S. Bedell, and G. E. Brown, *Phys. Rev. Lett.* **57**, 1955 (1986).

165. D. W. Hess, J. Low Temp. Phys. **68**, 311 (1987).

166. T. Holstein, R. E. Norton, and P. Pincus, *Phys. Rev. B* **8**, 2649 (1973).

167. M. Yu. Reizer, *Phys. Rev. B* **39**, 1602 (1989).

168. J. Solyom, *Adv. Phys.* **28**, 201 (1979).

169. C. M. Varma, P. B. Littlewood, S. Schmitt-Rink, E. Abrahams, and A. E. Ruckenstein, *Phys. Rev. Lett.* **63**, 1996 (1989).

INDEX